DISCARDED

# *Plant Parasitic Nematodes*

*VOLUME II*

## Cytogenetics, Host–Parasite Interactions, and Physiology

## Contributors

ALAN F. BIRD
C. E. CASTRO
A. J. CLARKE
A. F. COOPER, JR.
K. H. DEUBERT
C. C. DONCASTER
BURTON Y. ENDO
C. D. GREEN
L. R. KRUSBERG
N. T. POWELL
R. A. ROHDE
AUDREY M. SHEPHERD
DIETER STURHAN
C. E. TAYLOR
I. J. THOMASON
A. C. TRIANTAPHYLLOU
S. D. VAN GUNDY
B. M. ZUCKERMAN

# *Plant Parasitic Nematodes*

*Edited by*

*B. M. ZUCKERMAN*
LABORATORY OF EXPERIMENTAL BIOLOGY
UNIVERSITY OF MASSACHUSETTS
EAST WAREHAM, MASSACHUSETTS

*W. F. MAI*
DEPARTMENT OF PLANT PATHOLOGY
CORNELL UNIVERSITY
ITHACA, NEW YORK

*and*

*R. A. ROHDE*
DEPARTMENT OF PLANT PATHOLOGY
UNIVERSITY OF MASSACHUSETTS
AMHERST, MASSACHUSETTS

## VOLUME II

Cytogenetics, Host–Parasite Interactions, and Physiology

1971

ACADEMIC PRESS New York and London

ACADEMIC PRESS, INC.
111 Fifth Avenue, New York, New York 10003

*United Kingdom Edition published by*
ACADEMIC PRESS, INC. (LONDON) LTD.
Berkeley Square House, London W1X 6BA

LIBRARY OF CONGRESS CATALOG CARD NUMBER: 78-127710

PRINTED IN THE UNITED STATES OF AMERICA

# Contents

LIST OF CONTRIBUTORS xi
PREFACE xiii
CONTENTS OF VOLUME I xv

## GENETICS AND CYTOLOGY

### 13. Genetics and Cytology
*A. C. Triantaphyllou*

I. Introduction—Historical Review 1
II. Gametogenesis and Cytological Features of Reproduction 3
III. The Chromosomes of Nematodes 13
IV. Sexuality 16
V. Hybridization among Nematodes 27
VI. Cytogenetic Aspects of Nematode Evolution 29
References 32

## HOST–PARASITE INTERACTIONS

### 14. Specialized Adaptations of Nematodes to Parasitism
*Alan F. Bird*

I. Introduction 35
II. Morphological and Physiological Adaptations 36
III. Ecological Adaptations: Response to Stress 47
IV. Summary 48
References 48

### 15. Biological Races
*Dieter Sturhan*

I. Introduction 51
II. Sibling Species 53

III. Intraspecific Variation 54
IV. Types of Physiological Variation 59
V. Causes of Variability 61
VI. Methods of Race Identification 62
VII. Genetics of Physiological Characters 63
VIII. Development and Maintenance of Physiological Diversity 65
IX. Terminology 67
X. Conclusion 68
References 69

**16. Nematode Enzymes**

*K. H. Deubert and R. A. Rohde*

I. Introduction 73
II. Techniques in Enzymic Analysis 74
III. Nematode Enzymes 77
IV. Summary 88
References 89

**17. Nematode-Induced Syncytia (Giant Cells). Host–Parasite Relationships of Heteroderidae**

*Burton Y. Endo*

I. Introduction 91
II. Nematode Penetration and Migration 92
III. Mechanism of Feeding in the Heteroderidae 93
IV. Stimulation of Galls 94
V. Formation of Syncytia 97
VI. Nature of Resistance 111
VII. Conclusions 114
References 115

**18. Interaction of Plant Parasitic Nematodes with Other Disease-Causing Agents**

*N. T. Powell*

I. Introduction 119
II. Nematode–Fungus Complexes 120
III. Nematode–Bacteria Interactions 127
IV. Nematode–Virus Relationships 129
V. Some Effects of Complexes on Nematode Populations 131

VI. The Nature of Complexes Involving Nematodes 132
VII. Conclusions 134
References 135

19. **Feeding in Plant Parasitic Nematodes: Mechanisms and Behavior**
*C. C. Doncaster*

I. Introduction 137
II. Behavior Leading to Feeding 139
III. Pharyngeal Gland Secretions 141
IV. Ingestion 143
References 156

20. **Gnotobiology**
*B. M. Zuckerman*

I. Introduction 159
II. Methodology 161
III. Applications 171
IV. Conclusions 180
References 180

21. **Nematodes as Vectors of Plant Viruses**
*C. E. Taylor*

I. Introduction 185
II. The Vectors 187
III. The Viruses 192
IV. Relationships between Viruses and Vector Nematodes 196
V. Ecology and Control 204
References 207

**BIOCHEMISTRY AND PHYSIOLOGY**

22. **Chemical Composition of Nematodes**
*L. R. Krusberg*

I. Introduction 213
II. Inorganic Substances 214

III. Carbohydrates 215
IV. Amino Acids and Proteins 216
V. Lipids 222
VI. Plant Growth Regulators 231
VII. Other 232
VIII. Conclusions 233
References 233

**23. Respiration**
*R. A. Rohde*

I. Introduction 235
II. Factors Influencing Respiration 237
III. Methods of Measuring Respiration 244
References 245

**24. Mating and Host Finding Behavior of Plant Nematodes**
*C. D. Green*

I. Introduction 247
II. Sources of Stimulants 248
III. Dissemination of Stimuli 253
IV. Responses to Stimuli 258
V. Discussion 263
References 264

**25. Molting and Hatching Stimuli**
*Audrey M. Shepherd and A. J. Clarke*

I. Molting 267
II. Hatching 271
References 284

**26. Mode of Action of Nematicides**
*C. E. Castro and I. J. Thomason*

I. The State of Knowledge 289
II. Gross Effects 290
III. Permeation Characteristics 292
IV. Model Systems 294
V. Hypotheses 295
References 296

**27. Senescence, Quiescence, and Cryptobiosis**
*A. F. Cooper, Jr. and S. D. Van Gundy*

I. Introduction 297
II. Senescence 299
III. Quiescence 304
IV. Cryptobiosis 310
V. Summary 314
References 315

AUTHOR INDEX 319
SUBJECT INDEX 331

# List of Contributors

Numbers in parentheses indicate the pages on which the authors' contributions begin.

ALAN F. BIRD (35), C.S.I.R.O., Division of Horticultural Research, Glen Osmond, South Australia

C. E. CASTRO (289), Department of Nematology, University of California, Riverside, California

A. J. CLARKE (267), Rothamsted Experimental Station, Harpenden, Herts., England

A. F. COOPER, JR. (297), Department of Nematology, University of California, Riverside, California

K. H. DEUBERT (73), Laboratory of Experimental Biology, University of Massachusetts, East Wareham, Massachusetts

C. C. DONCASTER (137), Rothamsted Experimental Station, Harpenden, Herts., England

BURTON Y. ENDO (91), Crops Research Division, Agricultural Research Service, U. S. Department of Agriculture, Beltsville, Maryland

C. D. GREEN (247), Rothamsted Experimental Station, Harpenden, Herts., England

L. R. KRUSBERG (213), Department of Botany, University of Maryland, College Park, Maryland

N. T. POWELL (119), Department of Plant Pathology, North Carolina State University, Raleigh, North Carolina

R. A. ROHDE (73, 235), Department of Plant Pathology, University of Massachusetts, Amherst, Massachusetts

AUDREY M. SHEPHERD (267), Rothamsted Experimental Station, Harpenden, Herts., England

DIETER STURHAN (51), Institute of Nematology, Münster, Westphalia, Germany

C. E. TAYLOR (185), Scottish Horticultural Research Institute, Invergowrie, Dundee, Scotland

I. J. THOMASON (289), Department of Nematology, University of California, Riverside, California

A. C. TRIANTAPHYLLOU (1), Department of Genetics, North Carolina State University, Raleigh, North Carolina

S. D. VAN GUNDY (297), Department of Nematology, University of California, Riverside, California

B. M. ZUCKERMAN (159), Laboratory of Experimental Biology, University of Massachusetts, East Wareham, Massachusetts

# Preface

This two-volume treatise was written to provide an up-to-date reference source for students, teachers, and research and extension workers in plant nematology and related fields. Nematological advancements made since the publication of a similar book approximately ten years ago are discussed. A high proportion of the available knowledge obtained during this time has been in such important areas of nematology as ultrastructure, enzymology, chemistry of body composition, culturing, virus transmission, biological races, and nature of plant resistance. Thus, this is the first comprehensive reference work in nematology to include information from these new areas as well as from traditional ones.

An attempt has been made to coordinate and evaluate the phenomenal amount of research data of these years. In order to include the best possible coverage of the many diverse and specialized topics, a number of authors were invited to contribute to the text; many are actively engaged in the field about which they have written. Although each chapter was edited, the data and opinions expressed are those of the contributors.

Volume I includes a discussion of the history of plant nematology, the current status of research, and information pertaining to professional societies and publications. It also deals with nematode morphology, anatomy, taxonomy, and ecology, emphasizing plant parasitic forms and, where pertinent, drawing examples from free-living and animal parasitic nematodes.

Volume II deals with plant parasitic nematode genetics and cytology, host–parasite interactions, biochemistry, and physiology. As in Volume I, useful information relating to free-living and animal parasitic nematodes is included.

We wish to thank the authors for the considerable time spent in preparing their contributions. Such comprehensive treatises of important areas of plant nematology are invaluable to progress in this biological discipline. In fact, without them it would be difficult or impossible for students to become familiar with and research workers to keep abreast of the knowledge in specific areas.

B. M. Zuckerman
W. F. Mai
R. A. Rohde

## Contents of Volume I

### MORPHOLOGY, ANATOMY, TAXONOMY, AND ECOLOGY

Introduction
*W. F. Mai*

**Morphology and Anatomy**

Comparative Morphology and Anatomy
*Hedwig Hirschmann*

Nemic Relationships and the Origins of Plant Nematodes
*A. R. Maggenti*

Form, Function, and Behavior
*H. D. Crofton*

**Taxonomy**

Taxonomy: The Science of Classification
*G. W. Bird*

Taxonomy of Heteroderidae
*Mary T. Franklin*

Taxonomy of the Dorylaimida
*Virginia R. Ferris*

Classification of the Genera and Higher Categories of the Order Tylenchida (Nematoda)
*A. Morgan Golden*

**Ecology**

Biotic Influences in Soil Environment
*Richard M. Sayre*

Abiotic Influences in the Soil Environment
*H. R. Wallace*

Diagnostic and Advisory Programs
*K. R. Barker and C. J. Nusbaum*

Population Dynamics
*C. J. Nusbaum and K. R. Barker*

Author Index–Subject Index

*Genetics and Cytology*

# CHAPTER 13

# Genetics and Cytology

A. C. TRIANTAPHYLLOU

*Department of Genetics, North Carolina State University, Raleigh, North Carolina*

I. Introduction—Historical Review . . . . . . . . . . 1
II. Gametogenesis and Cytological Features of Reproduction . . . 3
A. Oogenesis . . . . . . . . . . . . . . 4
B. Spermatogenesis . . . . . . . . . . . . 9
C. Cytological Features of Meiotic Parthenogenesis . . . . 10
D. Cytological Features of Mitotic Parthenogenesis . . . . 11
E. Cytological Features of Pseudogamy . . . . . . . 12
F. Hermaphroditism . . . . . . . . . . . . 12
III. The Chromosomes of Nematodes . . . . . . . . . 13
A. General Characteristics . . . . . . . . . . 13
B. Chromosome Numbers . . . . . . . . . . 16
IV. Sexuality . . . . . . . . . . . . . . 16
A. Chromosomal Mechanisms of Sex Determination . . . . 16
B. Environmental Effect on Sex Differentiation . . . . . 17
V. Hybridization among Nematodes . . . . . . . . . 27
VI. Cytogenetic Aspects of Nematode Evolution . . . . . . 29
References . . . . . . . . . . . . . . 32

## I. INTRODUCTION—HISTORICAL REVIEW

Most of the early cytogenetic work in nematodes involves animal parasitic and free-living forms and was conducted in the late part of the nineteenth century and the early part of this century. Some of this work is still considered as classic because it elucidated certain basic cytological and biological phenomena that were incomprehensible until that time. Thus, Bütschli (1873) observed that two nuclei are present in fertilized eggs of *Caenorhabditis dolichura* which unite in the center of the egg to produce the nucleus of the first cell of the future embryo. This observation was one of the first steps toward understanding the "role of fertilization in reproduction," which was completely elucidated 2 years later by Hertwig in eggs of the sea urchin. Probably, the most significant contribution of nematode research in the field of cytology and heredity came a few years later when van Beneden (1883) discovered the process of "meiosis" occurring during maturation of the eggs of *Parascaris equorum* (*Ascaris megalocephala*). He demonstrated that during the formation of the polar bodies the chromosome number of the egg is reduced to one-half (meiosis) and that this number is doubled again in the cleavage nucleus which is formed by the fusion of the egg and sperm pronuclei (fertilization). Van Beneden (1883) and Boveri (1888) also introduced in cytology for the first time the concept of individuality and physical continuity of the chromosomes by demonstrating that *Parascaris* chromosomes persist during interphase. They observed that blastomere nuclei of *Parascaris equorum* eggs show a number of finger-shaped lobes which are formed at telophase by the free ends of their V-shaped chromosomes. During prophase of the following division the chromosomes reappear with their free ends lying in these lobes, indicating that chromosomal basic organization does not change during interphase when the chromosomes are not visible.

The processes of chromosome fragmentation and chromatin diminution were first discovered in nematodes. Boveri (1887) observed that the early cleavage divisions in *Parascaris equorum* eggs were unusual in several respects. The zygote nucleus of this nematode has four long chromosomes, each with a thin (euchromatic) central part, and two thicker (heterochromatic) club-shaped ends. During the first cleavage the chromosomes divide normally, mitotically. During the second cleavage, one of the blastomeres—$P_1$ or propagation cell—divides mitotically, but the other—$S_1$ or somatic cell—undergoes a peculiar division. Each of its chromosomes fragments into the two club-shaped ends and a number of small chromosomes that are derived from the thin central part (chromosome fragmentation). The small chromosomes then divide mitotically and the sister chromosomes migrate toward the spindle poles. The club-shaped ends, however, remain in the middle of the spindle and eventually are left in the cytoplasm of the daughter blastomeres where they degenerate (chromatin diminution). The process of chromosome fragmentation and chromatin diminution is repeated in one of the progeny of the P-cell line till the fifth cleavage. Eventually, all the cells of the embryo are diminished, except the last two P cells that will be enclosed in the genital primordium of the larva and will later give rise to all the gonial cells of that individual.

Because animal parasitic nematodes were recognized early as favorable material for cytological work, they were the second animal group after the insects to be studied extensively with regard to chromosomal mechanisms of sex determination. The X-O and multiple X chromosome situations were discovered in various nematodes as early as 1910 by Boveri, Gulick, Edwards, and others.

Nematodes provided also the first example of "gynogenesis" (pseudogamy or pseudofertilization) as a normal, nonpathological method of reproduction in animals. Krüger (1913) described this peculiar method of reproduction in the hermaphroditic *Rhabditis aberrans*. She observed that the sperm penetrates and thus activates the oocytes for further development but subsequently degenerates in the cytoplasm without fusing with the egg nucleus. Gynogenesis was confirmed later and demonstrated beyond doubt in a mutant line of *Rhabditis pellio* by Hertwig (1920) and by various workers in other nematodes and other animals.

Although these classic studies indicated that nematodes are favorable for karyological research, relatively little work followed. Most of this work involved studies on gametogenesis and the chromosomes of animal parasitic nematodes (Walton, 1924) and studies on gametogenesis, reproduction, and sexuality of free-living soil forms (Nigon, 1949).

The first attempt to study gametogenesis and the chromosomes of plant parasitic nematodes was made by Mulvey (1955). Although part of the early work is not very accurate, it indicated the potential usefulness of such studies in demonstrating cytogenetic differences among related nematodes. Cytogenetic work by various workers since 1960 has added much information regarding gametogenesis, chromosome numbers, mode of reproduction, and sexuality of plant parasitic nematodes. Purely genetic studies apparently have been hindered by difficulties involved in breeding experiments with plant parasitic nematodes and the almost complete absence of distinct morphological and physiological characters that could be used as genetic markers for genetic analysis. Therefore, this chapter is limited to an analysis of gametogenesis, reproduction, sexuality, and the cytogenetic aspects of evolution of plant parasitic nematodes. Some relevant information about common free-living soil nematodes is also included.

## II. GAMETOGENESIS AND CYTOLOGICAL FEATURES OF REPRODUCTION

Most species of soil and plant parasitic nematodes are bisexual, i.e., have males and females easily recognizable by primary and secondary sex characters. Species in which males and females appear in approximately equal numbers usually reproduce by amphimixis (cross-fertilization). In some species males appear in relatively small numbers, are rare, or absent. Reproduction in such species is usually by parthenogenesis. A number of species are hermaphroditic and reproduce by automixis (self-fertilization). However, most hermaphroditic species, besides the hermaphrodites, also have a variable number of males (incomplete or unbalanced hermaphroditism), which may or may not be functional. When the males are functional, reproduction is partially by automixis and partially by amphimixis. Some hermaphroditic and some bisexual species are pseudogamous (see cytological features of pseudogamy, Section II, E). Detailed accounts of gametogenesis with emphasis on free-living soil nematodes can be found in Nigon's (1965) extensive treatise. Reproduction of plant and soil nematodes has been reviewed by Triantaphyllou and Hirschmann (1964). Also, reviews of gametogenesis, particularly of animal parasitic nematodes, have been presented by Walton (1940, 1959). In the following, a general account of gametogenesis in free-living and especially in plant parasitic nematodes will be given. *Anguina tritici* will be used as the main example since it is amphimictic and the only plant parasitic nematode in which both oogenesis and spermatogenesis have been studied in detail (Triantaphyllou and Hirschmann, 1966).

### A. Oogenesis

Oogenesis in soil and plant parasitic nematodes follows the same general pattern known in most animals, with slight deviations in regard to the behavior of the chromosomes at synapsis and during meiotic prophase.

Multiplication of oogonia occurs in the apical, "germinal" zone of the ovary (Fig. 1). Oogonial cells are straight line descendents of the "propagation," P cells of the embryo and appear to be set aside from the remaining somatic cells during the early cleavage divisions. Oogonial divisions usually start in third-stage larvae and continue up to the early adult stage. Most divisions, however, occur during the fourth larval stage and the fourth molt. They are normal mitotic divisions and result in the production of a large number of oogonia, all of which have the somatic ($2n$) chromosome number. Oogonial divisions appear to be synchronized in some species (*Anguina tritici*, Fig. 1A), i.e., all oogonia of a region of the germinal zone may undergo division at the same time but occur with no particular order in others (*Meloidogyne* and *Heterodera*). The chromosomes in oogonial divisions of most soil and plant parasitic nematodes usually are not discrete (*Heterodera, Pratylenchus, Caenorhabditis,* and

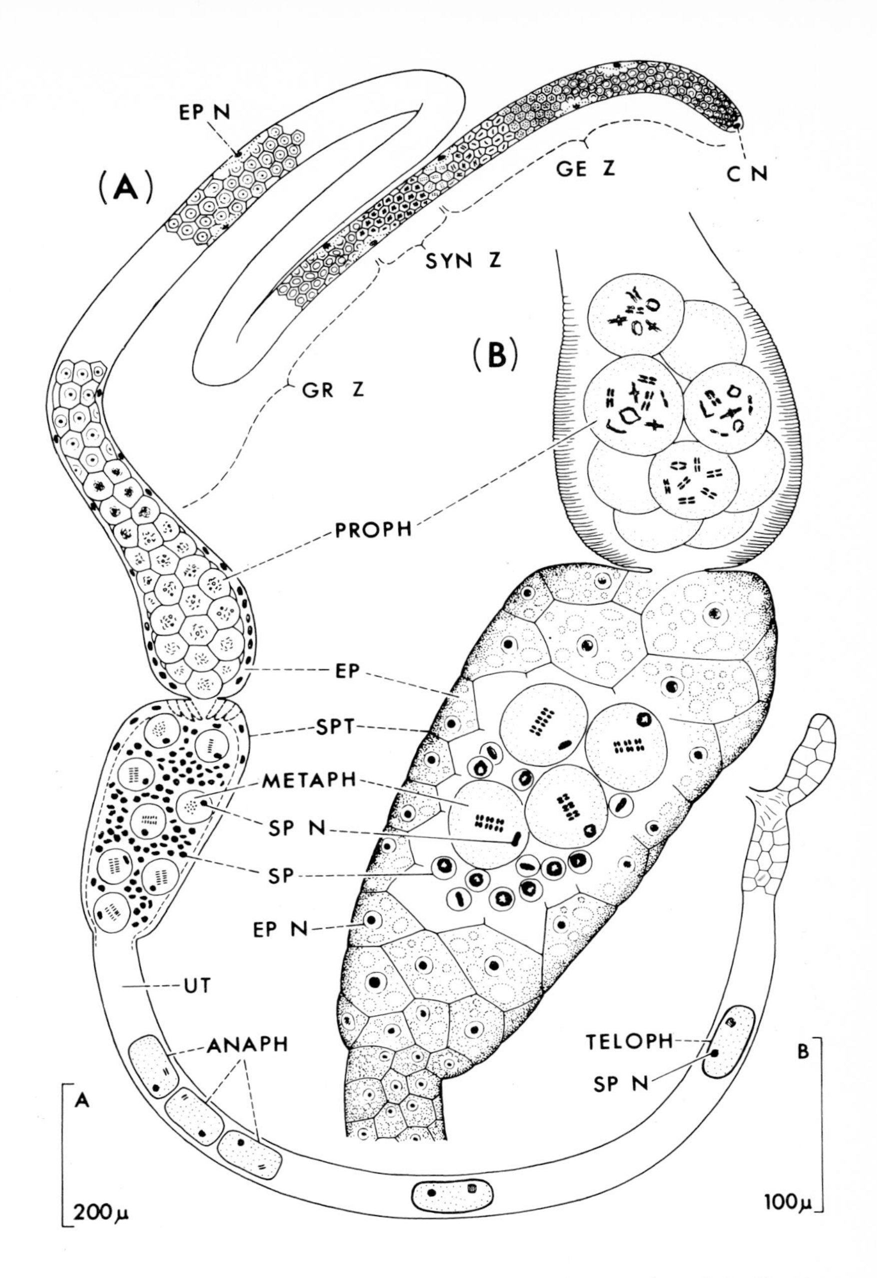


FIG. 1. (A) Diagrammatic presentation of the process of oogenesis and fertilization in *Anguina tritici*. (B) Enlargement of the spermatotheca region showing stages of maturation of the oocytes. ANAPH, oocytes in anaphase I; CN, cap cell nucleus; EP, epithelium; EP N, epithelial cell nucleus; GE Z, germinal zone of ovary; GR Z, growth zone of ovary; METAPH, oocytes in metaphase I; PROPH, oocytes in late prophase I; SP, spermatozoa; SP N, sperm nucleus in oocytes; SPT, spermatotheca; SYN Z, synapsis zone of ovary; TELOPH, oocytes in telophase I; UT, uterus. (After Triantaphyllou and Hirschmann, 1966.)

*Seinura*). However, the chromosome number can be determined and some information about the morphology and relative size of the chromosomes can be obtained from metaphase figures of some nematodes (*Anguina* and *Meloidogyne*) (Figs. 2 and 11).

The end of the germinal zone of the ovary coincides with the region in which oogonial divisions cease and the chromatin of the nuclei condenses in a compact network or mass, heavily staining with orcein, fuchsin, or other nuclear dyes (Fig. 3). This peculiar behavior of the chromatin indicates the occurrence of synapsis (pairing of homologous chromosomes) in the cells of that short region (zone of synapsis, Fig. 1A). The process of synapsis, including leptotene, zygotene, and pachytene stages, described in detail in other animals, cannot be clearly observed in soil and plant parasitic nematodes. Following the zone of synapsis the young oocytes enter the "growth zone" of the ovary and start increasing in size (growth period) with proportional increase of the size of their nuclei (Fig. 1A). One or more nucleoli appear in each nucleus at first, and the dense chromatic network slowly resolves into elongated double chromatin threads that probably represent the late pachytene or early diplotene stage (Fig. 4). The oocytes migrate down the growth zone of the ovary in a single file, or in two to three rows, but in some species they are arranged around a rachis in many rows (*Anguina tritici* and *Rhabditis anomala*). Their cytoplasm becomes progressively more granular through the accumulation of lipid droplets, refringent bodies, and various other storage materials. When the oocytes are midway down the growth zone, the double chromatin threads in their nuclei become diffuse, loose their stainability, and eventually may disappear completely (diffuse state). The large nucleus appears as a light sphere of uniform consistency, embedded in the darker, granular cytoplasm. High nuclear activity of DNA and RNA synthesis apparently takes place at this stage.

As the oocytes migrate farther down the ovary, they continue to increase in size, particularly those approaching the oviduct. The chromatin in their nuclei condenses progressively so that chromosomal configurations can be observed again (Figs. 5 and 6). The chromosomes at this stage are at early or advanced diakinesis. In the most advanced oocytes, the two homologs of each bivalent are usually associated end to end (Fig. 1B). In some cases the association may be subterminal giving the bivalent the appearance of a cross, or the homologs may be associated at both ends and the bivalent may appear like a double ring (Fig. 1B). This behavior of diakinetic chromosomes can be interpreted on the basis of the classic type of chiasmate association of the homologs observed in other animals, but some deviations may occur particularly in some rhabditids (Nigon and Brun, 1955). In oocytes located close to the

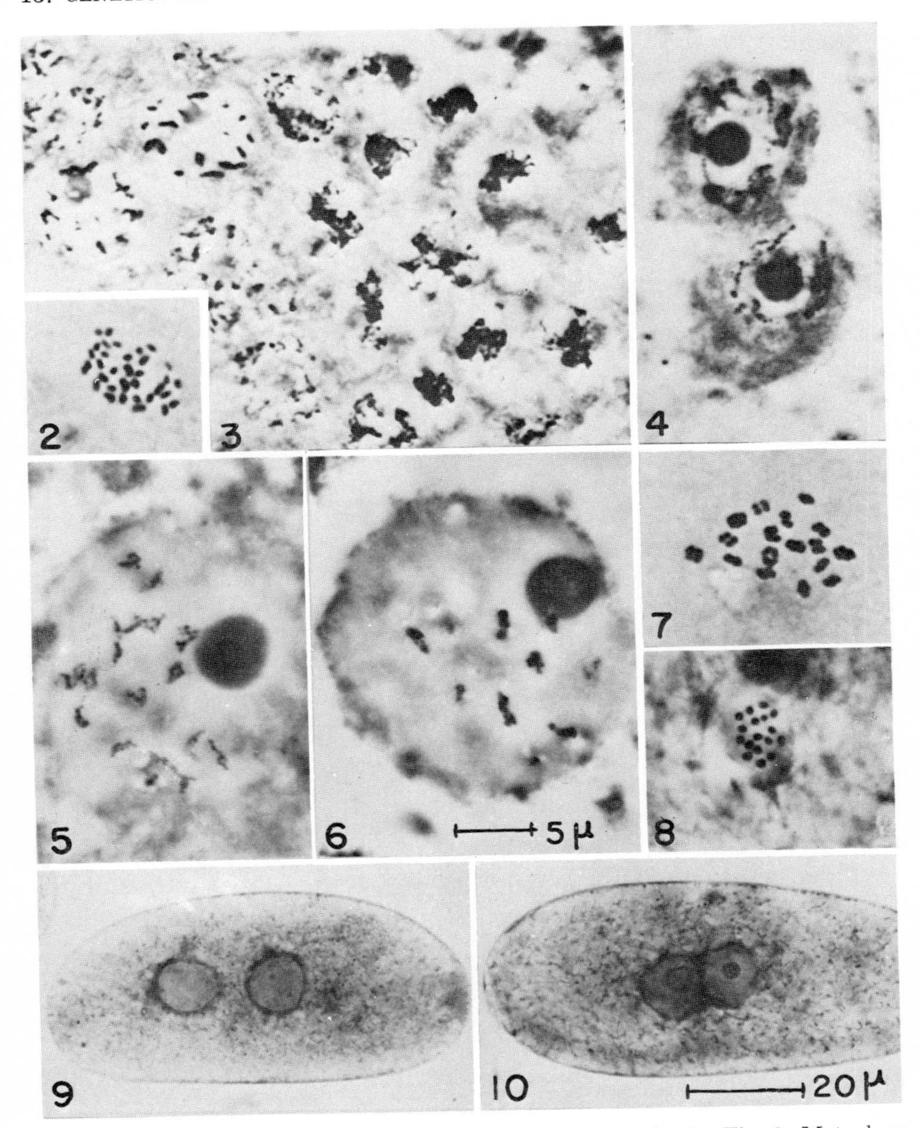


FIGS. 2–10. Oogenesis and fertilization in *Meloidogyne hapla*. Fig. 2. Metaphase of oogonial division. Fig. 3. End part of the germinal zone of the ovary (left half of the picture) and the zone of synapsis (right half)—some oogonia are at prophase and others at interphase. In the zone of synapsis the chromatin forms a dense network. Fig. 4. Oocytes in the beginning of the growth zone. Figs. 5 and 6. The diakinetic chromosomes of two oocytes approaching the end of the growth zone. Figs. 7 and 8. Prometaphase I and telophase II chromosomes. Figs. 9 and 10. Progressive stages toward fusion of the sperm and egg pronuclei. (After Triantaphyllou, 1966.)

oviduct* and those migrating down the oviduct into the spermatotheca, the nuclear membrane disappears and the diakinetic chromosomes contract further forming compact bivalent chromosomes. The two chromatids of each homolog are clearly separated so that each bivalent appears as a tetrad (Figs. 7 and 12). At this stage (prometaphase I) the oocytes enter the spermatotheca.

In noninseminated females a few oocytes may pass into the uterus, but they do not develop and remain at prometaphase or metaphase I. They also fail to develop an egg shell. Growth and development of oocytes in the growth zone ceases, and the entire gonad remains in a static state. In *Rotylenchulus reniformis* unfertilized eggs may proceed with development and undergo some cleavage divisions, but they never develop beyond the blastula stage (Nakasono, 1966).

In inseminated females one spermatozoon enters each oocyte close to the anterior or the posterior pole. Sperm entrance activates the oocyte which then advances to metaphase and passes into the uterus (in most other animals, the sperm does not enter the oocyte until the latter has undergone both maturation divisions). Soon after sperm entrance, the exterior membrane of the oocyte (vitelline membrane) becomes progressively thicker, apparently through impregnation with and deposition of materials condensed from the cytoplasm. A hard egg shell is thus formed, which is soon separated from the egg cytoplasm and leaves underneath it a perivitellous space. Later, another thin membrane is separated from the cytoplasm and is clearly visible in the perivitellous space, particularly at the poles of the egg. This thin membrane has been referred to as *lipoid membrane* by some investigators and *vitelline membrane* by others. The use of the latter term has caused considerable confusion because it also implies the original vitelline membrane which now has become the hard egg shell.

Anaphase I and telophase I follow rapidly, and one polar nucleus is formed close to the surface, usually in the middle of the oocyte. The polar nucleus is extruded as a polar body in most amphimictic nematodes or may be retained in the cytoplasm close to the surface in parthenogenetic nematodes (*Meloidogyne* spp. and *Helicotylenchus dihystera*). Following completion of the first maturation division, the egg chromosomes enter directly the prophase stage of the second maturation division without an intermediate interphase stage. Metaphase, anaphase, and telophase II occur rapidly (Figs. 8 and 19), and a second polar body is extruded or a second polar nucleus is formed close to the surface of the oocyte. In the

* *Anguina tritici* has no true oviduct. The dilated, thick-walled end part of the ovary directly joins the large ovoid and sperm-filled spermatotheca through a distinct narrowing or sphincter. Many other nematodes however have a distinct oviduct (*Meloidogyne* and *Pratylenchus*), usually a long narrow tube.

meantime the first polar nucleus usually divides into two or three bodies, usually through uneven distribution of its chromosomes. After the second maturation division, the egg chromosomes migrate close to the center of the egg where the egg pronucleus is formed. The pronucleus appears as a light, uniform sphere delimited from the granular cytoplasm by a thin nuclear membrane. No chromosomes can be observed in the egg pronucleus at the time of, or shortly after, its formation.

Since its entrance into the oocyte, the sperm nucleus has remained close to the entrance point, at the periphery of the egg, with the chromatin condensed in a single globular body or a number of interconnected smaller bodies that often resemble individual chromosomes. At the time the egg pronucleus is formed, the sperm nucleus is also transformed into a pronucleus, similar in appearance with the egg pronucleus but different in size in some cases. In *Pratylenchus* spp. one of the pronuclei, probably the sperm pronucleus, is always larger than the other.

Soon the two pronuclei approach each other and fuse to form the zygote nucleus (Figs. 9 and 10). The chromatin may condense to form distinct chromosomes shortly before fusion of the pronuclei (*Pelodera strongyloides;* Nigon, 1949) or as soon as the zygote nucleus is formed. The chromosomes at prophase of the first cleavage are elongated and usually entangled with each other. Later they contract into short or elongated rod-shaped bodies. The individual chromatids of each chromosome are not visible until early anaphase. The metaphase plate of the first cleavage division is always perpendicular to the long axis of the egg. All the chromosomes may be arranged on the same metaphase plate or the chromosomes of the egg and sperm pronuclei may be arranged in two separate plates on the same equatorial plane. The first cleavage division follows.

In most plant parasitic nematodes, eggs are deposited before maturation has been completed and, therefore, only one or a few eggs may be present in the short uterus of actively reproducing females (*Pratylenchus*). In old females, however, eggs may remain in the uterus for a longer period, i.e., until maturation has been completed and, occasionally, until embryonation has advanced. In the genus *Heterodera* most or all the eggs are maintained in the elongated uteri, where they undergo embryogenesis. In most free-living soil nematodes, the eggs are retained in the uterus until fertilization and several cleavage divisions have occurred.

## B. Spermatogenesis

Spermatogonial divisions occurring in the germinal zone of the testis, synaptic phenomena, and maturation of spermatocytes in the growth zone of the testis follow the same pattern as described for oogenesis. In fact, it is usually difficult to distinguish the testes from the ovaries within the same nematode species on the basis of cytological characteristics alone, and often even on a morphological basis. The only striking difference is the relatively small size of the primary spermatocytes located at the end of the testis as compared to the large oocytes of the end part of the ovary. The spermatocytes remain relatively small, because their cytoplasm does not increase excessively during the last part of the growth period. Fully grown spermatocytes in the end part of the testis are at late diakinesis or prometaphase I. As soon as they pass into the gonoduct (seminal vesicle or vas deferens depending on gonad structure), they advance to metaphase I and proceed to complete the two maturation divisions in a classic manner (Figs. 13–16). Four spermatids, each with the haploid chromosome complement, are thus produced from each spermatocyte. Details about cytokinesis during or following the maturation divisions are not available for plant parasitic nematodes. In *Pelodera strongyloides* cytokinesis does not take place until both maturation divisions have been completed (Nigon, 1949). During cytokinesis the four spermatids are separated in such a way that a protoplasmic mass is cut off and is not included in the spermatids. Transformation of spermatids into spermatozoa may take place shortly afterwards, much later in the vas deferens, or after the spermatids have been transferred into the female gonoducts (*Parascaris*).

Spermatozoa of plant parasitic nematodes are nonflagellate, usually spherical, amoeboid, conical, rod- or spindle-shaped bodies consisting of a central or eccentric nucleus and a thick, granular cytoplasmic covering. Within the nucleus the chromatin remains condensed in one or a small number of globular bodies, staining heavily with nuclear dyes (*Heterodera, Meloidogyne,* and *Helicotylenchus*). In few nematodes (*Seinura tenuicaudata, Diplogaster robustus, Mesodiplogaster lheritieri,* and *Panagrolaimus rigidus*), however, the chromosomes remain compact and discrete in the nuclei of the spermatozoa. The haploid chromosome number can thus be observed in mature spermatozoa present in the male or female gonoducts and even after penetration into oocytes.

### C. Cytological Features of Meiotic Parthenogenesis

Meiotic parthenogenesis occurs in *Meloidogyne hapla, M. graminicola, M. naasi, Heterodera betulae, Pratylenchus scribneri, Aphelenchus avenae,* and probably in many more plant parasitic nematodes. Oogenesis and spermatogenesis are similar to those of amphimictic species described earlier. Two divisions take place during maturation of the oocytes and two polar bodies or polar nuclei are produced. Mature eggs contain the haploid ($n$) number of chromosomes. Activation of oocytes is not triggered by the entrance of a spermatozoon but by some other agent, probably by substances produced in the glandular part of the uterus (Paramonov, 1962), or is unnecessary altogether. In the absence of actual fertilization, the somatic chromosome number is reestablished in the mature egg by various means. In *Meloidogyne hapla* the second polar nucleus, which is not extruded from the cytoplasm, appears to fuse with the egg pronucleus (Triantaphyllou, 1966). In *Heterodera betulae* a kind of endomitotic division takes place during prophase of the first cleavage division which results in the doubling of the chromosome number from $n = 12$ to $2n = 24$ before the first cleavage division occurs.

Meiotic parthenogenesis is "facultative" in the various *Meloidogyne* species, i.e., reproduction is by meiotic parthenogenesis in noninseminated females and by amphimixis in inseminated females. In general, it appears that meiotic parthenogenesis prevails under environmental conditions favorable for rapid development and reproduction when males are absent or rare, and amphimixis is common under adverse conditions when males are more abundant. Cytological observations have shown that in one and the same inseminated female some eggs may develop by amphimixis following penetration by a spermatozoon and others by meiotic parthenogenesis (Triantaphyllou, 1966).

Meiotic parthenogenesis, apparently, is "obligatory" in *Pratylenchus scribneri* and *Aphelenchus avenae* in which males are absent or very rare.

## D. Cytological Features of Mitotic Parthenogenesis

Mitotic parthenogenesis occurs in several species of the genera *Meloidogyne, Heterodera,* and *Pratylenchus* and in *Meloidodera floridensis*. Probably it is more common than meiotic parthenogenesis among plant parasitic nematodes. Although the behavior of the chromatin in the "zone of synapsis" of mitotically parthenogenetic nematodes appears to be the same as in amphimictic nematodes, the chromosomes reappearing in oocytes at the end of the growth zone of the ovary are not bivalent at diakinesis but univalent and in the diploid number. This indicates that synapsis of homologous chromosomes does not occur, or if it does, no chiasmata are formed and the homologs fall apart soon thereafter.

Only one maturation division occurs which is a regular mitotic division and results in the formation of one polar nucleus and the egg pronucleus, both with the somatic chromosome number. No sperm is needed for activation of the oocytes. In inseminated females of *Meloidogyne javanica*, one spermatozoon may enter each oocyte when the latter is at prometaphase of the single maturation division, but it does not fuse with the egg pronucleus and appears to degenerate in the cytoplasm while the egg nucleus undergoes cleavage (Triantaphyllou, 1962). It is evident that occasional fertilization of unreduced eggs in similar instances would result in progeny with higher chromosome numbers. This may be one way by which various polyploid forms have developed in the genus *Meloidogyne* and, probably, in other genera.

## E. Cytological Features of Pseudogamy

Pseudogamy has been reported in the hermaphroditic species *Rhabditis aberrans* (Krüger, 1913) and *R. anomala* (Hertwig, 1922) and in the bisexual species *Mesorhabditis belari* (Bělař, 1923), *Rhabditis maupasi*, *R. longicaudata* (Hertwig, 1922), and a mutant strain of *R. pellio* (Hertwig, 1920). Sperm penetration is necessary for activation of the oocytes for further development. The sperm nucleus, however, does not fuse with the egg pronucleus but degenerates in the cytoplasm. Maturation of the oocytes in most pseudogamous nematodes consists of a single mitotic division with no pairing of homologous chromosomes. From a genetic viewpoint, therefore, pseudogamy in most nematodes is equivalent to mitotic parthenogenesis. In *Mesorhabditis belari*, however, synapsis takes place and the reduced number of bivalent chromosomes appears at metaphase I in primary oocytes. After this stage, oocytes of the same female may develop in two different ways. Some undergo regular meiosis with two maturation divisions and then develop by amphimixis. Others undergo a modified type of meiosis. During anaphase I the chromosomes of both anaphase plates divide once so that at telophase I both telophase plates have the somatic ($2n$) chromosome number. No second maturation division takes place, and the egg proceeds towards the first cleavage without actual fertilization. Pseudogamy in this case is genetically equivalent to meiotic parthenogenesis.

## F. Hermaphroditism

Hermaphrodites have been observed in several rhabditids, diplogasterids, rhabdiasids, and aphelenchs, and probably occur in some criconematids. Most hermaphroditic nematodes are protandric, i.e., their gonads function first as testes and produce a number of spermatozoa. They later function as ovaries and produce oocytes which are fertilized by the previously produced sperm (automixis). Usually, maturation of the oocytes and spermatocytes in hermaphrodites is by regular meiosis as in bisexual amphimictic species.

In many hermaphroditic species the protandrically produced sperm are sufficient to fertilize only a small number of the eggs produced by the same animal. Eggs produced after the sperm are exhausted remain unfertilized and therefore cannot develop (*Seinura steineri, S. oxura, Rhabditis seychellensis,* and *Mesodiplogaster maupasi*). This reproductive inefficiency indicates that hermaphroditism in these nematodes has been recently acquired and has not been perfected or evolved to a final state. In advanced cases of hermaphroditism, sperm can be produced for a second and a third time in alternating zones of the ovotestis after the first sperm supply is exhausted (*Rhabditis guerneyi*).

## III. THE CHROMOSOMES OF NEMATODES

### A. General Characteristics

Studies of nematode chromosomes have been done only in connection with gametogenesis. Therefore, information is limited primarily to chromosomes of gonial cells, oogonia, and spermatogonia and those of oocytes and spermatocytes. On some occasions the chromosomes of the blastomeres during the early cleavage divisions have been studied. The chromosomes of later cleavage divisions and divisions of other somatic cells (hypodermis, epithelium of gonad) during postembryogenesis are very small and nondiscrete and have never been observed clearly. Figures 11–21 show the chromosomes of various plant parasitic nematodes.

At prometaphase and metaphase of gonial divisions of some plant parasitic nematodes, the chromosomes may be discrete, 0.5–1.5 $\mu$ in length and 0.4–0.6 $\mu$ in width. They are spherical or rod-shaped, and the daughter chromatids are not distinct until late metaphase or early anaphase (Figs. 2 and 11). The chromosomes of the zygote nucleus are similar but considerably larger (1–4 $\mu$ by 0.5–0.7 $\mu$). With each successive cleavage division, the chromosomes become smaller, and by the end of the blastula stage no discrete chromosomes can be distinguished in the very small metaphase plates of such divisions.

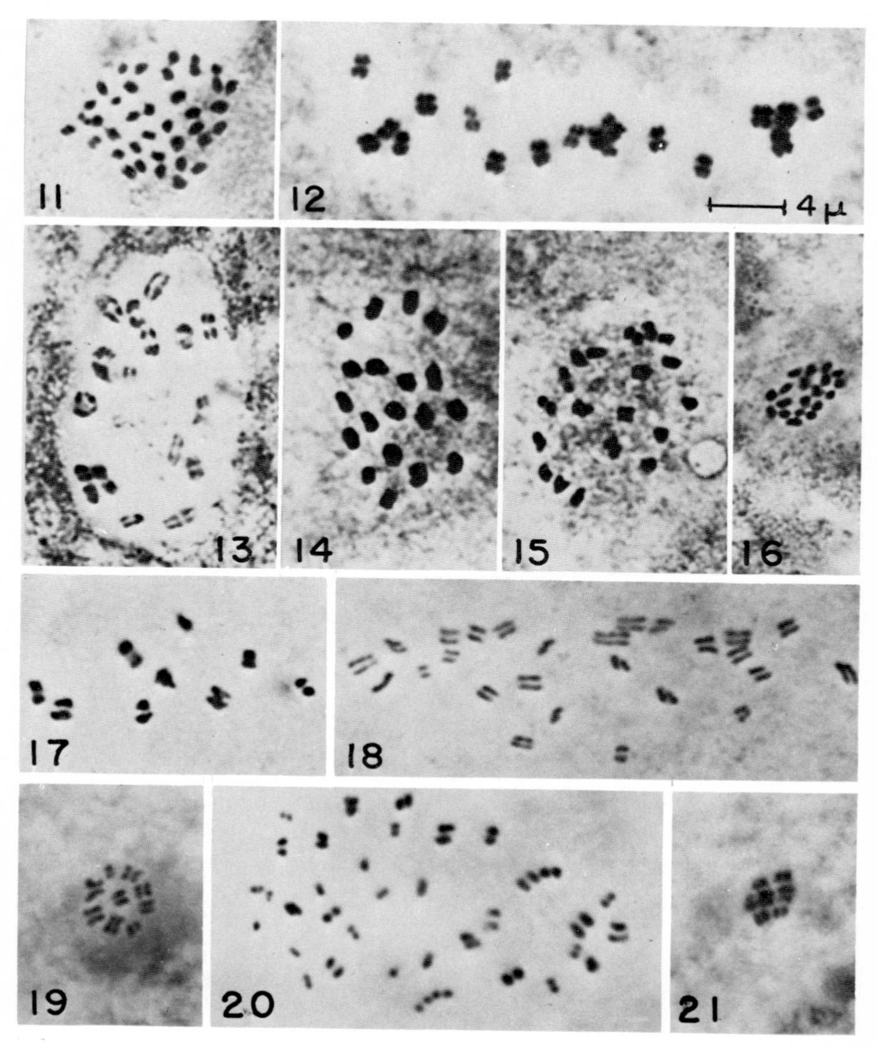


FIGS. 11–21. The chromosomes of various plant parasitic nematodes. Fig. 11. Metaphase of oogonial division of *Anguina tritici*. Fig. 12. Metaphase I in a primary oocyte of *A. tritici*. Figs. 13–16. Late prophase, metaphase I, metaphase II, and telophase II in spermatocytes of *A. tritici*. Fig. 17. Prometaphase I in a primary oocyte of *Heterodera glycines*. Fig. 18. Prometaphase I in an oocyte of *H. trifolii*. Fig. 19. Telophase I in an oocyte of *H. cruciferae*. Fig. 20. Prometaphase I in an oocyte of *Meloidogyne arenaria*. Fig. 21. Metaphase II in a secondary oocyte of *Meloidogyne spartinae*. (Figures 11–16 after Triantaphyllou and Hirschmann, 1966; Fig. 17 after Triantaphyllou and Hirschmann, 1962; Figs. 18, 19, and 21 are original; and Fig. 20 after Triantaphyllou, 1963.)

The meiotic chromosomes observed during maturation of the oocytes and spermatocytes are distinctly different from the chromosomes of gonial and somatic cells. At late diakinesis or prometaphase I, each bivalent is clearly tetrapartite in nature, consisting of two homologous chromosomes, each subdivided into two distinct chromatids (Figs. 7, 12, and 13). The two homologs are associated end to end, and the chromatids are parallel to each other.

The prometaphase I chromosomes of mitotic parthenogenetic species are univalent but appear as dyads. This is because the two chromatids of each chromosome separate during early prophase and become arranged parallel to each other at a distance of 0.4–0.8 $\mu$ with no visible centromeric connection between them (Figs. 18 and 20). Some type of physical connection or attraction must exist, however, because the chromatids never separate completely. In some nematodes (*Rhabditis aberrans*), the univalent metaphase I chromosomes during oogenesis and spermatogenesis of the hermaphrodites appear as short, single rods (Krüger, 1913).

Chromosomes within the same karyotype may differ considerably in dimensions and slightly in various other morphological features such as general shape and the presence of constrictions or heterochromatic segments. Variation in chromosome morphology from cell to cell and at different stages during division, however, has excluded a precise description of the karyotype of any plant parasitic nematode. Such karyotype descriptions may be possible in the future with improved staining procedures.

The parallel arrangement of the chromatids of each chromosome during maturation of the oocytes, particularly in mitotic parthenogenetic nematodes, the absence of a visible localized kinetochore, and the fact that the separating chromatids at anaphase remain parallel to each other as they migrate "broad side on" toward the poles suggest that the chromosomes of these nematodes have a diffuse type of kinetochore like the hemipteran insects and some other animals and plants. Experimental evidence about the nature of the kinetochore of nematode chromosomes, however, is lacking. The chromosomes of many ascarids have a multiple kinetochore (multicentric) as verified by the behavior of the small chromosomes of *Parascaris equorum* during mitotic anaphase of blastomeres undergoing chromosome fragmentation and chromatin diminution (see Section I). If the large *Parascaris* chromosomes had a localized kinetochore, only one of the small chromosomes, the one carrying the kinetochore, would divide normally. The others would not migrate to the poles during anaphase as do the heterochromatic ends which lack a spindle attachment, and they would be lost in the cytoplasm.

### B. Chromosome Numbers

The chromosome numbers observed by various workers in free-living soil and plant parasitic nematodes (Table I) vary from the lowest possible, $2n = 2$ in *Diploscapter coronata,* to a high of $2n = 54$ in *Meloidogyne arenaria* or $2n = 54$–$56$ in a giant form of *Ditylenchus dipsaci.* The basic chromosome number of most rhabditids and tylenchids, however, appears to be between $x = 6$ and $x = 9$. Deviations from the basic number are common in most genera. Thus, although the basic chromosome number of the genus *Heterodera* is $x = 9$, an undescribed *Heterodera* species has $n = 10$, and *H. betulae* has $n = 12$ and 13. In the genus *Pratylenchus* haploid numbers of $n = 5$, 6, and 7 have been detected in various species. Variation in the basic number has also been observed in the genus *Seinura* in which species with $n = 3$ and $n = 6$ have been found. Much greater variation is observed in groups of nematodes in which polyploidy occurs in association with a parthenogenetic mode of reproduction. Such polyploid forms, or aneuploid forms derived from polyploid forms, exist in the genera *Heterodera, Meloidogyne, Ditylenchus, Helicotylenchus,* and *Pratylenchus* and usually represent a triploid or a tetraploid condition. It is not clear whether polyploid forms exist among bisexual amphimictic nematodes. The high chromosome numbers of amphimictic species of *Meloidogyne* and of *Anguina tritici* may not have been derived through polyploidization but through other mechanisms of chromosome number increase.

## IV. SEXUALITY

### A. Chromosomal Mechanisms of Sex Determination

Very little is known about sex determination in plant parasitic nematodes. In most bisexual species the chromosomal situation of the male has not been studied in detail. In *Anguina tritici* males and females have the same chromosome number ($2n = 38$) and no sex chromosomes can be recognized. If specialized sex chromosomes exist, they are probably of the XX-XY type with the male as the heterogametic sex. Limited observations have suggested that a similar situation exists in the genus *Heterodera* ($2n = 18$ in both males and females), in *Meloidogyne hapla, M. graminis, M. spartinae, Pratylenchus vulnus,* and *P. coffeae. Seinura tenuicaudata* probably has an XX♀–XO♂ sex mechanism (Hechler, 1963). Among the rhabditids, most bisexual amphimictic species have an XX♀–XO♂ chromosomal mechanism of sex determination. In some species males and females have the same chromosome number and no sex chromosomes can be recognized. This is interpreted as an evolution of the original (XX–XO) sex mechanism toward the XX♀–XY♂ condition. This condition may have resulted through a transfer of the genetically active part of the X chromosome to an autosome which thus became the neo-X chromosome, whereas the original autosome became the Y chromosome. The small size of the chromosomes and the absence of detailed karyotypic analyses have precluded morphological recognition of the neo-X and Y chromosomes. On the other hand, evolution of the karyotype from the XY♂ to the XO♂ condition, through a loss of the Y chromosome, has been considered as another possibility in the animal parasitic nematode *Oswaldocruzia filiformis* (John, 1957).

No sex chromosomes have been observed in hermaphrodites which produce sperm and eggs with the same chromosomal complement. However, in the hermaphroditic species *Caenorhabditis dolichura* (Nigon, 1949), division of one of the bivalents during spermatogenesis may be delayed as compared to other bivalents, indicating that this bivalent is a sex chromosome (XX type). Males of the same species have an unpaired sex chromosome (XO condition) which undergoes only one mitotic division, either during the first or the second maturation division. Males are produced either as a result of amphimixis following fertilization of an oocyte by a spermatozoon without an X chromosome or as a result of automixis in the following manner. Hermaphrodites may produce a few gametes without an X chromosome, probably following meiotic nondisjunction. When participating in fertilization, such gametes produce male individuals (XO type). A similar mechanism apparently operates also in *Caenorhabditis elegans* and *C. briggsae*.

## B. Environmental Effect on Sex Differentiation

Nematodes have provided several examples of environmentally controlled sex expression. Sex differentiation of the larvae of various insect parasitic nematodes of the family Mermithidae (*Paramermis contorta* and *Mermis subnigrescens*) appears to be influenced by the environment provided by their insect hosts. Thus when few nematode larvae develop in the body of an insect, they usually differentiate as females. When many larvae develop in an insect, most or all of them differentiate as males (Caullery and Comas, 1928; Christie, 1929). Apparently, this

TABLE I
CHROMOSOME NUMBERS AND MODE OF REPRODUCTION OF SOME SOIL AND PLANT PARASITIC NEMATODES

| Family | Chromosome number[b] | | Mode of reproduction | Reference |
|---|---|---|---|---|
| Subfamily—species[a] | Haploid | Diploid | | |
| Tylenchidae | | | | |
| Tylenchinae | | | | |
| *Anguina tritici* | 19 ♀ (I, II); 19 ♂ (I, II) | 38 o, s, m | Amphimixis | Triantaphyllou and Hirschmann (1966) |
| *Ditylenchus dipsaci* | 12 ♀ (I); 12 ♂ (I, II) | | Amphimixis | Triantaphyllou (unpublished) |
| *Ditylenchus dipsaci* (from *Plantago maritima*) | | >50 | | D. Sturhan (personal communication) |
| *Ditylenchus dipsaci* (giant form from *Vicia faba*) | | (54–56) | | D. Sturhan (personal communication) |
| *Ditylenchus triformis* | | (30–40) m | | Triantaphyllou (unpublished) |
| Tylenchorhynchinae | | | | |
| *Tylenchorhynchus claytoni* | 8 ♀ (I) | | Amphimixis | Triantaphyllou (unpublished) |
| Pratylenchinae | | | | |
| *Pratylenchus penetrans* | 5 ♀ (I) | 10 m | Amphimixis | Hung and Jenkins (1969) Roman and Triantaphyllou (1969) |
| *Pratylenchus penetrans* | 6 ♀ (I) | | Amphimixis | Thistlethwayte (1970) |
| *Pratylenchus vulnus* | 6 ♀ (I); 6 ♂ (I) | | Amphimixis | Roman and Triantaphyllou (1969) |
| *Pratylenchus coffeae* | 7 ♀ (I); 7 ♂ (I) | | Amphimixis | Roman and Triantaphyllou (1969) |

| | | | | |
|---|---|---|---|---|
| *Pratylenchus scribneri* | 6 ♀ (I) | 12 m | Parthenogenesis (meiotic) | Roman and Triantaphyllou (1969) |
| *Pratylenchus scribneri* | | (25–26) ♀ (I) | Parthenogenesis (mitotic) | Roman and Triantaphyllou (1969) |
| *Pratylenchus zeae* | | 26 | Parthenogenesis | Hung and Jenkins (1969) |
| *Pratylenchus zeae* | | (21–26) ♀ (I) | Parthenogenesis (mitotic) | Roman and Triantaphyllou (1969) |
| *Pratylenchus neglectus* | | (20) ♀ (I) | Parthenogenesis (mitotic) | Roman and Triantaphyllou (1969) |
| *Pratylenchus brachyurus* | | (30–32) ♀ (I) | Parthenogenesis (mitotic) | Roman and Triantaphyllou (1969) |
| Hoplolaiminae | | | | |
| *Helicotylenchus dihystera* | | 30, 34, 38 ♀ (I) | Parthenogenesis (mitotic) | Triantaphyllou and Hirschmann (1967) |
| *Helicotylenchus erythrinae* | 5 ♀ (I, II) | 10 o | Amphimixis | Triantaphyllou and Hirschmann (1967) |
| *Rotylenchus buxophilus* | 8 ♀ (I) | | | Triantaphyllou (unpublished) |
| *Hoplolaimus galeatus* | 10 ♀ (I, II); 10 ♂ (I) | (20) o | Amphimixis | Triantaphyllou (unpublished) |
| Rotylenchulinae | | | | |
| *Rotylenchulus reniformis* | 9 ♀ (I) | | Amphimixis | Nakasono (1966) |
| *Rotylenchulus reniformis* | 9 ♀ (I) | 18 o, m | Amphimixis | Triantaphyllou (unpublished) |
| Heteroderidae | | | | |
| *Heterodera avenae* | 9 ♀ (I) | | Amphimixis | Mulvey (1960) |
| *Heterodera avenae* | 9 ♀ (I) | 19 ♀ (I) rarely | Amphimixis | Cotten (1965) |
| *Heterodera betulae* | 12, 13 ♀ (I) | 24 m | Parthenogenesis (meiotic) | Triantaphyllou (unpublished) |
| *Heterodera carotae* | 9 ♀ (I, II) | | Amphimixis | Triantaphyllou (unpublished) |
| *Heterodera cruciferae* | 9 ♀ (I) | | | Mulvey (1960) |

TABLE I (*Continued*)

| Family | Chromosome number[b] | | Mode of reproduction | Reference |
|---|---|---|---|---|
| Subfamily—species[a] | Haploid | Diploid | | |
| *Heterodera cruciferae* | 9, 10 ♀ (I) | | Amphimixis | Cotten (1965) |
| *Heterodera cruciferae* | 9 ♀ (I, II) | | Amphimixis | Triantaphyllou (unpublished) |
| *Heterodera galeopsidis* | | 32 ♀ (I) | Parthenogenesis (mitotic) | Hirschmann and Triantaphyllou (1965) |
| *Heterodera glycines* | 9 ♀ (I) | (18) m | Amphimixis | Triantaphyllou and Hirschmann (1962) |
| *Heterodera goettingiana* | 9 ♀ (I) | | | Mulvey (1960), Cotten (1965) |
| *Heterodera lespedezae* | | 27 ♀ (I) | Parthenogenesis (mitotic) | Hirschmann and Triantaphyllou (1965) |
| *Heterodera mexicana* | 9 ♀ (I, II) | | Amphimixis | Triantaphyllou (unpublished) |
| *Heterodera oryzae* | 9 ♀ (I) | | Amphimixis | Netscher (1969) |
| *Heterodera rostochiensis* | 9 ♀ (I) | | | Riley and Chapman (1957) |
| *Heterodera rostochiensis* | 9 ♀ (I); also 10, 11, 23–24 ♀ (I) | 18, 18–24 | | Cotten (1959, 1960) |
| *Heterodera sacchari* | | 27 ♀ (I) | Parthenogenesis (mitotic) | Netscher (1969) |
| *Heterodera schachtii* | 9 ♀ (I) | 18 m | Amphimixis | Mulvey (1957, 1960) |
| *Heterodera schachtii* | 9 ♀ (I) | 19 ♀ (I) in one egg | Amphimixis | Cotten (1965) |
| *Heterodera schachtii* | 9 ♀ (I, II) | | Amphimixis | Triantaphyllou (unpublished) |
| *Heterodera tabacum* | 9 ♀ (I, II) | | Amphimixis | Triantaphyllou (unpublished) |

| | | | | |
|---|---|---|---|---|
| *Heterodera trifolii* | | >24, 27 ♀ (I) | Parthenogenesis (mitotic) | Mulvey (1958a, 1960) |
| *Heterodera trifolii* | | 26, 27, 34 ♀ (I) | Parthenogenesis (mitotic) | Hirschmann and Triantaphyllou (1965) |
| *Heterodera virginiae* | 9 ♀ (I, II) | | Amphimixis | Triantaphyllou (unpublished) |
| *Heterodera weissi* | 9 ♀ (I, II) | | Amphimixis | Triantaphyllou (unpublished) |
| *Heterodera* sp. (from *Rumex crispus*) | | 24 ♀ (I) | Parthenogensis (mitotic) | Hirschmann and Triantaphyllou (1965) |
| *Heterodera* sp. (Osborne's cyst nematode) | 9 ♀ (I, II) | | Amphimixis | Triantaphyllou (unpublished) |
| *Meloidodera floridensis* | | 26, 27 ♀ (I), o | | Triantaphyllou (unpublished) |
| *Meloidogyne arenaria* (2*n* form) | | 36(34–37) ♀ (I) | Parthenogenesis (mitotic) | Triantaphyllou (1963) |
| *Meloidogyne arenaria* (3*n* form) | | 51–54 ♀ (I) | Parthenogenesis (mitotic) | Triantaphyllou (1963) |
| *Meloidogyne exigua* | 18 ♀ (I) | | | Triantaphyllou (unpublished) |
| *Meloidogyne graminicola* | 18 ♀ (I, II); 18 ♂ (I, II) | 36 o, m | Amphimixis and parthenogenesis (meiotic) | Triantaphyllou (1969) |
| *Meloidogyne graminis* | 18 ♀ (I); 18 ♂ (I) | 36 o, m | Amphimixis and parthenogenesis (meiotic) | Triantaphyllou (unpublished) |
| *Meloidogyne hapla* (race A) | 15, 16, 17 ♀ (I); (15, 16, 17) ♂ (I, II) | | Amphimixis and parthenogenesis (meiotic) | Triantaphyllou (1966) |
| *Meloidogyne hapla* (race B) | | 45 ♀ (I); (45) ♂ (I) | Parthenogenesis (mitotic) | Triantaphyllou (1966) |

TABLE I (*Continued*)

| Family / Subfamily—species[a] | Chromosome number[b]: Haploid | Chromosome number[b]: Diploid | Mode of reproduction | Reference |
|---|---|---|---|---|
| *Meloidogyne incognita* | | 41–44 ♀ (I), o | Parthenogenesis (mitotic) | Triantaphyllou (unpublished) |
| *Meloidogyne javanica* | | 43, 44, 46, 48 ♀ (I), o | Parthenogenesis (mitotic) | Triantaphyllou (1962) |
| *Meloidogyne naasi* | 18 ♀ (I, II); 18 ♂ (I, II) | (36) o | Parthenogenesis (meiotic) and amphimixis | Triantaphyllou (1969) |
| *Meloidogyne ottersoni* | 18 ♀ (I) | 36 o, m | Amphimixis and parthenogenesis (meiotic) | Triantaphyllou (unpublished) |
| *Meloidogyne spartinae* | 7 | | | Fassuliotis and Rau (1966) |
| *Meloidogyne spartinae* | 7 ♀ (I); 7 ♂ (I) | | Amphimixis | Triantaphyllou (unpublished) |
| Criconematidae | | | | |
| Criconematinae | | | | |
| *Hemicriconemoides* sp. | 5 ♀ (II) | | Amphimixis | Triantaphyllou (unpublished) |
| Aphelenchidae | | | | |
| *Aphelenchus avenae* | 8 ♀ (I) | | Parthenogenesis (meiotic) | Triantaphyllou (unpublished) |
| Aphelenchoididae | | | | |
| *Aphelenchoides composticola* | 3 ♀ (I); 3 ♂ (I); 2–3 ♂ (II) | | Amphimixis | Younes (1968) |
| *Seinura celeris* | 3 ♀ (I) | | Amphimixis | Hechler and Taylor (1966) |
| *Seinura oliveirae* | 3 ♀ (I) | | Amphimixis | Hechler and Taylor (1966) |
| *Seinura oxura* | 6 ⚥ (I) | | Automixis | Hechler and Taylor (1966) |
| *Seinura steineri* | 6 ⚥ (I) | | Automixis | Hechler and Taylor (1966) |
| *Seinura tenuicaudata* | 6 ♀ (I, II); 6 ♂ (I); 5?–6 ♂ (II) | | Amphimixis | Hechler (1963) |

| | | | | |
|---|---|---|---|---|
| Diplogasteridae | | | | |
| Diplogasterinae | | | | |
| *Diplogaster* sp. | 6 ♀ (I, II); 6(sperm) | | | Mulvey (1955) |
| *Mononchoides changi* | 7 ♀ (I, II); 7 ♂ (I); 6–7 ♂ (II) | | Amphimixis | Hechler (1970) |
| Rhabditidae | | | | |
| Rhabditinae | | | | |
| *Rhabditis aberrans* | Spermatogenesis 8 + 2 univalent ⚥ (I); 9 ⚥ (II) | Oogenesis 18 ⚥ (I); 18 ⚥ m | Pseudogamy? Parthenogenesis (mitotic-facultative) | Krüger (1913) |
| *Rhabditis aberrans* | 9 ⚥ (I); 8–9 ♂ (II) | | | Nigon (1965) |
| *Rhabditis guerneyi* | 5 ♀ (I, II) | 10 o; 10 m | | Bělař (1923) |
| *Rhabditis pellio* | 7 ♀ (I, II) | (14) m | | Hertwig (1920) |
| *Rhabditis pellio* | 7 ♀ (I, II); 7 ♂ (I); 6–7 ♂ (II) | 14 o; 13 s; 13–14 m | | Kröning (1923) |
| *Rhabditis pellio* (mutant) | | 14 ♀ (I) | Pseudogamy | Hertwig (1920) |
| *Rhabditis terricola* | 7 ♀ (I, II); 7 ♂ (I); 6–7 ♂ (II) | 14 o; 13 s 13–14 m | | Kröning (1923) |
| *Rhabditis* "XX" | Spermatogenesis Variable No. bivalent + univalent ⚥ (I, II) | Oogenesis 24 ⚥ (I) | Parthenogenesis (mitotic) pseudogamy? | Bělař (1923) |
| *Rhabditis* "XIX" | 5 ♀ (I) | 10 ♀ (telophase I) | Parthenogenesis (pseudomeiotic) | Bělař (1923) |
| *Pelodera strongyloides* | 11 ♀ (I, II); 10 + 1 univalent ♂ (I, II) | 22 ♀ o; 21 ♂ s | Amphimixis | Nigon (1949) |
| *Caenorhabditis briggsae* | | 12 ⚥; 11 ♂ s | Automixis; Amphimixis (rarely) | Nigon and Dougherty (1949) |
| *Caenorhabditis dolichura* | 6 ⚥ (I) | | Automixis | Honda (1925) |
| *Caenorhabditis dolichura* | 6 ⚥ (I, II); 5 + 1 univalent ♂ (I) | 12 ⚥ o, m; 11 ♂ s | Automixis; Amphimixis | Nigon (1949) |
| *Caenorhabditis elegans* | 6 ⚥ (I, II) | | Automixis | Honda (1925) |

TABLE I (*Continued*)

| Family | Chromosome number[b] | | | |
|---|---|---|---|---|
| Subfamily—species[a] | Haploid | Diploid | Method of reproduction | Reference |
| *Caenorhabditis elegans* | 6⚥(I), also higher numbers | 12⚥; 11♂ | Automixis | Nigon (1949) |
| *Caenorhabditis elegans* (polyploid) | 12⚥(I); 12 + 1 univalent ⚥(I); 11♂(I) | 24⚥ o; 23⚥ o; 22♂ s | | Nigon (1951) |
| *Mesorhabditis belari* | 10♀(I); 10♂(I) | 20♀(metaphase or telophase I), o, m | Pseudogamy (pseudomeiotic); Amphimixis (rarely) | Bělař (1923), Nigon (1949) |
| Diploscapterinae | | | | |
| *Diploscapter coronata* | | 2♀(I), m | Parthenogenesis (mitotic) | Hechler (1968) |
| Panagrolaimidae | | | | |
| Panagrolaiminae | | | | |
| *Panagrolaimus rigidus* | 4♀(I, II); 4♂(I, II) | 8 o, m | Amphimixis | Nigon (1949) |
| Dorylaimidae | | | | |
| Tylencholaiminae | | | | |
| *Xiphinema index* | 10♀(I) | 20 o, m | Parthenogenesis (meiotic) | Dalmasso and Younes (1969) |
| *Xiphinema mediterraneum* | 5♀(I) | | | Dalmasso and Younes (1970) |

[a] Classification of the Tylenchida according to Allen and Sher (1967). Classification of the Rhabditida and designation of species according to Goodey (1963).

[b] Numbers in parentheses are not considered as certain. ♀(I), primary oocyte; ♀(II), secondary oocyte; ♂(I), primary spermatocyte; ♂(II), secondary spermatocyte; ⚥(I), primary oocyte or spermatocyte of hermaphrodites; ⚥(II), secondary oocyte or spermatocyte of hermaphrodites; m, somatic cell; o, oogonium; and s, spermatogonium.

change in the direction of sex differentiation occurs after the larvae have entered the insect and therefore at a period when their genetic constitution could not be modified. It may result from the crowded conditions that create a shortage of space and food and probably alter the physiology and biochemistry of the insect host. It may also result from the concentration of substances secreted or excreted by the nematodes. In most of these species intersexual individuals also appear at varying frequencies. The cause of intersexuality in the mermithids is not known. Steiner (1923) suggested a mechanism similar to that of *Lymantria dispar* in which intersexuality results from hybridization of closely related genotypes. Christie (1929), however, believed that intersexes develop from female individuals which are under the influence of some environmental force tending to stimulate maleness.

Sex differentiation of larvae of various *Meloidogyne* species is also affected by the environment of the host plant (Triantaphyllou, 1960; McClure and Viglierchio, 1966; Davide and Triantaphyllou, 1968). Sex in these nematodes becomes morphologically recognizable in half-developed second-stage larvae in which the genital primordium, through cell divisions, becomes rod-shaped in male larvae and V-shaped in female larvae. Under conditions favorable for rapid growth and development practically all the larvae differentiate as female larvae and become adult females with two ovaries. Under adverse conditions most larvae differentiate as males with one testis. When conditions change from favorable to unfavorable shortly after initiation of sex differentiation, female larvae undergo sex reversal and become males with two testes, apparently corresponding to the two gonads of the female larvae from which they develop. Intersexual individuals are produced in *M. javanica* under similar conditions as a result of incomplete sex reversal (Triantaphyllou, 1960). Conditions that induce maleness are crowding of the larvae, low host suitability, changes of the physiology and biochemistry of the host plant induced by the application of plant growth regulators like maleic hydrazide or following decapitation of the host plant, incubation of infected roots, and probably high or low temperatures. Obviously, these general conditions exercise their effect indirectly. A more direct factor common to all these general conditions may be the quantity and quality of food available to the second-stage larvae before and shortly after initiation of sex differentiation. Food, in turn, may affect the physiology of the larvae, their hormonal balance, and various biochemical processes responsible for sex differentiation.

*Meloidodera floridensis* provides another example of environmentally controlled differentiation of sex (unpublished data). Second-stage larvae of this nematode can develop to adulthood without feeding if they are kept in a physiological salt solution for 1–3 months. Under such conditions 75–95% of the larvae develop into adult males. The remaining larvae die in the second larval stage. No females develop under such conditions. Contrary to this, larvae inoculated to pine seedlings develop into adult females. No males have ever been found in roots of pine plants. Apparently the same larvae differentiate as females after feeding and as males under conditions of starvation. Therefore, sex differentiation proceeds independently of the genetic constitution of the larvae and is exclusively controlled by the environment.

Wide fluctuation in the male-to-female ratio in various *Heterodera* species is attributed to the inability of the female larvae to develop to maturity under conditions of stress. Thus recent work by Kerstan (1969) showed that equal numbers of males and females of *H. schachtii* develop in plants infected by single larvae and that the increased male-to-female ratio obtained under crowded or other adverse conditions is caused by the reduction of the number of developing females rather than the increase of the number of males. A similar conclusion has been also reached by Johnson and Viglierchio (1969) for *H. schachtii* and by C. Koliopanos (personal communication) for *H. glycines*. It appears that sex in these *Heterodera* species is genetically controlled, and sex expression is not modified by environmental influences—only the sex ratio is. The only possible exception may be the case of *H. rostochiensis* in which single larva inoculations give rise to very few males and many females (den Ouden, 1960; Trudgill, 1967), whereas crowded conditions, unsuitable host plants, and other environmental conditions unfavorable for the development of the larvae result in a high male-to-female ratio (Ellenby, 1954; Trudgill, 1967).

Work related to the inheritance of the ability of *H. rostochiensis* to develop on potato varieties carrying genes for resistance from *Solanum tuberosum* s. sp. *andigena* and *S. multidissectum* provides a reasonable explanation of the cause of the unbalanced sex ratios in this nematode (Trudgill *et al.*, 1967; Jones *et al.*, 1967; Trudgill and Parrott, 1969). However, it does not add direct support to the hypothesis that sex differentiation is environmentally controlled. It appears that larvae with the proper genotype (*aa* homozygous recessive for overcoming resistance) can induce the formation of a sizable group of giant cells in the tissues of a resistant plant and obtain sufficient nourishment to develop into adult females and probably males (there is no evidence that all *aa* genotype larvae become females). All other larvae (*aA* and *AA* genotypes) cannot induce the formation of proper giant cells, their nourishment is limited, and they can develop only into males or die. What percent of these larvae actually develops into males is not known, and therefore it cannot be argued that female larvae, or larvae genetically undetermined with regard to sex, differentiate into males. It is possible that only male larvae develop into males whereas female larvae die.

Rationally, diploid amphimictic species of *Heterodera* would be expected to have a well developed genetic and possibly chromosomal mechanism of sex determination as most other nematode species do. Environmental influence on sex expression in the genera *Meloidogyne* and *Meloidodera* is easier to comprehend because of the polyploid and predominantly parthenogenetic nature of these organisms. Polyploidy and aneuploidy may have weakened the genetic mechanism of sex determination in *Meloidogyne* and *Meloidodera* equalizing the male and female factors in their chromosomal complement. The situation may be similar in most mermithids which also reproduce by parthenogenesis.

## V. HYBRIDIZATION AMONG NEMATODES

The question of potential or actual hybridization among related species of nematodes becomes very important when the biological species concept is to be applied in nematode taxonomy. A classic example is *Ditylenchus dipsaci* which occurs in many biotypes (host races). Critical morphological studies suggested that several biotypes could be described as separate species (Steiner, 1956). Subsequent hybridization tests, however, demonstrated that these biotypes can interbreed freely and produce viable progeny capable of further, continuous reproduction (Eriksson, 1965; Sturhan, 1964, 1966; Webster, 1967). Therefore, their distinction as separate species would not be justified. Similarly, populations of *Ditylenchus destructor* from various hosts interbred freely in experimental crosses indicating that they belong to the same species (Wu, 1960; Smart and Darling, 1963). Hybridization between two host races of *Heterodera avenae* also was successful, and analysis of $F_1$ and $F_2$ progeny partially elucidated the genetic mechanism involved in their capacity to infect two varieties of barley (Andersen, 1965).

Some degree of incompatibility apparently exists among various biotypes (pathotypes) of *Heterodera rostochiensis* in which single male–female matings are more successful within than between biotypes (Parrott, 1968). This may indicate the first step toward genetic isolation of these biotypes and eventual phylogenetic branching of the group into a number of distinct species. Similarly, crosses between various biotypes of *Heterodera glycines* with different host range and pathogenicity on certain soybean varieties are possible and produce progeny capable of further reproduction but with an increased percent of larval abnormalities (Koliopanos, personal communication).

Interspecific hybridization tests in nematodes were conducted early in this century by Maupas (1900, 1919), who attempted to cross several related species of rhabditids and diplogasterids. Similar tests within the same groups of organisms were conducted later by Nigon and Dougherty (1949), Hirschmann (1951), and Osche (1952, 1954). Related species were found to be reproductively isolated in practically all the cases.

Interspecific crosses among plant parasitic nematodes have been attempted on only a few occasions. Crosses between *Heterodera glycines* and the closely related species *H. schachtii* have been partially successful and have given progeny capable of further, but probably limited, reproduction (Potter and Fox, 1965). Similarly, experimental crosses between various pathotypes of *H. rostochiensis,* on one hand, and *H. tabacum, H. virginiae, H. mexicana,* and Osborne's cyst nematode on the other, have been quite successful (Green and Miller, 1969). These results indicate the close phylogenetic relationship of the species tested, but at the same time question the species status in these organisms since natural interbreeding among them may be possible. It appears, however, that some degree of incompatibility exists in most cases, and future studies should be directed toward determining the effectiveness of the existing degree of reproductive isolation in keeping the species biologically distinct, particularly under natural conditions.

Attempted crosses between *Heterodera glycines, H. carotae,* and Osborne's cyst nematode have been unsuccessful, primarily because of failure of the females of each species to attract males of the other species (Fox, 1967). Males of *Heterodera schachtii* can inseminate females of the triploid parthenogenetic *H. trifolii,* and this indicates a close relationship of these two organisms (Mulvey, 1958b). Whether actual fertilization takes place is rather questionable in spite of the observation that giant and morphologically abnormal larvae are present among the progeny of impregnated females. Such larvae are commonly found even in pure cultures of *H. trifolii* and probably represent cytological anomalies occurring during oogenesis.

Attempted crosses between *Ditylenchus destructor* and *D. myceliophagus* were not successful according to Smart and Darling (1963). Therefore, these closely related species are biologically distinct. If further studies confirm that these species are morphologically indistinguishable from each other, they should still be considered as different species (sibling species) on the basis of their reproductive isolation.

Undoubtedly, more work of this type will be done in the future with many related nematode species in order to learn more about the degree of their relationship and to clarify existing taxonomic uncertainties.

## VI. CYTOGENETIC ASPECTS OF NEMATODE EVOLUTION

Phylogenetic diversification of nematodes appears to have been influenced extensively by the evolution of their karyotype and mode of reproduction. Of course, changes in karyotype and reproduction are part of the over-all evolution and therefore should not be regarded as its cause. It is only because of the strong impact these two biological phenomena have on the rate of genetic evolution that they are often presented as the cause rather than the result or the expression of evolution. Changes in chromosome number or in mode of reproduction, for example, automatically set the basis for further genetic diversification which eventually may lead to speciation. Our present knowledge, although limited, clearly supports the view that karyotypic variation and variation in mode of reproduction are closely associated with and probably the cause of extensive speciation in several nematode groups.

Karyotypic changes in nematodes usually include modifications of the basic chromosome number, polyploidy, and aneuploidy. Extensive karyotypic variation resulting from various types of chromosomal rearrangements undoubtedly occurs, but evidence is lacking because of the complete absence of detailed cytogenetic studies in nematodes.

Changes in mode of reproduction include the establishment of meiotic and mitotic parthenogenesis in practically every nematode group studied and protandric hermaphroditism in some groups. Studies of the karyotype and mode of reproduction in plant parasitic nematodes have been confined primarily to the genera *Heterodera, Meloidogyne,* and *Pratylenchus.*

In *Heterodera* most species are diploid ($n = 9$) amphimictic with quite similar karyotypes (Fig. 17). Cytogenetic information, therefore, has not helped in the interpretation of the phylogenetic relationships of these species. Extensive changes of the karyotype and the mode of reproduction have occurred, however, in several *Heterodera* species whose phylogenetic derivation can be determined or assumed (Triantaphyllou, 1969, 1970). Thus, *H. trifolii, H. galeopsidis, H. lespedezae,* and an undescribed species from *Rumex crispus* are polyploid or polyploid derivatives with chromosome numbers ranging from $2n = 24$ to 34, and reproduce by mitotic parthenogenesis. Undoubtedly, these forms have been derived secondarily from diploid amphimictic ancestors, probably relatives of *H. schachtii* to which they are morphologically related. *Heterodera sacchari* has $2n = 27$ chromosomes and reproduces by mitotic parthenogenesis. It probably has evolved from the morphologically similar *H. oryzae* or another diploid amphimictic relative as a triploid parthenogenetic form. *H. leuceilyma* may have evolved along the same parthenogenetic line as *H. sacchari* for the same reasons.

As can be seen from these two examples, evolution in these nematodes is closely associated with the establishment of polyploidy and parthenogenesis. This type of evolution appears to be common in nematodes and particularly in plant parasitic forms. Related polyploid parthenogenetic forms of the same phylogenetic origin such as *H. trifolii, H. galeopsidis,* and *H. lespedezae* may differ slightly in several characters—most often in host range. For practical purposes such variants have been described as different species. Biologically, however, they belong to the same parthenogenetic species complex, and this should be taken into consideration in their future taxonomic treatment.

A third line of parthenogenetic evolution in *Heterodera* involves *H. betulae* which has a chromosome number of $n = 12$ and 13 and reproduces by meiotic parthenogenesis. No polyploidy is involved in this case. The chromosome number has increased from the basic 9 to 12 and 13, and reproduction has changed from amphimixis to meiotic parthenogenesis. The exact phylogenetic derivation of this species is not clear, but its morphological relationship with the *H. cacti* species group suggests that it may have evolved from this group.

The cytogenetic picture of the genus *Meloidogyne* is even more complex and difficult to interpret than *Heterodera.* The genus is predominantly parthenogenetic and shows extensive variation in chromosome numbers. *Meloidogyne carolinensis,* the only amphimictic species, appears to have $n = 18$ chromosomes, i.e., the same number as *M. graminicola, M. naasi, M. graminis,* and *M. ottersoni,* all of which can reproduce by both amphimixis and meiotic parthenogenesis (Table I). Consequently, 18 has been recognized as the basic chromosome number for the genus. *Meloidogyne hapla* populations with $n = 17$, 16, or 15 chromosomes and amphimictic or meiotic parthenogenetic reproduction have probably evolved from ancestors with $n = 18$ chromosomes through various mechanisms of chromosome number reduction. *Meloidogyne hapla* and *M. arenaria* populations with somatic chromosome numbers of 36–54 and mitotic parthenogenetic reproduction must have been derived secondarily from ancestral types of *M. hapla* that were still undergoing meiosis. Diploid forms developed directly through suppression of meiosis and establishment of mitotic parthenogenesis as the sole method of reproduction. Forms with higher chromosome numbers may have been derived through occasional fertilization of unreduced eggs with reduced or unreduced sperm and the simultaneous or subsequent establishment of mitotic parthenogenesis. Polyploid forms in *Meloidogyne,* therefore, must have been derived at different occasions from a variety of ancestral types. This may explain the wide diversity of forms present in each parthenogenetic species.

*Meloidogyne incognita* and *M. javanica,* two biologically highly successful species, are also polyploid reproducing by mitotic parthenogenesis, but their closest ancestors are not yet known. Undoubtedly, more *Meloidogyne* species need to be studied before the cytogenetics of the entire genus can be interpreted satisfactorily.

*Meloidogyne spartinae* with $n = 7$ chromosomes (Fig. 21) and amphimictic reproduction is cytogenetically different from all other *Meloidogyne* species and apparently belongs to a different phylogenetic group. Also, its relationship to the other genera of the family Heteroderidae is not clear.

The numerical relationship (1:2) of the basic chromosome numbers of *Heterodera* ($n = 9$) and *Meloidogyne* ($n = 18$) and the predominantly parthenogenetic nature of *Meloidogyne* suggested originally that the karyotype of the latter genus represents the tetraploid state of an ancestral *Heterodera*-type karyotype. More recent studies, however, have shown that the *Meloidogyne* chromosomes behave like those of a diploid organism during maturation of gonial cells and that the amount of DNA present in hypodermal nuclei of *Meloidogyne* larvae is less than half that of nuclei of *Heterodera* larvae (Lapp and Triantaphyllou, 1969). This information does not support the hypothesis of polyploidy and introduces the following alternative possibilities: (a) the *Meloidogyne* chromosomes have been derived from an ancestral, *Heterodera*-type karyotype, through various mechanisms of chromosome number increase, other than polyploidization; (b) the *Heterodera* karyotype has been derived from an ancestral, *Meloidogyne*-type karyotype through centric fusions or other mechanisms of chromosome number reduction; and (c) the karyotypes of the two genera are not closely related and, possibly, are of different or remote phylogenetic origin. The last possibility is favored at present.

Since *Meloidodera floridensis* is triploid parthenogenetic, it cannot be regarded as a direct phylogenetic link of *Meloidogyne* and *Heterodera*. *Meloidodera floridensis* is cytologically related to *Heterodera* and exhibits some primitive morphological characteristics such as the subequatorial position of the vulva indicative of the ancestral organism from which both *Meloidodera* and *Heterodera* may have evolved.

In the genus *Pratylenchus* variation in chromosome numbers has been detected among diploid amphimictic species ($n = 5$ in *P. penetrans,* 6 in *P. vulnus,* and 7 in *P. coffeae*) indicating that extensive karyotypic changes have occurred during the evolution of the genus. It has not been possible to identify the original, "basic" chromosome number and the direction of karyotype evolution. Further diversification of the genus occurred with the establishment of meiotic parthenogenesis and also the establishment of various degrees of polyploidy in conjunction with mitotic parthenogenesis (Table I).

In the genus *Seinura* the evolution of the karyotype, sexuality, and reproduction provide some guides for speculative interpretation of the phylogenetic development of the group. It is possible that *S. tenuicaudata,* a diploid ($n = 6$) amphimictic species with a sex ratio of approximately 1:1 and probably an XX♀–XO♂ chromosomal mechanism of sex determination, is a basic species representative of the ancestor of the genus. Evolution of the group may have followed two different directions. *Seinura steineri* and *S. oxura* evolved toward hermaphroditism with a progressive decrease of the proportion of males and a change (functional) of the females into protandric hermaphrodites without noticeable changes of the basic karyotype. On the other hand, *S. celeris* and *S. oliveirae* maintained the amphimictic mode of reproduction but underwent a decrease of the haploid chromosome number from the basic $x = 6$ to $n = 3$. Such a decrease of the chromosome number would be expected to occur in phylogenetically advanced species. *Seinura celeris* and *S. oliveirae* can be considered as advanced species on the basis of their molting pattern (Hechler and Taylor, 1966). They have only three molts instead of the normal four of *S. tenuicaudata* and other nematodes. The production of only 15–40% males in *S. celeris* and 5–20% in *S. oliveirae* also indicates a tendency toward a reduction of the number of males which could eventually lead to hermaphroditism or parthenogenesis. *Aphelenchoides compositicola* with $n = 3$ chromosomes and amphimictic reproduction may have evolved karyotypically along similar lines. An alternative interpretation that evolution has occurred from a basic chromosome number of $x = 3$ to $n = 6$ cannot be excluded altogether.

## REFERENCES

Allen, M. W., and Sher, S. A. (1967). *Annu. Rev. Phytopathol.* **5,** 247–264.
Andersen, S. (1965). *Nematologica* **11,** 121–124.
Bělař, K. (1923). *Biol. Zentrbl.* **43,** 513–518.
Boveri, T. (1887). *Anat. Anz.* **2,** 688–693.

Boveri, T. (1888). *Jena. Z. Naturwiss.* **22,** 685–882.
Bütschli, O. (1873). *Nova Acta Acad. Leopold. Carol.* **36**(5), 1–144.
Caullery, M., and Comas, M. (1928). *C. R. Acad. Sci.* **186,** 646–648.
Christie, J. R. (1929). *J. Exp. Zool.* **53,** 59–76.
Cotten, J. (1959). *Nature (London)* **183,** 128.
Cotten, J. (1960). *Nematologica* **5,** Suppl II, 123–126.
Cotten, J. (1965). *Nematologica* **11,** 337–342.
Dalmasso, A., and Younes, T. (1969). *Ann. Zool. Ecol. Anim.* **1,** 265–272.
Dalmasso, A., and Younes, T. (1970). *Nematologica* **16,** 51–54.
Davide, R. G., and Triantaphyllou, A. C. (1968). *Nematologica* **14,** 37–46.
Den Ouden, H. (1960). *Nematologica* **5,** 215–216.
Ellenby, C. (1954). *Nature (London)* **174,** 1016–1017.
Eriksson, K. B. (1965). *Nematologica* **11,** 244–248.
Fassuliotis, G., and Rau, G. J. (1966). *Nematologica* **12,** 90. (Abstr.)
Fox, J. A. (1967). *Nematologica* **13,** 143–144. (Abstr.)
Goodey, T. (1963). "Soil and Freshwater Nematodes" (rev. by J. B. Goodey from 1951 Ed.), 2nd Ed., 544 pp. Wiley, New York.
Green, C. D., and Miller, L. I. (1969). *Rep. Rothamsted Exp. Sta.* 1968, pp. 153–155.
Hechler, H. C. (1963). *Proc. Helminthol. Soc. Wash.* **30,** 182–195.
Hechler, H. C. (1968). *Proc. Helminthol. Soc. Wash.* **35,** 24–30.
Hechler, H. C. (1970). *J. Nematol.* **2,** 125–130.
Hechler, H. C., and Taylor, D. P. (1966). *Proc. Helminthol. Soc. Wash.* **33,** 71–83.
Hertwig, P. (1920). *Arch. Mikrosk. Anat.* **94,** 303–337.
Hertwig, P. (1922). *Z. Wiss. Zool.* **119,** 539–558.
Hirschmann, H. (1951). *Zool. Jahrb. (Syst.)* **80,** 132–170.
Hirschmann, H., and Triantaphyllou, A. C. (1965). *Phytopathology* **55,** 1061. (Abstr.)
Honda, H. (1925). *J. Morphol. Physiol.* **40,** 191–233.
Hung, Chia-Ling, and Jenkins, W. R. (1969). *J. Nematol.* **1,** 352–356.
John, B. (1957). *Chromosoma* **9,** 61–68.
Johnson, R. N., and Viglierchio, D. R. (1969). *Nematologica* **15,** 144–152.
Jones, F. G. W., Parrott, D. M., and Ross, G. J. S. (1967). *Ann. Appl. Biol.* **60,** 151–171.
Kerstan, U. (1969). *Nematologica* **15,** 210–228.
Kröning, F. (1923). *Arch. Zellforch.* **17,** 63–85.
Krüger, E. (1913). *Z. Wiss. Zool.* **105,** 87–124.
Lapp, N. A., and Triantaphyllou, A. C. (1969). *J. Nematol.* **1,** 296. (Abstr.)
McClure, M. A., and Viglierchio, D. R. (1966). *Nematologica* **12,** 248–258.
Maupas, E. (1900). *Arch. Zool. Exp. Gen.* **8,** 463–624.
Maupas, E. (1919). *Bull. Biol. Fr. Belg.* **52,** 466–498.
Mulvey, R. H. (1955). *Can. J. Zool.* **33,** 295–310.
Mulvey, R. H. (1957). *Nature (London)* **180,** 1212–1213.
Mulvey, R. H. (1958a). *Can. J. Zool.* **36,** 91–93.
Mulvey, R. H. (1958b). *Can. J. Zool.* **36,** 839–841.
Mulvey, R. H. (1960). *In* "Nematology: Fundamentals and Recent Advances with Emphasis on Plant Parasitic & Soil Forms" (J. N. Sasser and W. R. Jenkins, eds.), pp. 323–330. Univ. of North Carolina Press, Chapel Hill, North Carolina.
Nakasono, K. (1966). *Appl. Entomol. Zool.* **1,** 203–205.
Netscher, C. (1969). *Nematologica* **15,** 10–14.
Nigon, V. (1949). *Ann. Sci. Natur. Zool.* **11,** 1–132.
Nigon, V. (1951). *Bull. Biol. Fr. Belg.* **85,** 187–225.

Nigon, V. (1965). *In* "Traité de Zoologie" (P. Grassé, ed.), Vol. 4, pp. 218–386. Masson, Paris.
Nigon, V., and Brun, J. (1955). *Chromosoma* **7,** 129–169.
Nigon, V., and Dougherty, E. C. (1949). *J. Exp. Zool.* **112,** 485–503.
Osche, G. (1952). *Zool. Jahrb.* (*Syst.*) **81,** 190–280.
Osche, G. (1954). *Zool. Jahrb.* (*Syst.*) **82,** 618–654.
Paramonov, A. A. (1962). *In* "Plant-Parasitic Nematodes"(K. Skrjabin, ed.), Vol. 1, 390 pp. Transl. from Russ. by Israel Program for Sci. Transl., Monson, Jerusalem, 1968.
Parrott, D. M. (1968). *Rep. Rothamsted Exp. Sta.* 1967, pp. 147–148.
Potter, J. W., and Fox, J. A. (1965). *Phytopathology* **55,** 800–801.
Riley, R., and Chapman, V. (1957). *Nature* (*London*) **180,** 662.
Roman, J., and Triantaphyllou, A. C. (1969). *J. Nematol.* **1,** 357–362.
Smart, G. C. Jr., and Darling, H. M. (1963). *Phytopathology* **53,** 374–381.
Steiner, G. (1923). *J. Hered.* **14,** 147–158.
Steiner, G. (1956). *Proc. 14th Int. Congr. Zool., Copenhagen 1953,* pp. 377–378.
Sturhan, D. (1964). *Nematologica* **10,** 328–334.
Sturhan, D. (1966). *Z. Pflanzenkr. Pflanzenschutz* **73,** 168–174.
Thistlethwayte, B. (1970). *J. Nematol.* **2,** 101–105.
Triantaphyllou, A. C. (1960). *Ann. Inst. Phytopathol. Benaki, N.S.* **3,** 12–31.
Triantaphyllou, A. C. (1962). *Nematologica* **7,** 105–113.
Triantaphyllou, A. C. (1963). *J. Morphol.* **113,** 489–499.
Triantaphyllou, A. C. (1966). *J. Morphol.* **118,** 403–413.
Triantaphyllou, A. C. (1969). *J. Nematol.* **1,** 62–71.
Triantaphyllou, A. C. (1970). *J. Nematol.* **2,** 26–32.
Triantaphyllou, A. C., and Hirschmann, H. (1962). *Nematologica* **7,** 235–241.
Triantaphyllou, A. C., and Hirschmann, H. (1964). *Annu. Rev. Phytopath.* **2,** 57–80.
Triantaphyllou, A. C., and Hirschmann, H. (1966). *Nematologica* **12,** 437–442.
Triantaphyllou, A. C., and Hirschmann, H. (1967). *Nematologica* **13,** 575–580.
Trudgill, D. L. (1967). *Nematologica* **13,** 263–272.
Trudgill, D. L., and Parrott, D. M. (1969). *Nematologica* **15,** 381–388.
Trudgill, D. L., Webster, J. M., and Parrott, D. M. (1967). *Ann. Appl. Biol.* **60,** 421–428.
Van Beneden, E. (1883). *Arch. Biol.* **4,** 265–641.
Walton, A. C. (1924). *Z. Zellen Gewebelehre* **1,** 167–239.
Walton, A. C. (1940). *In* "An Introduction to Nematology" (G. B. Chitwood and M. B. Chitwood, eds.), Sect. II, Pt. I, pp. 205–215. Monumental Printing, Co., Baltimore, Maryland.
Walton, A. C. (1959). *J. Parasitol.* **45,** 1–20.
Webster, J. M. (1967). *Ann. Appl. Biol.* **59,** 77–83.
Wu, Liang-Yu. (1960). *Can. J. Zool.* **38,** 1175–1187.
Younes, T. (1968). Thése, pp. 1–37. Faculté Sci., Univ. Lyon, Lyon, France.

*Host–Parasite Interactions*

# CHAPTER 14

# Specialized Adaptations of Nematodes to Parasitism

ALAN F. BIRD

*C.S.I.R.O., Division of Horticultural Research,*
*Glen Osmond, South Australia*

I. Introduction . . . . . . . . . . . . . . . . . 35
II. Morphological and Physiological Adaptations . . . . . . . 36
A. The Mouth Spear or Buccal Stylet . . . . . . . . 36
B. Cuticle . . . . . . . . . . . . . . . . 39
C. Muscle . . . . . . . . . . . . . . . . 39
D. Glands . . . . . . . . . . . . . . . . 41
III. Ecological Adaptations: Response to Stress . . . . . . . 47
A. Growth . . . . . . . . . . . . . . . . 47
B. Production of Males . . . . . . . . . . . . 47
IV. Summary . . . . . . . . . . . . . . . . . 48
Acknowledgments . . . . . . . . . . . . . . 48
References . . . . . . . . . . . . . . . . 48

## I. INTRODUCTION

Reviews on nematode–host interactions are usually more concerned with the responses of the host to the parasite than with the influence that the host has on the parasite. The host's influence has, however, had a marked effect on plant parasitic nematodes and has led to a number of highly specialized adaptations. Those parasites which can adapt themselves in some advantageous way to their rather specialized environment have a better chance of surviving in competition with their less fortunate neighbors. Parasitic adaptation may be morphological, physiological, or ecological, but this chapter deals primarily with morphological adaptations. Also, endoparasitic nematodes will be chiefly discussed because in these forms the transition from the external environment to that of the host is probably much greater than it is in the ectoparasitic forms leading to much more specialized adaptations to parasitism.

## II. MORPHOLOGICAL AND PHYSIOLOGICAL ADAPTATIONS

### A. The Mouth Spear or Buccal Stylet

The most obvious adaptation to parasitism in phytoparasitic nematodes is their universal possession of a buccal stylet. Occasionally this structure is absent in degenerate males, but it is always present in the larval forms of these species. Two types of stylets are found in plant parasitic nematodes, and these are not only structurally different but also originate from different localities in the anterior region of the nematode. In the Dorylaimida the anterior part of the stylet is developed from a special cell in the anterior part of the esophagus from which it moves into place during and after each molt. This type of stylet is known as an onchiostyle in contrast to the stomatostyle of the Tylenchida which is thought (Allen, 1960) to have developed, at least in part, from a collapse and fusion of the walls of the stoma or buccal cavity. Both types are used in feeding and, as such, represent an obvious adaptation to parasitism although they may have functions other than feeding. For instance, Doncaster and Shepherd (1967) have shown that the larval spear in *Heterodera rostochiensis* responds to root exudations, prior to emerging from the egg, by cutting a slit in the shell. This is done by using the tip of the spear to make a line of close perforations through the shell which merge to make the cut. The process, which has been filmed, has been shown to be remarkably precise and to be dependent on a high degree of coordination. It is initiated by host root exudates and is an interesting example of parasitic adaptation.

The dimensions of the mouth spears of plant parasitic nematodes vary even within the same species. In *Meloidogyne javanica* for instance, the buccal stylet of the preadult larva (Fig. 1A) is much slighter than in the adults of either the male (Fig. 1B) or the female (Figs. 1C and D). The stylet of members of the Tylenchida is rather like a hypodermic needle. As a result of observations with conventional and scanning electron microscopes (Bird, 1969; Ellenby and Wilson, 1969), we now know that this structure (Fig. 2) in some of the endoparasitic nematodes is about a micron in width with a lumen which is less than 0.5 $\mu$ and that the whole

Fig. 1 Stylets of living specimens of *M. javanica*. (A) Infective larva viewed under phase contrast. (B) Male viewed under normal transmitted light. (C) and (D) Head region of female also viewed under normal transmitted light. ×2500. Note the increase in size of the stylet (sp) in the adult forms, the location of the terminal duct (td) of the dorsal esophageal gland, and the well developed ampulla (a) of this gland in the endoparasitic sedentary female (from Bird, 1968b).

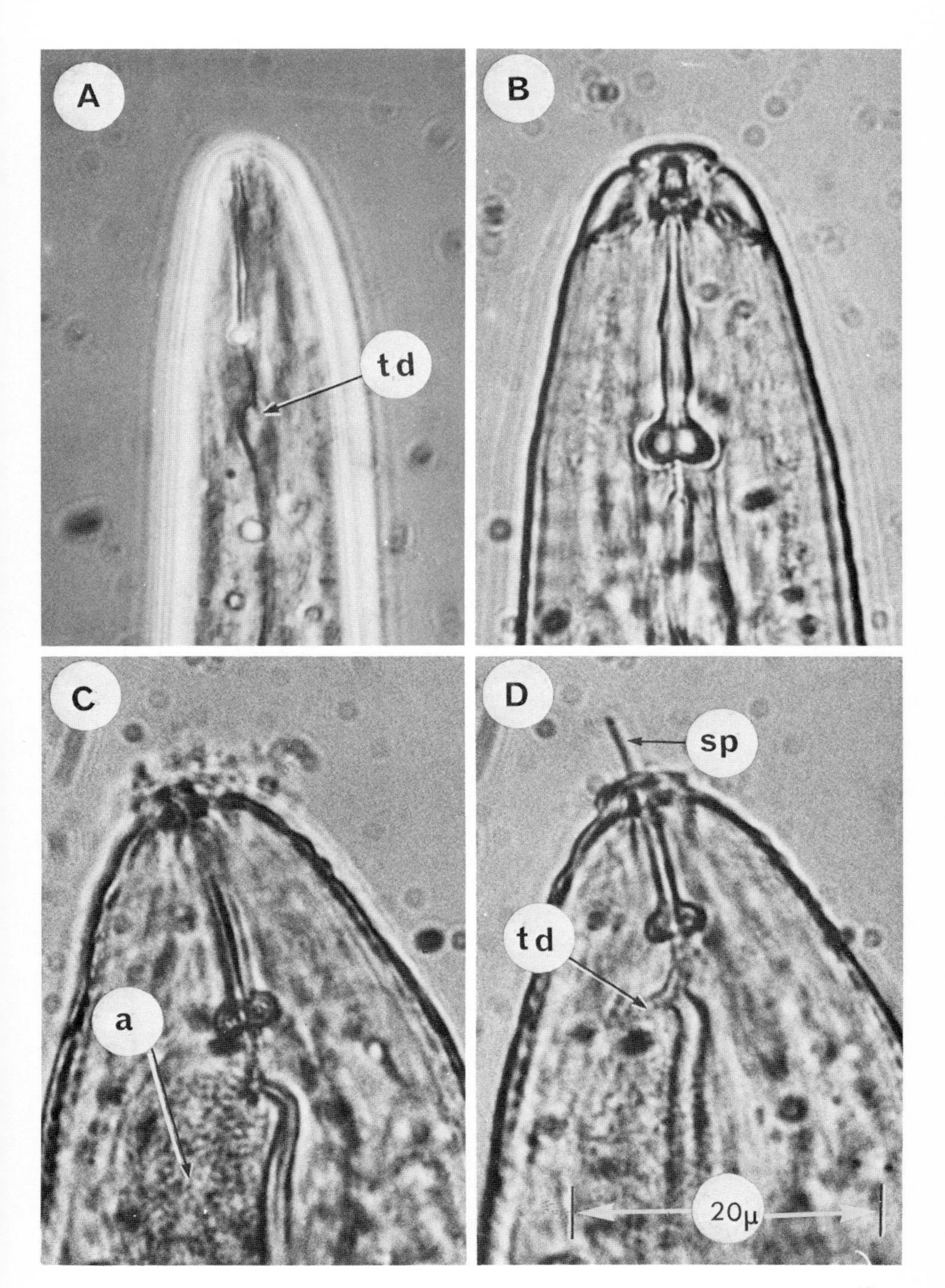
A
td
B
C
a
D
sp
td
20μ

stylet tapers to a diameter of just over 0.25 $\mu$ at its tip. Thus it acts as an efficient bacterial filter while still permitting the injection of substances which induce and maintain "giant cell" formation, as discussed elsewhere in this book, without damaging these particular cells.

This initiation and development of a giant cell or syncytial system on

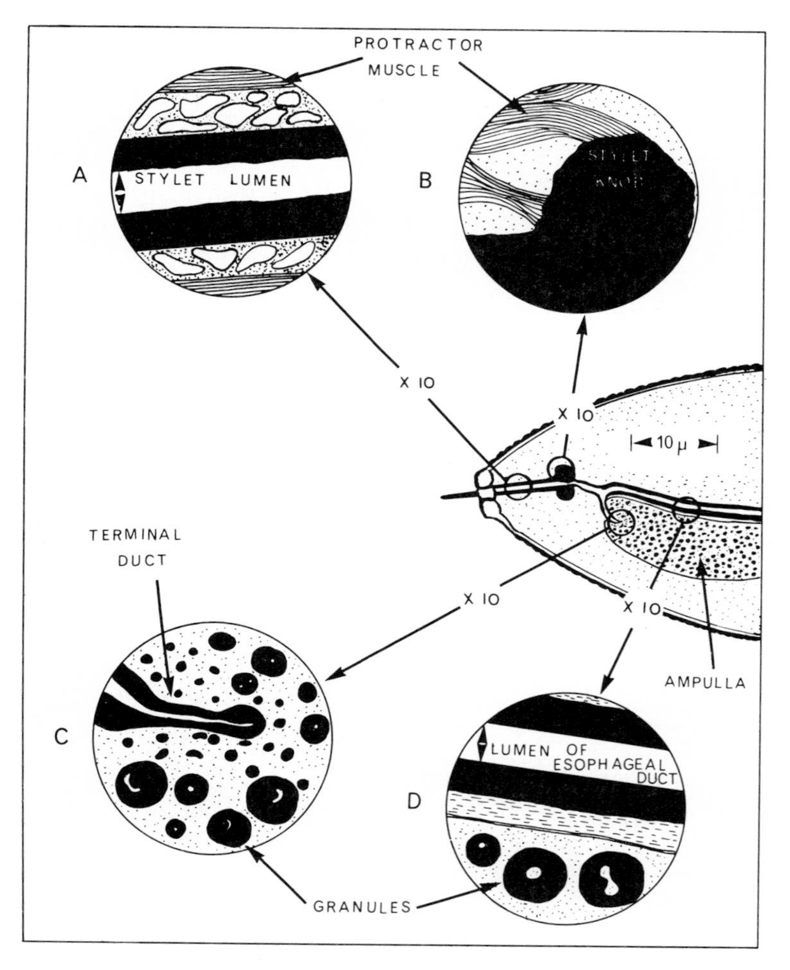


FIG. 2. Scale diagram of head region of an adult, female *M. javanica* showing spear and ampulla of dorsal esophageal gland. The insets represent scale electron micrographs of sections cut through various regions. (A) Longitudinal section through center of stylet. (B) Longitudinal section through base of stylet showing attachment of muscles to one of the knobs. (C) Section through region of terminal duct of dorsal esophageal gland showing breakdown of granules. (D) Longitudinal section through part of ampulla and center of esophageal duct showing some of the granules.

which the nematode feeds in various host plants is characteristic of the family Heteroderidae and probably represents the pinnacle of evolutionary adaptation to parasitism in phytoparasitic nematodes. It is interesting to examine the changes which take place in one of these phytoparasites as it adapts itself to its new mode of life within the plant root. These processes have been examined in *Meloidogyne javanica* (Bird, 1967, 1968a,b, 1969; Bird and Saurer, 1967). Changes occur within a few days of the nematode's entering the host plant. One of the most impressive of these, which was revealed by high resolution electron microscope observations, is the change in structure of the cuticle.

## B. Cuticle

The cuticle of the preparasitic larvae of various nematodes, including the plant parasitic species *Ditylenchus dipsaci, Meloidogyne hapla, Meloidogyne javanica, Heterodera rostochiensis,* and *Tylenchulus semipenetrans* (Yuen, 1967; Ibrahim and Hollis, 1967; Bird, 1968a; Wisse and Daems, 1968), contains regularly arranged vertical rods or striations in the basal layer. These striations have a regular periodicity, which tends to vary a little depending on methods of fixation, but which is slightly greater transversely than longitudinally. In the larval cuticle of *M. javanica* it averages about 20 nm (Fig. 3). These regular spacings, which are thought to represent a protein with close linkages between the molecules giving rise to a resistant layer that protects the larva from fluctuations in the environment, are lost within a week of the larva's entering its host's root (Figs. 3B and C). They are not found in any of the subsequent endoparasitic stages, either larval or female, although they are found in the cuticle of the male, which can, of course, move out of the root and is thus exposed to a similar environment to the preparasitic larva. Since a relatively structureless cuticle would be more adaptable to the increases in width which occur with the onset of parasitism, this change in cuticle structure is probably therefore an adaptation associated with the endoparasitic mode of life of *M. javanica.*

## C. Muscle

Another adaptation to parasitism in endoparasitic nematodes is the rapid loss of movement suggestive of somatic musculature atrophy. This can be detected by measuring the rate of movement of larvae through 1 cm of sand of particle size 150–250 $\mu$ in a tube of 0.5 cm diam at hourly intervals. This is plotted as the percentage of the total number of larvae put in the tube. Within a day of entry into the host there is a significant

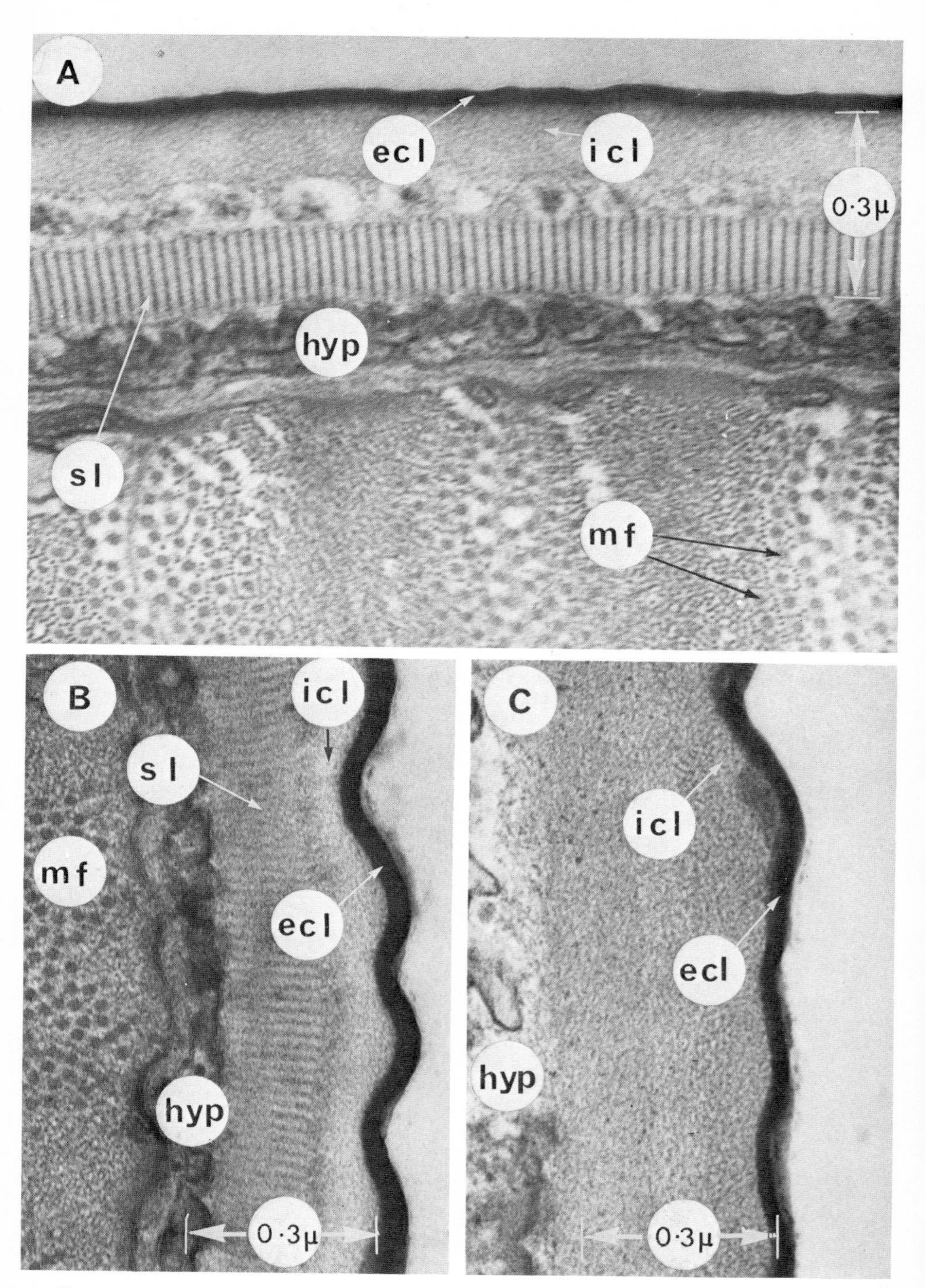
A
ecl
icl
0·3μ
hyp
sl
mf
B
icl
sl
mf
ecl
hyp
0·3μ
C
icl
ecl
hyp
0·3μ

drop in the mobility of *M. javanica* larvae, and this is even more pronounced in the case of larvae that have been in the root for 2 days (Fig. 4). Thus, while 99% of freshly hatched preparasitic larvae have moved through the sand column within 6 hr, only 12% of 1–2-day-old parasitic larvae have done this in the same length of time (Fig. 4). Atrophy of somatic musculature continues until at the time of molting the nematode is incapable of movement except in the head region where some movement is required in order to feed from the giant cells. This partial atrophy of the parasitic nematode's musculature is a most interesting example of adaptation to the endoparasitic mode of life. That part of the musculature which is necessary for the establishment of a sound host–parasite relationship is retained, while that part which is associated with movement of the whole nematode is reabsorbed. The conversion of unused muscle in these sedentary forms into a source of energy for the growing nematode would obviously be more advantageous than the maintenance of muscles which are no longer functional. All of the females of the sedentary endoparasitic forms are saccate and highly adapted for reproduction. This ability to produce large numbers of eggs is characteristic of parasitic organisms of all types. Among the sedentary endoparasitic nematodes it has resulted in the enlargement and modifications of a number of glands.

## D. Glands

### 1. Esophageal Glands

By means of a variety of techniques which includes the use of an anesthetic coupled with high resolution phase contrast microscopy to facilitate observations on living unstained preparasitic and parasitic larvae, as well as histochemical and microspectrophotometric methods (Bird, 1967, 1968a,b; Bird and Saurer, 1967), it can be shown that the esophageal glands of *M. javanica* undergo marked changes with the onset of the parasitic mode of life. It can be seen, for instance, that the ducts of the

---

Fig. 3. Electron micrographs of transverse sections through cuticles of second-stage larvae of *M. javanica* ×100,000 showing changes which take place during the onset of the parasitic mode of life. (A) Preparasitic larva showing external cortical layer (ecl), internal cortical layer (icl), striped layer (sl), hypodermis (hyp), and two types of muscle fibers (mf). (B) Two- to three-day-old parasitic larva showing external cortical layer, internal cortical layer, partly atrophied striped layer, hypodermis, and muscle fibers. (C) One-week-old parasitic larva showing external cortical layer, internal cortical layer, hypodermis, and absence of striped layer (from Bird, 1968a). Abbreviations for (B) and (C) as in (A).

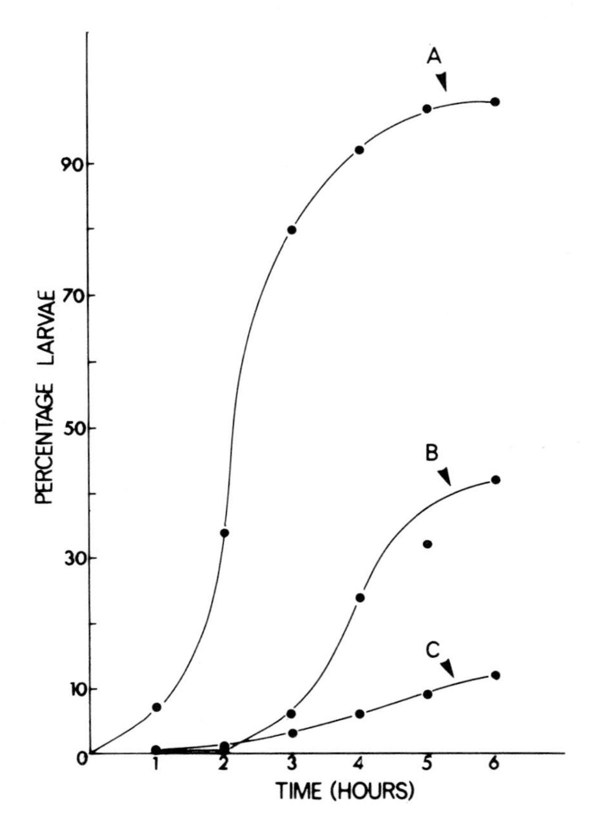


FIG. 4. Decrease in mobility which takes place as larvae become more established as parasites. This is measured by the percentage of larvae which migrate through a 1-cm column of sand over a given period of time. (A) Freshly hatched infective (preparasitic) larvae, (B) parasitic larvae 0–1 day old, and (C) parasitic larvae 1–2 days old (from Bird, 1967).

subventral esophageal glands in preparasitic larvae contain granules (Fig. 5) which are approximately 0.6–0.8 $\mu$ in diameter and are somewhat irregular in shape. These observations have been confirmed by high resolution electron microscopy. Both morphological and chemical changes take place in these glands within 2 days of penetration into the host root. Morphologically, it appears, at the level of resolution of the light microscope, that there are no longer any granules in the ducts of the subventral glands. However, when viewed under the electron microscope (Fig. 5), it can be seen that the irregular granules of the preparasitic larvae which have a distinct outer membrane, are replaced in parasitic larvae by smaller granules which have indistinct membranes and a speckled ap-

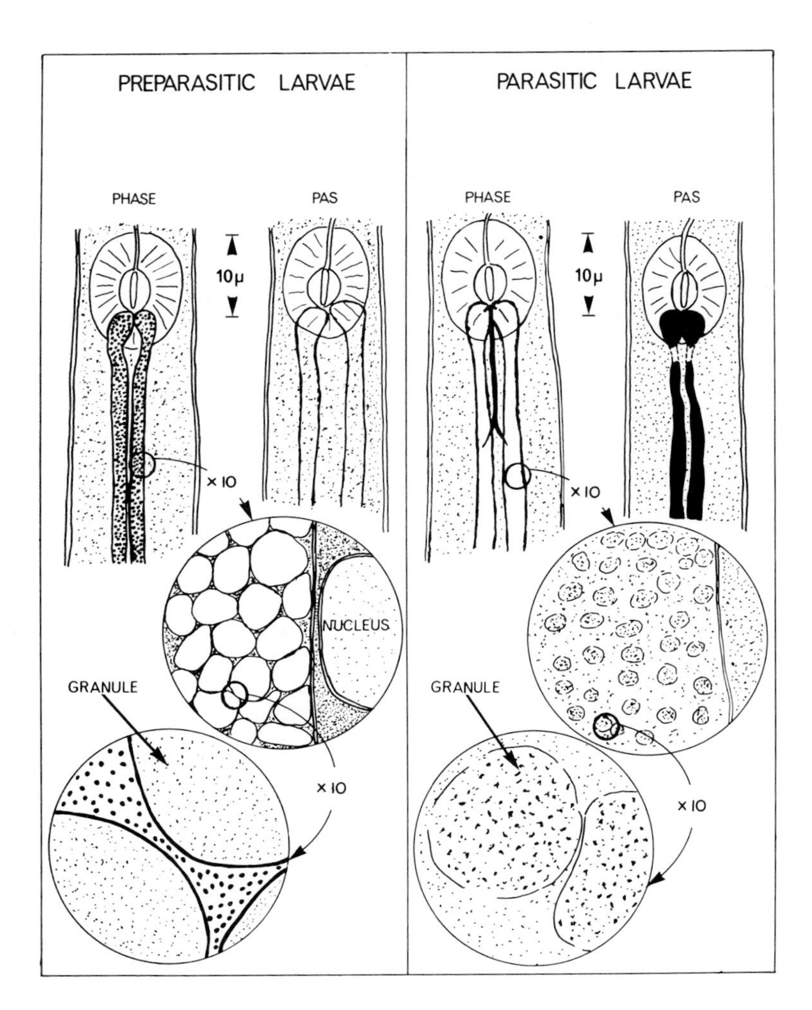


FIG. 5. Scale diagrams showing how the changes which occur in the subventral esophageal glands of *M. javanica* larvae as they become parasitic are revealed by phase contrast microscopy and histochemical staining by the periodic acid Schiff (PAS) technique. The insets represent scale electron micrographs of sections cut through the ducts of these glands and show the ultrastructural changes which occur in their granular contents.

pearance. There is also a change in the chemical composition of the contents of these ducts with the onset of the parasitic mode of life. In preparasitic larvae there is little or no staining of the ducts when stained with the periodic acid Schiff (PAS) technique, but when parasitic larvae are treated in a similar fashion (Fig. 5) there is a strong positive response, which as a result of this and other histochemical tests indicates that these ducts are now filled with mucopolysaccharide material. It is thought that this material has a role in the initiation of the giant cell system.

As these parasites grow within their host's roots the dorsal esophageal gland appears to take over the role of giant cell stimulation, and after molting this gland becomes larger and more active than the two subventrals. The role of this gland in feeding was first demonstrated by Linford (1937) in a root-knot nematode, and its role in this regard has subsequently been demonstrated in various other phytoparasites as described later in this book in the chapter on the mechanics of feeding (Doncaster, Chapter 19).

A characteristic feature of the dorsal esophageal gland contents of various phytoparasitic nematodes is the presence of granules. In *M. javanica* these are synthesized in the dorsal esophageal gland behind the median bulb in cytoplasm that is rich in rough endoplasmic reticulum and Golgi bodies (Bird, 1969). The granules move forward from this position and collect in an ampulla close to the terminal duct (Figs. 1B and C and Fig. 2).

The granules in the dorsal esophageal gland of females differ from those found in the subventral esophageal glands of infective larvae described above in that they do not appear to be bounded by a membrane, they have invaginations which give rise to irregularities in their internal structure (Fig. 2) and they are somewhat smaller having an average diameter of 0.5–0.6 $\mu$. These granules do not contain nucleic acid and consist principally of protein. They appear to break down before being extruded in the exudate, which consists of basic proteins with histochemical properties similar to the histones. It is tempting to speculate that the female *M. javanica* injects these histonelike proteins into the cytoplasm of the giant cells and that these substances are responsible for the control of nucleic acid and protein synthesis in these areas. It is certain that the nematode injects something into the giant cells which is responsible for their growth and maintenance (Bird, 1962), but it is not known for certain whether this exudation into the cytoplasm *in vivo* is similar in composition to that which these nematodes exude *in vitro*, although this seems highly likely since the esophageal glands of adult female *M. javanica* feeding on well developed giant cells are rich in basic proteins.

If care is taken, the exudate produced by these females *in vitro* can be sectioned and its structure examined under the electron microscope. It has been shown, by means of this technique (Bird, 1969) that the exudate is particulate, being made up of structures of similar dimensions to ribosomes. These are not as discrete as ribosomes, however, and form a meshwork which coalesces at the periphery of the exudate and becomes granular. For various reasons (Bird, 1969) this is not thought to be regurgitated food but to emanate from the dorsal esophageal gland, being derived from a breakdown of the granules in the ampulla of this gland which is located at the base of the stylet as described above (Fig. 2). A normal active female *M. javanica* is capable of producing up to 1000 $\mu^3$ of exudate in 2 hr. In terms of normal plant cells this is a huge volume, as indeed it is when one considers the size of the nematode producing it; but, of course, giant cells are not normal plant cells and have a relatively enormous volume. It has been calculated that the average fully developed nematode-induced giant cell in a tomato plant has a volume of about 6,237,000 $\mu^3$; since there are at least three or four of these per nematode, then this parasite, by means of its dorsal esophageal gland exudation, is controlling the metabolism of about 20 million $\mu^3$ of its host's cytoplasm.

### 2. Rectal Glands

The females of several genera of nematodes endoparasitic in plants envelop their eggs in a gelatinous matrix. This matrix is a specialized adaptation to the nematode's parasitic mode of life since it protects the large number of eggs that are produced by these endoparasitic forms from predators, and more importantly it has been shown (Wallace, 1968) that at least in *M. javanica* it acts as an efficient barrier to water loss for the eggs contained inside. The cyst wall of *Heterodera rostochiensis* has also been shown to have a similar function (Ellenby, 1968), but here it is the cuticle which has become adapted for this particular purpose rather than a secretion from glands. The adaptation of cells in endoparasitic nematodes to function as synthesizing areas for the gelatinous matrix is an example of convergent evolution toward the parasitic mode of life as this material is produced by the excretory cell in *Tylenchulus* (Maggenti, 1962), the uterine wall in *Heterodera* (Mackintosh, 1960), and the rectal glands in *Meloidogyne* (Maggenti and Allen, 1960). It has been most highly developed in *Meloidogyne,* where it is a most interesting example of specialized development.

In many nematodes the number of rectal gland cells varies from three to six, depending on the sex of the adult. In *M. javanica* females they are six in number and have developed into huge cells (Bird and Rogers, 1965) which are easily detected in sections of this nematode viewed under the light microscope. They measure approximately 100 $\mu$ in length and 40 $\mu$ in width (Fig. 6), and each has a large irregularly shaped nucleus which is about 25 $\mu$ in diameter and contains a large nucleolus which is about 10 $\mu$ in diameter. The exudation of the gelatinous matrix in *Meloidogyne* has been observed and described by Maggenti and Allen (1960). It is asso-

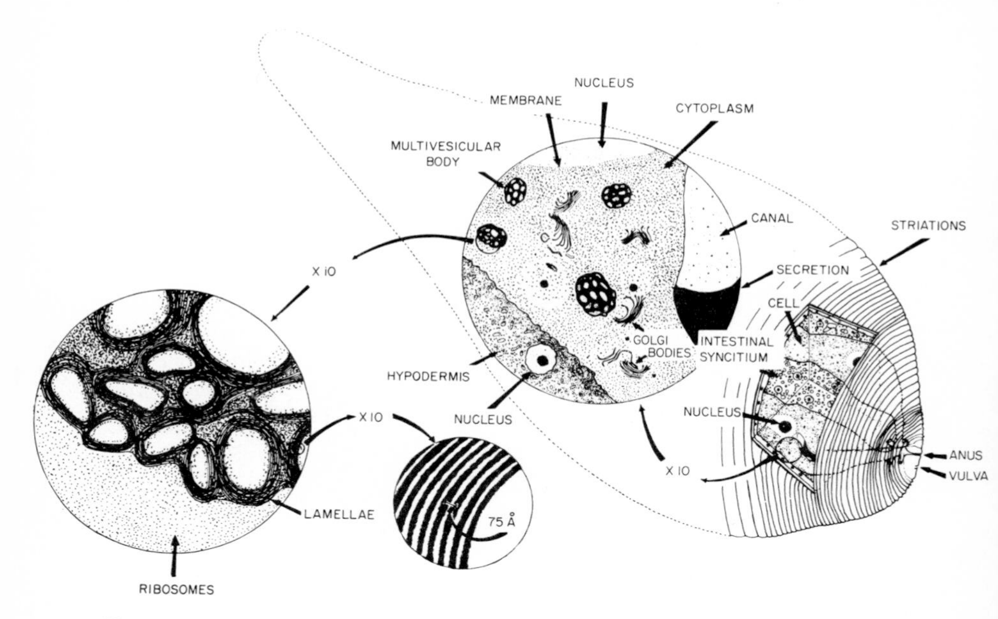


FIG. 6. Scale diagrams showing two of the six rectal gland cells in relation to the external orifices. The insets represent both low and high power electron micrographs of sections cut through these cells and show their characteristic dense cytoplasm. (From Bird and Rogers, 1965.)

ciated with a rhythmic contraction of the dorsal depressor ani muscle. In the female of *Meloidogyne* the connections between the anterior part of the alimentary tract and the anus have atrophied, and the anus functions simply as an orifice through which the gelatinous matrix is exuded. The amount of matrix exuded is copious, and these six rectal gland cells between them produce a mass of material which is frequently larger than the nematode itself. This material has been shown by histochemical tests and chromatographic techniques to contain traces of some enzymes and lipid but to consist predominantly of protein and mucopolysaccharide (Bird, 1958; Bird and Rogers, 1965). When sections of these cells are examined under the electron microscope (Fig. 6) they show structures that are always found in cells associated with active protein synthesis. These include a dense cytoplasm with an anastamosing endoplasmic reticulum in which are embedded many ribosomes and numerous Golgi bodies that are closely related to large multivesicular lamellar bodies. Both of these structures are thought to be associated with the production of mucopolysaccharide.

These multivesicular lamellar bodies are thought to act as frameworks on which enzymes of many different types may be supported. Structures similar to these have been described in some other organisms where they too are thought to play an important part in the synthesis of mucopolysaccharide.

## III. ECOLOGICAL ADAPTATIONS: RESPONSE TO STRESS

By ecological adaptations the author means those characteristics having survival value which occur in nematodes as a result of changes in their host environment. Here again *M. javanica* is a good example since its responses to some of these fluctuations have been measured.

### A. Growth

It has recently been suggested (Bird, 1970) that the rate of growth of *M. javanica* is related to the degree of stress to which it is subjected.

It has been shown that in *M. javanica,* when the nematode/plant weight ratio is high, a stress is induced which results in a decrease in the growth rate of the nematode. As this ratio of number of nematodes to plant weight decreases so does the stress until a stage is reached when the rate of growth of the nematode is at a maximum. As this ratio further decreases the rate of growth of the nematode becomes slightly but significantly less.

This growth rate response of the nematode to a particular level of stress is regarded as an adaptation of this nematode to its host environment, because this acceleration in growth of the parasite at a time when its host is declining permits it to survive when a more slowly developing nematode would perish.

### B. Production of Males

Another ecological adaptation to stress is the relationship between the numbers of males produced and the degree of stress to which the parasitic larva is subjected. This observation first described by Tyler (1933), who noticed that the percentage of males in cultures of *Meloidogyne* in tomato increased with increasing adverse conditions of nutrition, has subsequently been verified by numerous authors in this and other endoparasitic nematodes. Thus males may be induced by the presence of other pathogens, increased numbers of nematodes, nutritional deficiency of the host plant, reduced rate of photosynthesis in the host, and a number of other factors all of which induce a degree of stress on the parasite. This particular adaptation leads to an increase in the ratio of motile males to sedentary females and enables some of the population to leave the area under stress and perhaps also to the production of more fertilized eggs which may stand a better chance of surviving.

## IV. SUMMARY

Among plant parasitic nematodes those that show the most specialized adaptations to the phytoparasitic mode of life are the sedentary endoparasitic forms. However, there are some adaptations to parasitism which are of universal occurrence in all plant parasitic nematodes. The most obvious of these is the possession in all of these forms of a mouth spear or buccal stylet.

Other adaptations associated with the onset of the parasitic mode of life, which have been demonstrated largely through the use of the electron microscope, are changes in the structure of the cuticle, atrophy of somatic musculature in sedentary females, accentuated development of various glands associated with feeding and reproduction, and physiological responses of these phytoparasites to changes in the host environment which induce changes in rates of growth and sex differentiation. The specialized morphological adaptations of plant parasitic nematodes described above are all associated with either feeding, growth, or reproduction. It is the evolution of structures which facilitate these processes in various hosts that leads to the development of the most competitive and successful parasitic forms.

## ACKNOWLEDGMENTS

The author is indebted to the *Journal of Parasitology* for permission to publish Figs. 1, 3, and 4 and to *Nematologica* for permission to publish Fig. 6.

## REFERENCES

Allen, M. W. (1960). *In* "Nematology: Fundamentals and Recent Advances with Emphasis on Plant Parasitic and Soil Forms" (J. N. Sasser and W. R. Jenkins, eds.), pp. 136–139. Univ. of North Carolina Press, Chapel Hill, North Carolina.

Bird, A. F. (1958). *Nematologica* **3**, 205–212.

Bird, A. F. (1962). *Nematologica* **8**, 1–10.

Bird, A. F. (1967). *J. Parasitol.* **53,** 768–776.
Bird, A. F. (1968a). *J. Parasitol.* **54,** 475–489.
Bird, A. F. (1968b). *J. Parasitol.* **54,** 879–890.
Bird, A. F. (1969). *J. Parasitol.* **55,** 337–345.
Bird, A. F. (1970). *Nematologica* **16,** 13–21.
Bird, A. F., and Rogers, G. E. (1965). *Nematologica* **11,** 231–238.
Bird, A. F., and Saurer, W. (1967). *J. Parasitol.* **53,** 1262–1269.
Doncaster, C. C., and Shepherd, A. M. (1967). *Nematologica* **13,** 476–478.
Ellenby, C. (1968). *Proc. Roy. Soc. Ser. B* **169,** 203–213.
Ellenby, C., and Wilson, E. M. (1969). *Nematologica* **15,** 290–291.
Ibrahim, I. K. A., and Hollis, J. P. (1967). *Proc. Helminthol. Soc. Wash.* **34,** 137–139.
Linford, M. B. (1937). *Phytopathology* **27,** 824–835.
Mackintosh G. McD. (1960). *Nematologica* **5,** 158–165.
Maggenti, A. R. (1962). *Proc. Helminthol. Soc. Wash.* **29,** 139–144.
Maggenti, A. R., and Allen, M. W. (1960). *Proc. Helminthol. Soc. Wash.* **27,** 4–10.
Tyler, J. (1933). *Hilgardia* **7,** 373–388.
Wallace, H. R. (1968). *Nematologica* **14,** 231–242.
Wisse, E., and Daems, W. T. (1968). *J. Ultrastruct. Res.* **24,** 210–231.
Yuen, P. H. (1967). *Can. J. Zool.* **45,** 1019–1033.

# CHAPTER 15

# Biological Races

DIETER STURHAN

*Institute of Nematology, Münster, Westphalia, Germany*

I. Introduction . . . . . . . . . . . . . . . . 51
II. Sibling Species . . . . . . . . . . . . . . . . 53
III. Intraspecific Variation . . . . . . . . . . . . . . 54
A. Variation between Populations . . . . . . . . . 54
B. Variation within Populations . . . . . . . . . 58
IV. Types of Physiological Variation . . . . . . . . . 59
V. Causes of Variability . . . . . . . . . . . . . 61
VI. Methods of Race Identification . . . . . . . . . . 62
VII. Genetics of Physiological Characters . . . . . . . . 63
A. Genetic Basis of Variation in Pathogenicity . . . . . 63
B. Inheritance of Pathogenicity . . . . . . . . . 64
VIII. Development and Maintenance of Physiological Diversity . . 65
A. Sources of Variability . . . . . . . . . . . 65
B. Formation of Biological Races . . . . . . . . . 66
IX. Terminology . . . . . . . . . . . . . . . . 67
X. Conclusion . . . . . . . . . . . . . . . . 68
References . . . . . . . . . . . . . . . . 69

## I. INTRODUCTION

Physiological specialization is a fundamental phenomenon of life. Differences in biology, physiology, and ecology of a single species are to be expected between populations of different geographical origin and from different habitats. Such variation occurs most frequently in parasites, commonly named *biological races,* which are generally host races, showing none or only small morphological differences but definite differences in biology such as food preferences.

Since the eighties of the last century, when Ritzema Bos (1888) carried out his experimental research with *Ditylenchus dipsaci* and observed that stem eelworms from different plants had different host preferences, variation in pathogenicity has been known to occur in plant nematodes. In the past 80 years knowledge of biological races in nematodes has greatly increased, and physiological variation has been observed in a considerable number of phytoparasitic species, designated as biological or physiological races, biotypes, pathotypes, biological strains, etc.

The category *biological race*, as commonly used, includes a heterogeneous mixture of biological phenomena, such as sibling species, geographical races, host races, and distinct phenotypes within single populations, of which only host races rightly deserve this term. Morphologically similar nematode species were often confused with biological races, and there are obviously still more separate species masked under this name. To differentiate between geographical races, which have developed different physiological characteristics under the environmental conditions in a certain region, and host races is usually impossible in plant parasitic nematodes since most of the species were widely distributed by man, and it is difficult to obtain information on the natural distribution. The term *race* is also incorrectly used to designate physiological polymorphism of nematode populations.

Because there are so many unanswered questions and our knowledge of these phenomena is still very limited, biological races in the proper sense will not be discussed here, but physiological variation in phytonematodes will be reviewed generally.

In the past, the problem of host specificity was discussed by several nematologists, among them Steiner (1925), who explained the origin of races by adaptation to new plants; Goodey (1931), who postulated a "food-memory hypothesis"; and De Bruyn Ouboter (1930), who considered selection from genetically different material and mutation as the only causes of the development of biological races. Most of the early theories on race genesis were based on observations on the former *Tylenchus dipsaci*, *Heterodera schachtii*, and *Heterodera marioni*, each of which was later found to comprise several distinct species. More recent general discussions of the problem were published by Bovien (1955), Wallace (1963), Hesling (1966), and Sturhan (1966b, 1969).

While early theories on race genesis had to be very hypothetical, accumulation of observational and experimental data through the years has provided a solid basis for a better understanding of the nature of biological races, their development, and maintenance. In *D. dipsaci*, which probably represents the species with the greatest extent of physiological variation among plant parasitic nematodes, the problem has been studied most extensively.

## II. SIBLING SPECIES

It has been established that certain single species, as they were previously conceived, are in fact groups of species. Many "biological races" were shown to be distinct species. Several of these are differentiated by marked, but previously overlooked, morphological differences. Others are morphologically very similar or even indistinguishable but possess distinct specific biological characteristics and are reproductively isolated: so-called sibling species.

Most workers in earlier days (as Steiner, 1925) regarded the beet cyst nematode, the potato cyst nematode, the oat cyst nematode, etc., as biological strains within the species *Heterodera schachtii*. Later it became evident that all are independent species. *Heterodera tabacum* was first considered a strain of *H. rostochiensis* but is now generally regarded a separate species.

What formerly had been called *Heterodera marioni* was shown to be a taxonomic complex of species. By 1949, when Chitwood described several *Meloidogyne* species, the existence of biological races had been postulated by several workers.

From *Ditylenchus dipsaci*, with its great number of biological races, *D. destructor*, which shows marked differences in morphological characters as well as in food preferences and other bionomics, was first classified as a separate species. *Ditylenchus myceliophagus*, closely related to *D. destructor* and difficult to separate morphologically from it, but not pathogenic to potato as is *D. destructor*, sometimes was regarded as a physiological race of this species. Crossbreeding experiments by Smart and Darling (1963) and by Webster and Hooper (1968), however, showed both species to be reproductively isolated.

Sturhan (1969), who found $2n = 24$ chromosomes in "normal" *D. dipsaci*, observed approximately double the chromosome number in a *Ditylenchus* population from *Plantago maritima* and in the "giant race" from *Vicia faba*, which he regards as tetraploid. Both of these forms, which each show only slight morphological differences from *D. dipsaci* and similarities in general bionomics and host range, are obviously sibling species, although crosses with "normal" stem eelworms resulted in $F_1$ progenies.

Probably there are more sibling species which are confused with biological races, because they cannot be or so far have not been separated by morphological features. Some examples to support this suggestion are found in the genera *Meloidogyne* and *Heterodera*. The possibility should be considered that sibling species may be involved even in connection with the well-recognized pathotypes of *H. rostochiensis* after additional separating characters have been described, such as differences in cyst color, reproductive rates, and geographical distribution (Guile, 1967), and indications of genetic incompatibility of pathotypes exist (Parrott, 1968).

On the other hand, it has been demonstrated that several nematode forms which were given species rank are merely biological or geographical races.

## III. INTRASPECIFIC VARIATION

### A. Variation between Populations

Populations from different locations frequently show differences in behavior. Plants attacked in one region are not attacked in another. These phenomena, explained by the assumption of the presence of different races or strains, have been observed in an increasing number of plant parasitic nematodes.

*Ditylenchus dipsaci* (Kühn). More than 20 biological races, differing in their host ranges and named after their preferred hosts or those plants on which they were first detected, have been described. Many races are polyphagous, attacking plants from many different families, while others appear to have a limited host range. The literature on physiological differentiations in *D. dipsaci* is extensive; recent discussions of the general problems of races in this species are published by Hesling (1966) and Sturhan (1969). Differences in host range have also been found between different isolates of one race (cf. Hesling, 1966), indicating that the biological races are not uniform in pathogenicity. Steiner (1956) attempted to demonstrate that many of the races are distinguishable by morphological characters and are true species. His suggestions were refuted by successful crossing among a great number of biological races (Sturhan, 1964, 1966a; Eriksson, 1965; Webster, 1967). According to Metlitski (1968) species names such as *D. fragariae* and *D. phloxidis* should be rejected. Separate species seem to be involved in only a few cases (cf. Section II).

*Ditylenchus destructor* Thorne. Differences in pathogenicity among populations from different hosts have been observed by Hastings *et al.* (1952) and by Smart and Darling (1963). All four isolates in Smart and Darling's experiments interbred, as did *D. destructor* populations from different hosts used by Wu (1960).

*Ditylenchus radicicola* (Greeff). s'Jacob (1962) demonstrated differential host preferences of populations from Scandinavia and the Netherlands. A population deviating in host range was reported from Rhode Island (Chiaravalle, 1964).

*Heterodera rostochiensis* Woll. The occurrence of a race capable of reproducing on resistant *Solanum andigenum* and its *S. tuberosum* hybrids was first recorded by Van der Laan and Huijsman (1957) and Dunnett (1957). Resistance-breaking pathotypes from many European countries and South America are now known. At present at least four pathotypes are differentiated by their ability to overcome resistance in potato varieties bred from *S. andigenum, S. kurtzianum, S. vernei,* and *S. multidissectum,* usually designated as pathotypes A, B, C, D or A, AB, ABC, ABCD on the European continent and A, B, E (C) in Britain. Because definition and naming are not yet uniform, direct comparison between results from different countries is difficult. Investigations by Jones and Pawelska (1963) revealed considerable variability in pathogenicity between populations and suggest the existence of many "biotypes." Ross and Huijsman (1969), who speak of A and B "race groups," found differences in host range among 11 European B populations and consequently call each a "race." Pathotypes were successfully crossed by Webster (1965), Parrott (1968), and Dunnett and Bedi (1970).

*Heterodera avenae* Woll. Since Duggan (1958) and Andersen (1959) observed variations in host preferences between populations of the oat cyst nematode, races or pathotypes which differ mainly in their pathogenicity to certain oat and barley varieties are also known to occur in this nematode species. On this basis two races have been distinguished in Denmark, Sweden, and Ireland; three in England; and four in the Netherlands and Germany (see Wålstedt and Bingefors, 1963; Kort *et al.*, 1964; Cotten, 1967; Neubert, 1967; and others). The races from the different countries are not completely identical but show certain similarities. Certain races prefer grasses as hosts, such as the Dutch pathotype B (Kort, 1964). Andersen (1965) successfully crossed the Danish races 1 and 2.

*Heterodera schachtii* Schmidt. Variations found in host range studies among populations from different localities suggest that biological races may exist (Raski, 1952; Shepherd, 1959; Steudel, unpublished). It needs to be checked whether or not the development of beet cyst nematodes on tomato is indicative of different strains.

*Heterodera glycines* Ichinohe. Ross and Brim (1957) first found some evidence of biological races. Later differences in host range and reaction to resistant soybean varieties were observed among isolates from several states in the United States (see Ross, 1962; Smart, 1964; and others).

Studies by Epps and Golden (1967) indicate that there are two, possibly three, biological races in the United States. Miller (1965, 1967) reports differential development of 11 American isolates on six legumes and *Beta vulgaris,* which indicates a strong physiological variation in this nematode species.

*Heterodera humuli* Filipjev. In comparative studies Merny (1968) detected only small morphological differences between *H. humuli* and *H. fici.* He is inclined to consider them as "hop race" and "ficus race" of *H. humuli.*

*Meloidogyne javanica* (Treub). Physiological differences have been found between a strain from Southern Rhodesia and strains from Maryland and Georgia (Martin, 1958; Daulton and Nusbaum, 1962). Colbran (1958) reported biological races from Queensland, and Goplen *et al.* (1959) differentiated two California biotypes according to a different reaction on alfalfa varieties. Strich-Harari and Minz (1961) observed a different Japanese race which attacked strawberries. Sasser (1966), however, could not detect differences in pathogenicity among 13 *M. javanica* populations from 10 countries using 10 host differentials.

*Meloidogyne incognita* (Kofoid & White). A large number of reports on biotypes or strains and populations, differentiated by their pathogenicity to various hosts, exists for this species (see Allen, 1952; Martin, 1954; Dropkin, 1959; Riggs and Winstead, 1959; Goplen *et al.,* 1959; Winstead and Riggs, 1963; Sasser, 1966; and others). Most of these observations, which suggest that the species is comprised of many physiologically different populations, are from the United States.

*Meloidogyne arenaria* (Neal). Differences in host range and pathogenicity among different populations were observed by Colbran (1958) in Queensland and by Riggs and Winstead (1959) and Minton (1963) in the United States. In a study of the reproductive potential on certain differential hosts, Sasser (1963) identified five biotypes in populations of different geographical origin and later (Sasser, 1966) recorded different behavior between many populations from various locations.

*Meloidogyne hapla* Chitwood. Observations by Colbran (1958) suggested the existence of different races in Australia. Goplen *et al.* (1959) found populations in California differing in their ability to attack a resistant alfalfa variety, and Gillard (1961) mentioned differences in pathogenicity to barley between a population from Belgium and another of American origin. Sasser (1966), studying 20 populations from six different countries, found strawberry to be a suitable host for 15 populations of *M. hapla,* whereas five were nonpathogenic.

*Pratylenchus penetrans* (Cobb). Slootweg (1956) suggested that this species is composed of several different races. Olthof (1968) showed that at least two distinct races occur in Ontario, differentiated by their reproductive potential and pathogenic behavior on tobacco and celery.

*Pratylenchus neglectus* (Rensch). According to Loof (1960) there are some indications that biological races exist in this species.

*Pratylenchus loosi* Loof. There is evidence that populations from different localities vary in pathogenicity on tea (Anonymous, 1965).

*Radopholus similis* (Cobb). Bally and Reydon (1931) first reported the possibility of physiological strains in this species. Two biological races, which are found in Florida, are separated by their ability to parasitize citrus and banana, the "banana race" attacking banana but not citrus, the "citrus race" attacking both. There is evidence of a third race in Florida (DuCharme and Birchfield, 1956). Different races may also exist in Australia, Puerto Rico, and Cuba.

*Belonolaimus longicaudatus* Rau. At least two and possibly three physiological strains differing in their ability to attack citrus and peanuts have been discovered in Florida (Perry and Smart, 1968).

*Rotylenchulus reniformis* Linford & Oliveira. Data on host preferences suggest physiological variations between populations of this polymorphic species. There is also evidence of a difference in host range between the bisexual amphimictic *R. reniformis* and the apparently parthenogenetic *R. nicotiana,* which was synonymized with *R. reniformis* by Dasgupta *et al.* (1968).

*Tylenchulus semipenetrans* Cobb. Certain differences in infectivity on citrus and some other hosts between populations from Californian orchards are interpreted by Baines *et al.* (1969) as evidence of the existence of two "biotypes." They suggest that differing "biotypes" occur in Japan. Different host races, which do not attack grape and olive such as those in California and elsewhere, are present in Israel and Spain (Cohn, 1966; Jiménez Millán, 1966). A "grass strain," which parasitizes *Andropogon rhizomatus* but failed to infect citrus, was reported by Stokes (1969).

*Aphelenchoides fragariae* (Ritzema Bos), *A. ritzemabosi* (Schwartz). Burckhardt (unpublished) observed variation in virulence and aggressiveness between many strains from different hosts and different localities in both species, which indicates great intraspecific heterogeneity. Observations by Junges (1938) already suggested variations in host range.

*Longidorus elongatus* (de Man). Van Hoof (1967), investigating transmission of tomato black ring virus by several Dutch populations of *L. elongatus,* demonstrated that populations may differ in their ability to transmit virus. Moreover, differences in host preferences have been observed between British and German populations (Wyss, 1969).

*Trichodorus pachydermus* Seinhorst. From virus transmission experiments by Van Hoof (1968) it is obvious that certain populations of this species are specific for transmitting certain isolates of tobacco rattle virus (TRV). The TRV strain has to "suit" the *Trichodorus* population.

*Trichodorus christiei* Allen. According to Bird and Mai (1967), who studied eight geographical isolates from the United States, physiological races, differentiated by reproductive potentials on selected hosts, exist within this species.

This brief review shows that in many plant parasitic nematode species variation in pathogenicity and other physiological characters has been found, mainly between populations of different geographical origin and also among different populations of a single biological race.

## B. Variation within Populations

Besides variation between different populations (group variation) there is an increasing number of observations on variation of physiological characters within populations (individual variation).

From numerous observations on resistance-breaking "races" in *Heterodera rostochiensis* and *H. avenae* it is well known that very often more than one pathotype occurs in one location. In the case of such mixtures it is difficult to calculate the percentage of each pathotype present or even to determine exactly which pathotypes are present (see Jones and Pawelska, 1963; Kort, 1966; and others). Cultivation of certain plants and plant varieties is changing the composition of such populations.

In *H. glycines* two isolates from single cysts from an infested field differed greatly in development of cysts on Lee soybean (Miller, 1966). Experimental data by Steele (1964) suggest that selective development of physiologically different strains, which were present in the original inoculum, may have occurred by prolonged association of *H. schachtii* with tomato. The possibility of selection and/or nutritionally adapted individuals from mixed populations is indicated by Baines *et al.* (1969), who found an increase in the infectivity during hosting of *Tylenchulus semipenetrans* on *Poncirus*, a resistant host.

Riggs and Winstead (1959) reported the selection of new "B biotypes" causing severe root knot on resistant tomato plants from "A biotype" greenhouse cultures of *Meloidogyne incognita* and *M. arenaria*. B biotypes survived in mixtures of A and B biotypes on susceptible tomato plants. Populations derived from single larvae of B populations showed no variation in virulence, which suggests that the B populations are genetically stable (Winstead and Riggs, 1963). Sturhan (1965) observed variation in host range between "pure" cultures of *Ditylenchus dipsaci*, each obtained from one fertilized female of the same field population and all regarded as belonging to the same biological race.

In a few instances variation in pathogenicity has been found even within the progeny of a single female. Thus, Sturhan (1966a) obtained spontaneously a high rate of reproduction of a red clover strain of *D. dipsaci* on sugar beet but could not obtain the same result with subsequent inoculations. Furthermore, in some cases he found stem eelworm larvae, which unexpectedly did not reach maturity in host plants of the "pure" strain from which they originated despite being present and apparently feeding there for more than 3 months (Sturhan, 1968).

Several similar examples are known from the genus *Meloidogyne*. Riggs and Winstead (1959) observed differences in virulence in the offspring of several *M. incognita* cultures of biotype A developed from single larvae. Likewise, considerable variation among individuals of single egg-mass isolates was found in the same species by Allen (1952), Martin (1954), and Triantaphyllou and Sasser (1960).

Kort (1966) found strong indications from his experiments with single cysts of *Heterodera rostochiensis* that the eggs inside the same cyst are heterogenous, and Howard (1968) obtained from one cyst of the potato cyst nematode two inbreeding lines with different fecundity.

These few examples show that variation in physiological characters is to be expected within populations and even within the descendants of a single female.

## IV. TYPES OF PHYSIOLOGICAL VARIATION

Different types of physiological variation have been recognized in plant parasitic nematodes. Biological races are generally detected by variation in host preference, and thus almost all observations on physiological differentiation refer to pathogenic behavior.

In most instances this is a variation in host range, i.e., in the ability to attack different plant species and plant varieties and to multiply on the host. Even resistance-breaking biotypes or pathotypes may show additional differences in host specificity on the level of plant species [cf. Ross and Huijsman (1969) for *Heterodera rostochiensis*]. Differences may be present in response to plant attractants, although apparently this has not yet been investigated, and in respect to penetration into the plant, as suggested by observations of Baines *et al.* (1969) for *Tylenchulus semi-*

*penetrans* and of Sturhan (1968) for *Ditylenchus dipsaci*. Moreover, such variation has been found among physiological strains that some may live and apparently feed in a nonhost for a long period without undergoing further development, whereas others are able to feed, to reach maturity, or even to produce some progeny on a nonhost, as indicated by various observations on *H. rostochiensis* and by similar results of Sturhan (1968, 1969) and others on *D. dipsaci*.

Differences in reproduction rate, speed of development, and degree of infestation among races or pathotypes on the same host are not unusual, e.g., in *H. rostochiensis* (see Kort, 1966; and others), *Meloidogyne incognita* and *M. arenaria* (see Minton, 1963; Sasser, 1966; and others), *T. semipenetrans* (Baines *et al.*, 1969), *D. dipsaci* (see Sturhan, 1968; and others), and *Trichodorus christiei* (Bird and Mai, 1967).

Differences in host plant reactions—in type as well as in degree—are often observed, such as the ability of aggressive *Heterodera* and *Meloidogyne* strains to form giant cells and of host races of *D. dipsaci* to dissolve the middle lamella. In the stem eelworm differential expression of symptoms on tulip by attack of the tulip race and a race from onion was reported (Kok *et al.*, 1963) and by attack of the tulip race and the narcissus race (Windrich, 1970). Howell (1966) mentioned different symptoms in pea infested with two different stem eelworm populations. Epps and Golden (1967) differentiated two populations of *Heterodera glycines* by foliage symptoms of heavily infested plants: One population caused yellowing of the foliage of susceptible soybean varieties, while the other population, even with heavy infestations, did not. Differences in the ability to transmit virus are reported for *Longidorus* and *Trichodorus* species (Van Hoof, 1967, 1968).

Except for variation in pathogenicity there are only a few observations on differences in bionomics and other physiological characters. Evidence that biological races differ in fecundity and reproduction potential is indicated by Olthof (1968) for *Pratylenchus penetrans*, from observations by several authors on the *H. rostochiensis* pathotypes, and from the findings by Howard (1968) on inbreeding lines in this species. In hatching tests with *H. rostochiensis* cysts, Guile (1967) found differences in hatching and diapause among the pathotypes. Marked variation in the tolerance to high and low soil moisture was observed by Daulton and Nusbaum (1962) between *Meloidogyne javanica* populations from Southern Rhodesia and Georgia (United States). Palo (1962) found differences in resistance to increased humidity between biological races of *D. dipsaci*. A curious case of intraspecific variation is reported by Olthof (1968), who observed that a *P. penetrans* strain developed significantly better in the roots of tobacco in the presence of *Thielaviopsis basicola* than in its absence, whereas the fungus had no effect on the development of another strain.

## V. CAUSES OF VARIABILITY

Research on the cause of physiological variation in plant parasitic nematodes has just begun. To a limited extent only, variation in pathogenicity and other physiological characters may be due to biological, ecological, or morphological differences between biological races and strains. At present, there are only a few indications for this such as differences in reproduction potentials and differential tolerance to soil moisture and desiccation (cf. Section IV).

Variation in pathogenicity as the principal distinguishing character of biological races is primarily due to physiological or biochemical differences within the species and probably largely a matter of enzymes, which determine the ability to invade plants, to feed on them, and to reproduce. These factors need not be correlated, as indicated by some experiments with stem eelworms (Sturhan, 1968), and probably depend on different enzyme systems. It is known from observations of *D. dipsaci*, *Heterodera*, and *Meloidogyne* species that highly pathogenic and less or nonpathogenic populations may invade tissues of certain plants in equal numbers.

So far, comparative biochemical studies on enzymes and other substances discharged by nematodes or required for their development have been confined mainly to the species level, except for a few investigations on *H. rostochiensis* and *D. dipsaci*. Wilski and Giebel (1966) found great activity of β-glucosidase in pathotype A of the potato cyst nematode but low enzymic activity in pathotype B. The hypothesis is proposed that in resistant potatoes, phenolic glucosides, which are hydrolyzed by β-glucosidase secreted by the larvae of pathotype A, are present in the root cells. As a result, free polyphenols are released, causing necrosis. Since the activity of β-glucosidase secreted by larvae of pathotype B is apparently low, no such hypersensitive necrotic reaction is incited.

Howell (1966) reported that a Raleigh, North Carolina, population of *D. dipsaci* caused conspicuous gall formation on pea seedlings of the Wando variety, whereas a Waynesville, North Carolina, population produced a necrotic reaction and death of the apical meristem. Muse and Williams (1969) found that pectolytic enzyme activity differed between the two nematode populations, whereas cellulolytic activity was similar.

## VI. METHODS OF RACE IDENTIFICATION

Since biological races are mainly differentiated by varying host preferences, determination of intraspecific physiological groups is accomplished by studying host range and pathogenicity. In species with named and more or less defined biological races, pathotypes, or biotypes, a number of host differentials are generally used, as proposed by Seinhorst (1957) for distinction of 11 biological races of the stem eelworm or by Kort *et al.* (1964) for identification of four pathotypes of the oat cyst nematode.

Many attempts have been made to distinguish biological races by morphological characters. Steiner (1956), De Bruyn Ouboter (1930), and others attempted to demonstrate that separate host strains in *D. dipsaci* are distinguished by morphological characters, but these suggestions have not been substantiated.

Research by Epps and Golden (1967) indicates that geographical isolates of *H. glycines* differ in morphology and biology. Miller and Duke (1967) observed morphological variation between 11 isolates from different sites in the United States, which are partially also physiologically differentiated, and considered each of the isolates a geographical race. On the other hand, Miller (1966) reported morphological differences even between two isolates obtained from the same field, which additionally showed variation in development on resistant Lee soybean. Guile (1967) demonstrated that the *H. rostochiensis* pathotypes in Britain can be distinguished by the color changes shown by the white cysts during their maturation. In *Meloidogyne* certain morphological variation among different isolates has also been observed. Riggs and Winstead (cf. Section III, B) even found some indications of morphological differences between parental populations and their descendant strains in *M. incognita* and *M. arenaria*. Observations by Bird and Mai (1968) indicate that physiologically differing populations of *Trichodorus christiei* were partly distinguished by morphological characters.

Attempts to distinguish biological races on morphological grounds will generally fail because morphology is often strongly influenced by the host plant, interbreeding of races when occurring sympatrically is possible, and morphological characters are not usually genetically linked with physiological characters. Distinct morphological differences between populations of different origin are more indicative of geographical races than of proper biological races.

New approaches using serological, biochemical, and genetic techniques may provide better methods for race identification. Preliminary results suggest that it may prove possible to use serological techniques for identifying strains of *H. rostochiensis* and biological races of *D. dipsaci* (Mabbott, 1968; Gibbins and Grandison, 1968; Webster and Hooper, 1968). Videgård and Andreasson (1968) attempted to use electrophoretic techniques for determining potato cyst nematode pathotypes. In certain cases, especially in parthenogenetic species, cytology may be useful for identifying physiological strains, but as yet the possible relationship between chromosome complement and physiological characters has not been studied.

## VII. GENETICS OF PHYSIOLOGICAL CHARACTERS

### A. Genetic Basis of Variation in Pathogenicity

To evaluate phenotypic variability, it is fundamentally important to know whether this is due to nongenetic modifications or genetic factors, because biological races should be designated only when heredity of the characterizing features has been proven or is suggested.

There are no obvious indications concerning direct adaptation to new plants or resistant varieties. It seems justifiable to assume that all, or almost all, physiological differences in plant parasitic nematodes are genetically controlled. Careful breeding tests, inbreeding, and studies on heredity have confirmed the genetic basis of pathogenicity. In certain cases only, extrachromosomal determination might be expected, as discussed by Brun (1966) for the saprozoic *Caenorhabditis elegans* adapted to increased temperatures, or conditioning to hosts without genetic change by, e.g., inducible enzyme synthesis, as observed in some other pathogenic organisms.

Observations on several nematode species (cf. Section III, B) have demonstrated that a considerable genetic variability of natural populations is to be expected, with a great number of genotypes. The physiological behavior of populations depends on their genetic composition and the genetic factors present. In several cases, what had been regarded as increased conditioning to special plants or specialization to new hosts was nothing other than selection of the best-adapted gene combinations.

The physiological potentials of a population may be assumed to be largely latent and will appear only under certain conditions and changing selection pressure, especially when characters are determined by several genes or a number of alleles exist. Variations observed in the progeny of single females indicate widespread occurrence of heterozygosity and recombinations—if a common appearance of new mutations is not assumed.

## B. Inheritance of Pathogenicity

Heredity studies are the only way to obtain decisive information on the genetic basis and the mechanism promoting physiological variability. But such exact studies, despite their outstanding importance and especially in reference to breeding for resistance, are still extremely scanty.

In *Heterodera rostochiensis* several attempts have been made to explain the genetics of pathotypes from analyses of results obtained from pot and field experiments. Jones and Parrott (1965) and Jones (1966) advanced the hypothesis that dominant genes for resistance to *H. rostochiensis* in potato hybrids are matched by recessive genes in nematodes able to multiply on them. They assumed that males of all genetic constitutions (AA, Aa, aa) can mature in the roots of resistant plants but only recessive females (aa) can mature. Trudgill *et al.* (1967) suggested that this ability may be based on a polygenic system. Results obtained by Dunnett and Bedi (1970) with hybrid progenies from controlled matings between pathotypes A and B using single larvae showed that specificity A appeared to be dominant over specificity B since hybrids did not produce cysts on potato plants with the resistance gene $H_1$ ex *Solanum andigenum.* They postulated that the characteristics A and B are governed by a pair of alleles. Ross and Huijsman (1969) concluded from their studies that a gene-for-gene connection in the sense of Flor (1959) seems nonexistent, as suggested before.

Andersen (1965) crossed the Danish *Heterodera avenae* races 1 and 2 and explained the results obtained in the $F_1$ and $F_2$ progenies by a theory of two dominant genes in race 2, which is able to produce cysts on the barley variety Drost, whereas race 1 is not.

In *Ditylenchus dipsaci* host range studies of hybrid populations produced from crosses between biological races were conducted by Eriksson (1965), Sturhan (1966a), and Webster (1967), and the genetic basis of pathogenicity and its hereditability was demonstrated. But since genetic segregation and backcrossing were to be expected in their experiments, only indications on the mode of inheritance of different pathogenic characters could be obtained. Later, Sturhan (1969), working with defined $F_1$ generations in *D. dipsaci,* demonstrated recessive inheritance of pathogenicity to red clover and alfalfa. He found that pathogenicity to field beans is due to at least two different genetic factors, possibly alleles at the same locus, one being recessive, the other dominant. Such examples of intraspecific variation may be explained by a similar genetic situation, where different races or pathotypes show differences in reproductive potentials or expression of symptoms on the same host species or variety, despite cultivation on this host for several generations.

## VIII. DEVELOPMENT AND MAINTENANCE OF PHYSIOLOGICAL DIVERSITY

### A. Sources of Variability

The origins of genetic variability are mutations, which generally will arise spontaneously but may even be induced in nematodes by influences of the host plant (Sturhan, 1967). Cytological studies on *D. dipsaci* showed that chromosome aberrations do not seem to be rare (Sturhan, unpublished). Whether or not accessory chromosomes or the absence of single chromosomes observed in many nematode species have any influence on the expression of physiological or morphological characters is still unknown; nor is it known whether or not there is any effect of multiplication of the whole chromosome complement found within several *Meloidogyne*, *Heterodera*, and *Pratylenchus* species and in *D. dipsaci*. Some evidence of differences in host preferences between bisexual and parthenogenetic populations exists for *Rotylenchulus reniformis* (cf. Section III, A).

The most important source of genetic variation in populations of sexually reproducing species is recombination. By such formation of new genotypes a population can produce a wide phenotypical variability over many generations without genetic additions by mutation or gene flow. No two individuals of sexually reproducing species are identical, and thus a totally uniform physiological behavior in such populations cannot be expected.

Gene flow, i.e., the exchange of genetic factors between populations, introduces new characters to the gene pool of a population. This is much favored by agricultural practices and extended transport and exchange of plant material. Studies by Eriksson (1965) indicate that interbreeding of *D. dipsaci* races may result in more polyphagous hybrid populations, and Sturhan (1966a) showed that hybrid races arise in this way combining characters of both parental races. A heterosis effect could be suggested from the results obtained by Webster (1967) in studies on the host plants of hybrid populations and their parental races because in some cases hybrids of *D. dipsaci* multiplied slowly on certain hosts, whereas their parental races did not multiply; in other cases hybrids propagated rapidly, but both parental strains only slowly. In rare cases an introduction of characters may even be expected from closely related species, which may be assumed from the successful crossings between *Heterodera schachtii* and *H. glycines* by Potter and Fox (1965).

Parthenogenetic populations represent a different situation since there is no gene flow and no recombination in species with ameiotic parthenogenesis; thus, the progeny of one individual, except for the appearance of genetic mutations or chromosomal aberrations, should have the same genotype. It is difficult to understand the mechanism by which clones of physiologically different isolates developed under the selection pressure of resistant host varieties as reported by Riggs and Winstead (1959) and Triantaphyllou and Sasser (1960) for *Meloidogyne* species with mitotic parthenogenesis (cf. Section III, B). The probability of phenotypic variation is generally regarded to be low in species with obligatory ameiotic parthenogenesis as well as in polyploid species, where each new allele has to compete with several wild-type alleles. However, this might be overcome, e.g., by increased variability in chromosome number.

### B. Formation of Biological Races

Excessive genetic variation is reduced mainly by selection, which modifies the genetic composition of populations, favoring special genes and gene combinations and suppressing others. Selection of new biological races, pathotypes, or biotypes is normally achieved by the cultivation of new plant varieties or certain plant species. However, the basis for selection may not only be pathogenic abilities but also other characters, viz., resistance to desiccation, temperature tolerance, fecundity, and reproduction potentials. Thus it has been demonstrated by Kort (1966) in *Heterodera rostochiensis* that promotion of resistance-breaking pathotypes is not only a matter of partial selection by resistant host plants but also related to the reproductive capacity of the pathotype, which differs among pathotypes on the same host plant and host variety.

Superiority of heterozygotes, which is also to be expected for nematodes and has been suggested for *H. rostochiensis*, prevents the elimination of recessive factors, even under strong and continuous selection pressure. Since only a few genetic factors may be exposed to selection, the remaining physiological potential will be cryptic and will appear only under changing selection pressure. Under natural conditions—without the influence of man—the probability of the formation of new distinct biological races is low. It is, however, strongly favored by particular husbandry methods, intensive cultivation of crop plants, continuous growing of the same host, the growing of perennial crops, and weed control.

Biological races may have developed in geographical isolation and been introduced to other regions later, where they now occur sympatrically with other races (such as suggested for *H. rostochiensis*), or they may have evolved in ecological and physiological isolation on a host plant. Selection of physiologically differentiated populations in the latter case is especially promoted in endoparasitic nematodes, particularly if mating takes place only within the host (presumably, e.g., in *D. dipsaci*) or is restricted to the root surface of the host (*Heterodera* and *Meloidogyne*). Since populations within plants are relatively isolated and often small, the chances are fairly high that favored genetic factors will increase. As yet, almost all known cases of intraspecific variation in pathogenicity are confined to endoparasitic species.

## IX. TERMINOLOGY

Various terms are applied to designate physiological variation and pathogenic specificity in nematodes. They are not always clearly defined and often incorrectly used, partly because of inadequate knowledge of actual variability of nematodes. The term *biological* (or *physiological*) *race* is most commonly used to designate groups of individuals which have several important host preferences and other biological characters in common. However, high intrapopulation variability, overlapping of host ranges, interbreeding, etc., make evident the difficulty of defining, naming, and fixing the limits of races—if one would not name every population differing in some respects a special race. The main problem is that such races may occur sympatrically. In the case of race mixtures, naming races is no more justifiable, because of the great extent of genetic exchange, with formation of numerous different combinations of characters.

A special genetic constitution, considered to characterize a race, may arise polytopically, i.e., independently at different sites and at different times. In such a case it seems no longer appropriate to speak of "uniform" races because the remaining genetic background may be totally different and no reproductive connection has ever existed. Moreover, biological races may show only phenotypical uniformity. As explained before, host specificity may be due to different genetic factors, and if new mutations appear independently one cannot assume that they are completely identical in different and geographically isolated populations.

The use of the term *pathotype* for pathogenic specificity in respect to one or only a few plant species or varieties may give rise to difficulties because the consideration of more plants often reveals the existence of more differentiating characters. Pathotypes—as well as biological races—may be heterogeneous genetically and merely phenotypically similar with respect to their pathogenic abilities, which may be due to different alleles, different genes, or different gene combinations.

Recently it has been proposed to use the term *trophotype* instead of pathotype, because differences in feeding and reproduction rather than differences in "disease" should be indicated. The meaning of the Greek "τροφή," however, does not include reproduction. But since physiological "races" may feed and even mature on nonhosts (cf. Section IV), this term likewise appears not totally appropriate.

According to definition (see Johannsen, 1903) the term *biotype* should be restricted to genetically identical individuals. Since no two individuals of a sexually reproducing species can be expected to be alike, this designation could correctly only be used in species with ameiotic parthenogenesis such as several *Meloidogyne* species.

## X. CONCLUSION

Physiological differentiation, expressed as group variation between populations and likewise as individual variation within populations, is common in plant parasitic nematodes. It is probable that if more populations of additional species are examined very carefully for physiological variation, many more races will be found than are now recognized. This applies especially to geographically isolated populations and endoparasitic species. Variation is principally reported for parasitic abilities, but other physiological and biological characters are also involved. Increasing evidence indicates that every local population has a considerable amount of genetic variability according to host specificity. Only with caution may results obtained with one population be transferred to other populations.

A deeper understanding of the real nature of the so-called biological or physiological races has been gained especially by the evaluation of genetic background, but there is still a very definite need for more information concerning the physiological differences exhibited by populations. Considerable additional research is needed in this important field. A knowledge of biological races in plant parasitic nematodes is essential and basic to studies of host–parasitic relationships and to the success of plant breeding programs for disease resistance. Genetic studies of nematodes are as important as similar studies in the host plant.

The probable number of existing biological races cannot be estimated; an increase in the kinds of test plants used nearly always results in the discovery of additional races. This necessitates the development of a system of nomenclature, using a limited number of the most important cultivated plants and plant varieties. Such a system of race identification will serve practical purposes but may scarcely indicate actual relationships.

## REFERENCES


Allen, M. W. (1952). *Proc. Helminthol. Soc. Wash.* **19,** 44–51.

Andersen, S. (1959). *Nematologica* **4,** 91–98.

Andersen, S. (1965). *Nematologica* **11,** 121–124.

Anonymous (1965). *Tea Res. Inst. Ceylon, Pt. II, Rep.* 1964, pp. 67–72.

Baines, R. C., Miyakawa, T., Cameron, J. W., and Small, R. H. (1969). *J. Nematol.* **1,** 150–159.

Bally, W., and Reydon, G. A. (1931). *Arch. Koffiecult. Ned.-Ind.* **5,** 23–216.

Bird, G. W., and Mai, W. F. (1967). *Phytopathology* **57,** 1368–1371.

Bird, G. W., and Mai, W. F. (1968). *Nematologica* **13,** 617–632.

Bovien, P. (1955). *Ann. Appl. Biol.* **42,** 382–390.

Brun, J.-L. (1966). *Ann. Biol. Anim. Biochim. Biophys.* **6,** 439–466.

Chiaravalle, P. D. (1964). *Diss. Abstr.* **25,** 746–747.

Chitwood, B. G. (1949). *Proc. Helminthol. Soc. Wash.* **16,** 90–104.

Cohn, E. (1966). *Nematologica* **11,** 593–600.

Colbran, R. C. (1958). *Queensl. J. Agr. Sci.* **15,** 101–136.

Cotten, J. (1967). *Plant Pathol.* **16,** 54–59.

Dasgupta, D. R., Raski, D. J., and Sher, S. A. (1968). *Proc. Helminthol. Soc. Wash.* **35,** 169–192.

Daulton, R. A. C., and Nusbaum, C. J. (1962). *Nematologica* **8,** 157–168.

De Bruyn Ouboter, M. P. (1930). *Tijdschr. Plantenziekten* **36,** 125–228.

Dropkin, V. H. (1959). *Phytopathology* **49,** 18–23.

DuCharme, E. P., and Birchfield, W. (1956). *Phytopathology* **46,** 615–616.

Duggan, J. J. (1958). *Econ. Proc. Roy. Dublin Soc.* **4,** 103–118.

Dunnett, J. M. (1957). *Euphytica* **6,** 77–89.

Dunnett, J. M., and Bedi, A. S. (1970). *Proc. 9th Int. Nematol. Symp. Warsaw, 1967* pp. 229–230. (Abstr.)

Epps, J. M., and Golden, A. M. (1967). *Nematologica* **13,** 141. (Abstr.)

Eriksson, K. B. (1965). *Nematologica* **11,** 244–248.

Flor, H. H. (1959). *In* "Plant Pathology, Problems and Progress" (C. S. Holton, G. W. Fischer, R. W. Fulton, H. Hart, and S. E. A. McCallan, eds.), Univ. of Wisconsin Press, Madison, Wisconsin.

Gibbins, L. N., and Grandison, G. S. (1968). *Nematologica* **14,** 184–188.

Gillard, A. (1961). *Meded. Landbouwhogesch. Opzoekingssta. Staat Gent* **26,** 515–646.

Goodey, T. (1931). *Ann. Appl. Biol.* **18,** 414–419.

Goplen, B. P., Stanford, E. H., and Allen, M. W. (1959). *Phytopathology* **49,** 653–656.

Guile, C. T. (1967). *Ann. Appl. Biol.* **60,** 411–419.

Hastings. R. J., Bosher, J. E., and Newton, W. (1952). *Sci. Agr.* **32**, 304–310.
Hesling, J. J. (1966). *Rep. Glasshouse Crops Res. Inst.* 1965, pp. 132–141.
Howard, H. W. (1968). *Ann. Appl. Biol.* **62**, 485–491.
Howell, R. K. (1966). *Phytopathology* **56**, 882. (Abstr.)
Jiménez Millán, F. (1966). *Boln. Real Soc. Espan. Hist. Natur., Secc. Biol.* **64**, 57–62.
Johannsen, W. (1903). "Über Erblichkeit in Populationen und reinen Linien." Fischer, Jena.
Jones, F. G. W. (1966). *Rep. Rothamsted Exp. Sta.* 1965, pp. 301–316.
Jones, F. G. W., and Parrott, D. M. (1965). *Ann. Appl. Biol.* **56**, 27–36.
Jones, F. G. W., and Pawelska, K. (1963). *Ann. Appl. Biol.* **51**, 277–294.
Junges, W. (1938). *Z. Parasitenk.* **10**, 559–607.
Kok, M. W. S., Seinhorst, J. W., and Kaai, C. (1963). *Meded. Dir. Tuinbouw (Neth.)* **26**, 494–497.
Kort, J. (1964). *Meded. Landbouwhogesch, Opzoekingssta. Staat Gent.* **29**, 783–787.
Kort, J. (1966). *Meded. Landbouwhogesch. Opzoekingssta. Staat Gent* **31**, 601–608.
Kort, J., Dantuma, G., and Van Essen, A. (1964). *Neth. J. Plant Pathol.* **70**, 9–17.
Loof, P. A. A. (1960). *Neth. J. Plant Pathol.* **66**, 29–90.
Mabbott, T. W. (1968). *Rep. 8th Int. Symp. Nematol., Antibes, 1965* p. 46. (Abstr.)
Martin, G. C. (1958). *Nematologica* **3**, 332–349.
Martin, W. J. (1954). *Plant Dis. Rep. Suppl.* **227**, 86–88.
Merny, G. (1968). *Rep. 8th Int. Symp. Nematol., Antibes, 1965* p. 28. (Abstr.)
Metlitski, O. Z. (1968). *Parazitologiya* **2**, 528–534.
Miller, L. I. (1965). *Phytopathology* **55**, 1068. (Abstr.)
Miller, L. I. (1966). *Phytopathology* **56**, 585. (Abstr.)
Miller, L. I. (1967). *Phytopathology* **57**, 647. (Abstr.)
Miller, L. I., and Duke, P. L. (1967). *Nematologica* **13**, 156. (Abstr.)
Minton, N. A. (1963). *Phytopathology* **53**, 79–81.
Muse, B. D., and Williams, A. S. (1969). *J Nematol.* **1**, 19–20. (Abstr.)
Neubert, E. (1967). *Nachrichtenbl. Deutsch. Pflanzenschutzd. (Berlin)* **21**, 66–68.
Olthof, T. H. A. (1968). *Nematologica* **14**, 482–488.
Palo, A. V. (1962). *Nematologica* **7**, 122–132.
Parrott, D. M. (1968). *Rep. Rothamsted Exp. Sta.* 1967, pp. 147–148.
Perry, V. G., and Smart, G. C. (1968). *Rep. Fla. Agr. Exp. Sta.* 1967, p. 69.
Potter, J. W., and Fox, J. A. (1965). *Phytopathology* **55**, 800–801.
Raski, D. J. (1952). *Plant Dis. Rep.* **36**, 5–7.
Riggs, R. D., and Winstead, N. N. (1959). *Phytopathology* **49**, 716–724.
Ritzema Bos, J. (1888). *Arch. Mus. Teyler, Ser. 2* **3**, 161–348, 545–588.
Ross, H., and Huijsman, C. A. (1969). *Theor. Appl. Genetics* **39**, 113–122.
Ross, J. P. (1962). *Plant Dis. Rep.* **46**, 766–769.
Ross, J. P., and Brim, C. A. (1957). *Plant Dis. Rep.* **41**, 923–924.
Sasser, J. N. (1963). *Phytopathology* **53**, 887–888. (Abstr.)
Sasser, J. N. (1966). *Nematologica* **12**, 97–98. (Abstr.)
Seinhorst, J. W. (1957). *Nematologica* **2**, Suppl., 355–361.
Shepherd, A M. (1959). *Nature (London)* **183**, 1141–1142.
s'Jacob, J. J. (1962). *Nematologica* **7**, 231–234.
Slootweg, A. F. G. (1956). *Nematologica* **1**, 192–201.
Smart, G. C., Jr. (1964). *Plant Dis. Rep.* **48**, 542–543.
Smart, G. C., Jr., and Darling, H. M. (1963). *Phytopathology* **53**, 374–381.
Steele, A. E. (1964). *J. Amer. Soc. Sugar Beet Technol.* **13**, 170–176.

Steiner, G. (1925). *Phytopathology* **15,** 499–534.
Steiner, G. (1956). *Proc. 14th Int. Zool. Congr., Copenhagen, 1953* pp. 377–379.
Stokes, D. E. (1969). *J. Nematol.* **1,** 306. (Abstr.)
Strich-Harari, D., and Minz, G. (1961). *Isr. J. Agr. Res.* **11,** 75–77.
Sturhan, D. (1964). *Nematologica* **10,** 328–334.
Sturhan, D. (1965). *Meded. Landbouwhogesch. Opzoekingssta. Staat Gent* **30,** 1468–1474.
Sturhan, D. (1966a). *Z. Pflanzenkr. (Pflanzenpathol.) Pflanzenschutz* **73,** 168–174.
Sturhan, D. (1966b). *Mitt. Biol. Bundesanst. Berlin-Dahlem* **118,** 40–53.
Sturhan, D. (1967). *Naturwissenschaften* **54,** 649–650.
Sturhan, D. (1968). *Meded. Rijksfak. Landbouwetensch. Gent* **33,** 679–685.
Sturhan, D. (1969). *Mitt. Biol. Bundesanst. Berlin-Dahlem* **136,** 87–98.
Triantaphyllou, A. C., and Sasser, J. N. (1960). *Phytopathology* **50,** 724–735.
Trudgill, D. L., Webster, J. M., and Parrott, D. M. (1967). *Ann. Appl. Biol.* **60,** 421–428.
Van der Laan, P. A., and Huijsman, C. A. (1957). *Tijdschr. Plantenziekten* **63,** 365–368.
Van Hoof, H. A. (1967). *Nematologica* **12,** 615–618.
Van Hoof, H. A. (1968). *Nematologica* **14,** 20–24.
Videgård, G., and Andreasson, B. (1968). Abstr. of paper read at 1st Int. Congr. Plant Pathol., London.
Wallace, H. R. (1963). "The Biology of Plant Parasitic Nematodes," pp. 186–189. Arnold, London.
Wålstedt, I., and Bingefors, S. (1963). *Recent Plant Breed. Res., Svalöf 1946–1961* pp. 222–232.
Webster, J. M. (1965). *Nematologica* **11,** 299–300.
Webster, J. M. (1967). *Ann. Appl. Biol.* **59,** 77–83.
Webster, J. M., and Hooper, D. J. (1968). *Parasitology* **58,** 879–891.
Wilski, A., and Giebel, J. (1966). *Nematologica* **12,** 219–224.
Windrich, W. A. (1970). *Neth. J. Plant Pathol.* **76,** 93–98.
Winstead, N. N., and Riggs, R. D. (1963). *Plant Dis. Rep.* **47,** 870–871.
Wu, L. Y. (1960). *Can. J. Zool.* **38,** 1175–1187.
Wyss, U. (1969). Dissertation, Techn. Univ. of Hannover.

## CHAPTER 16

# Nematode Enzymes

K. H. Deubert and R. A. Rohde

*Laboratory of Experimental Biology, University of Massachusetts, East Wareham, Massachusetts,* and
*Department of Plant Pathology, University of Massachusetts, Amherst, Massachusetts*

I. Introduction . . . . . 73
II. Techniques in Enzymic Analysis . . . . . 74
A. Detection in Exudates . . . . . 75
B. Detection in Homogenates . . . . . 75
C. Histochemical Techniques . . . . . 76
III. Nematode Enzymes . . . . . 77
A. Oxidoreductases . . . . . 77
B. Transferases . . . . . 80
C. Hydrolases . . . . . 80
D. Lyases . . . . . 87
E. Ligases . . . . . 88
IV. Summary . . . . . 88
References . . . . . 89

## I. INTRODUCTION

This chapter is concerned with nematode enzymes and with some of the physiological processes catalyzed by these enzymes. Processes such as biological oxidation, respiration, and certain metabolic sequences are discussed by Rohde, Chapter 23. In many instances, our knowledge about enzymes in nematodes is still too fragmentary for detailed discussion, and errors of fact or interpretation are inevitable. This is a compilation, nevertheless, of an impressive volume of information accumulated in about one decade by a small group of workers.

The rapid growth of enzymology has led to difficulties in nomenclature. The rules for trivial nomenclature according to the Recommendations of the International Union of Biochemistry (I.U.B.) are followed herein, and where applicable, the names used in the original publication follow in parentheses. The enzyme classes are grouped according to the Recommendations of the I.U.B. However, the relatively small number of enzymes described for nematodes did not allow following this classification in all cases; therefore, the enzymes belonging to each class are discussed according to their importance in nematology.

## II. TECHNIQUES IN ENZYMIC ANALYSIS

It is beyond the scope of this chapter to discuss methods of enzymic analysis applicable to plant parasitic nematodes. In most cases it is not the choice of an analytical method which limits research efforts but the availability of sufficient quantities of nematodes if enzymic activity is to be detected in homogenates. Negative results are not proof of the absence of an enzyme.

The study of plant parasitic nematode enzymes started only recently; therefore, specific sources of error are unknown. Proper handling of reagents and nematodes is of the utmost importance, even if quantitative evaluation is not intended. Thoroughly clean glassware and sterility of nematodes and homogenates are imperative. Glassware cleaned with detergents should be carefully rinsed with hot tap water, redistilled acetone, distilled and then doubly distilled water. Despite these precautions, a check should be made to determine if detergent residues interfere with the enzyme reaction.

Several species of plant parasitic nematode can be cultured aseptically on plant tissue; therefore, it is not difficult to obtain sufficient quantities of these nematodes for analysis. In other cases surface sterilization with an antimicrobial agent followed by several washings with sterile water may be satisfactory. When studying free-living nematodes, axenic culture can provide enough nematodes for comparative studies. Buffer solutions and substrates should be kept sterile since they are ideal media for microorganisms.

Enzymes are very sensitive to analytical reagents, and the stability especially of dilute enzyme solutions varies. Traces of heavy metals or high temperatures may cause loss of activity. Nematode extracts, homogenates, or enzyme solutions should be stored, if necessary, at 2°–4°C and protected from light.

Nematode enzymes can be detected when discharged into the surrounding medium, after wet homogenization of the nematodes, or enzyme sites can be located by histochemical methods.

## A. Detection in Exudates

Plant parasitic nematodes are known to release substances into cells they feed on (see Doncaster, Chap. 19). Therefore, it would appear simple to hold nematodes in a medium and later analyze the medium for discharged enzymes. In fact, this method was applied to demonstrate the presence of hydrolytic enzymes in plant parasitic nematodes for the first time. Myuge (1957a,b) and Zinoviev (1957) detected amylase and protease activity in water in which *Ditylenchus allii, D. destructor,* and *Meloidogyne* spp. had been incubated. Goffart and Heiling (1962) used a commercial pectin preparation and oatmeal as a medium and demonstrated the activity of invertase, pectinase, and amylase after incubation of *Ditylenchus destructor, D. dipsaci, Heterodera rostochiensis,* and *H. schachtii.*

The major drawback of these methods is that very small amounts of enzyme are diluted in large volumes of medium. The use of large numbers of nematodes seems to be appropriate in this case, but a number of nematodes may die during incubation and lysis could release enzymes into the medium. Generally, it is probable that enzymic activity in the medium is below the sensitivity of the analytical method. On the other hand, there is always the question of whether or not a feeding stimulus, e.g., cell surface, is necessary for secretion. This is important in drawing conclusions about parasitism.

Sterility of nematodes and medium throughout the incubation is imperative because populations of microorganisms may build up rapidly and give reactions falsely attributed to nematode enzymes. Myuge (1957a,b) and Zinoviev (1957) make no mention of experimental conditions; therefore, one must assume that sterility was not maintained throughout their experiments. Goffart and Heiling (1962) reported microbial growth on a medium after several days of incubation. Bird (1966a) analyzed exudates obtained under sterile conditions from *Meloidogyne* larvae and found very low amylase activity as compared to the findings of Goffart and Heiling (1962).

## B. Detection in Homogenates

Homogenates have been used in most studies aimed at detecting enzymes in plant nematodes. Although raising sufficient nematodes aseptically is time consuming, at present this is the only way to obtain the amounts of tissue necessary for enzyme analysis.

Unfortunately, analytical results obtained from homogenates provide a distorted picture of what actually occurs within the living nematodes for homogenization changes the physicochemical state and the concentration of metabolites. Therefore, results of analyses are influenced by the methods used and by the properties of the enzyme tested.

It must also be considered that metabolites may be unstable during homogenization. Therefore, temperature, time, pH and ionic strength during homogenization must be compatible with the decay constants. It is apparent that one method alone cannot provide optimum conditions for the analysis of every enzyme. The heterogeneity of different nematode tissues and various developmental stages in a nematode population require a complementary method of extraction if the result of one analysis is dubious or negative.

Wet homogenates can be prepared with the aid of a glass homogenizer, ultrasonic treatment, or the French pressure cell. Electrophoretic techniques have become valuable tools in the analysis of homogenates. For the application of these methods to nematology see Benton and Myers (1966) and Gysels (1968).

### C. Histochemical Techniques

There are a number of advantages in the use of histochemical methods to detect enzymic activity. First, individual nematodes, intact or sectioned, can be stained with little or no interference from other organisms. Also, the reaction may be localized within specific structures and organs using these techniques (e.g., demonstration of acetylcholinesterase in phasmids) and many enzymes which would be so diluted in homogenates that their detection would be very unlikely, even if enough nematodes for a homogenate were available, could be detected histochemically.

Histochemical methods have been useful in the study of host–parasite interactions. Reactions can be observed in living plant tissues under "natural" conditions, and localization of reactions permits conclusions about the particular cells involved. Here again, many enzymes and substrates occur in such minute quantities that standard techniques using homogenates are not practicable. An additional advantage is that necessary cofactors and physical conditions are apt to be present.

Unfortunately, few histochemical tests are very specific and even fewer are quantitative. When they are applicable, however, they give information not obtainable by other techniques.

Many reactions are more likely to be detectable after nematodes have been killed by freezing. Nematodes are frozen at —20° to —30°C in a small volume of water, and after 10–15 minutes the ice is allowed to melt in cold incubating medium. If no reaction occurs the nematode can be ruptured mechanically to expose the organs directly to the medium.

## III. NEMATODE ENZYMES

### A. Oxidoreductases

Metabolites obtained from carbohydrates, lipids, and proteins provide the energy which is necessary for the cell to perform its functions. Biological oxidation of the metabolites is the final step in this complex process.

The oxidoreductases detected in nematodes will be discussed in connection with the pathways of intermediary metabolism in which they occur. It should be mentioned that the mere presence of the enzymes of a cycle does not necessarily mean the cycle is functional. The significance of these cycles is discussed further by Krusberg, Chapter 22.

#### 1. Embden-Meyerhof Cycle

In anaerobes the Embden-Meyerhof, or glycolytic, cycle is the primary energy source. This cycle is primarily a means of conversion of glucose and its isomers to pyruvic acid and to acetyl-CoA.

Phosphoglyceraldehyde dehydrogenase converts, in three separate steps, glyceraldehyde 3-phosphate (a triose) into 1,3-diphosphoglyceric acid, an energy-rich compound. This is the first oxidation in the glycolytic sequence. Morikawa (1965) detected phosphoglyceraldehyde dehydrogenase (triosephosphate dehydrogenase) in *Panagrellus redivivus*.

Krusberg (1960) found lactate dehydrogenase (lactic dehydrogenase) in homogenates of *Ditylenchus triformis* and *D. dipsaci*. This enzyme catalyzes pyruvic acid to lactic acid under anaerobic conditions.

#### 2. Krebs Cycle

The Krebs cycle accounts for the major part of aerobic respiration and may be considered as the main producer of adenosine triphosphate (ATP). Oxidoreductases present in this cycle have been described for nematodes.

Oxidative decarboxylation of isocitric acid to $\alpha$-ketoglutaric acid is catalyzed by isocitric dehydrogenase. However, there seems to be two different enzymes, one specific for nicotinamide-adenine dinucleotide (NAD) and the other for nicotinamide-adenine dinucleotide phosphate (NADP). The NADP-isocitric dehydrogenase is capable of converting added oxalosuccinic acid, which is considered an intermediate in the Krebs cycle, to $\alpha$-ketoglutaric acid. Krusberg (1960) detected NADP-isocitric dehydrogenase activity in *D. triformis* and *D. dipsaci.* No isocitric dehydrogenase was detected in *Turbatrix aceti* (Ells and Read, 1961).

Enzymes involved in the oxidation of $\alpha$-ketoglutaric acid to succinyl-CoA were not detectable in *D. triformis* (Krusberg, 1960).

Succinic dehydrogenase, catalyzing the conversion of succinic acid to fumaric acid (trans-isomer) was found in *Panagrellus* (Read, 1960). Malic dehydrogenase, which converts malic acid to oxalacetic acid, was detected in *D. triformis* and *D. dipsaci* (Krusberg, 1960), and *T. aceti* (Ells and Read, 1961).

### 3. Pentose Shunt

The pentose shunt, also known as the sedoheptulose monophosphate or hexose monophosphate shunt, is another pathway by which glucose can be metabolized. This pathway is unique in the fact that it is an aerobic mechanism. Enzymes which catalyze the first reactions of this pathway, the conversions of glucose-6-phosphate to 6-phosphogluconic acid and the 6-phosphogluconic acid to the 5-carbon sugar ester ribulose-5-phosphate were found in *D. triformis* and *D. dipsaci* by Krusberg (1960).

### 4. Electron Transport Chain

Pyrimidine nucleotides, flavoproteins, coenzyme Q, and the cytochromes are members of the electron transport chain. Krusberg (1960) reported NADH-cytochrome c reductase activity (DPNH cytochrome c reductase) was present in homogenates of *D. triformis* and *D. dipsaci.* The enzyme catalyzes the transfer of electrons or hydrogen from NADH to ferrocytochrome.

Diaphorase, a flavoprotein with relatively negative redox potential, was found in *D. triformis* and *D. dipsaci* (Krusberg, 1960). The enzyme is believed to mediate the transfer of hydrogen from NADH (DPNH) to various dyes, and it catalyzes the transfer of hydrogen from reduced lipoate to NAD.

Cytochrome oxidase (composed of cytochromes $a + a_3$) is the terminal member of the electron transport chain and transfers electrons to oxygen. Pennoit-de Cooman (1950) reported the presence of this enzyme in

*Panagrellus silusiae*. First evidence for the presence of cytochrome oxidase in a plant parasitic nematode (*D. triformis*) was given by Krusberg (1960). Deubert and Zuckerman (1968) detected cytochrome oxidase histochemically in *Pratylenchus scribneri, Caenorhabditis briggsae,* and *Panagrellus redivivus*. The ubiquitous nature of the enzyme makes it unlikely that there are nematodes without it. Negative results observed by Bird and Rogers (1965) from histochemical tests on *Meloidogyne javanica* may have been owing to inactivation by fixation.

There are indications that nematodes may be partially independent of cytochrome oxidase. Bryant *et al.* (1967) showed that $10^{-4}$ *M* cyanide for 1 week was not fatal to *C. briggsae*. *Panagrellus redivivus* move considerably slower in an aqueous cyanide solution of the same concentration, but they do not die (Deubert, unpublished results). After transfer from the cyanide solution to distilled water, the cyanide inhibition in *P. redivivus* seems to diminish gradually. Perhaps this is because of the high dissociation constants of the cytochrome oxidase·CN compound; thus, the enzyme may not be completely inhibited.

### 5. Miscellaneous Oxidoreductases

Formation of hydrogen peroxide results from the transfer of hydrogen to oxygen by aerobic dehydrogenases or oxidases. This compound is toxic. Catalase or peroxidase catalyze the breakdown of hydrogen peroxide, utilizing it to oxidize substances which combine with oxygen only with difficulty. Attempts to detect peroxidase activity in *M. javanica* by Bird and Rogers (1965) were not successful.

Acedo (1968) showed increased peroxidase activity in frozen sections of *Pratylenchus penetrans*–infected cabbage. It was assumed that the enzyme was of plant origin, but its production was stimulated by the nematode.

Polyphenoloxidase catalyzes the oxidation of certain phenols by molecular oxygen. This enzyme complex is also known as phenolase, cresolase, diphenol oxidase, catecholase, and tyrosinase. When polyphenoloxidase acts upon a monophenol such as tyrosine, the *ortho*-diphenol, dihydroxyphenylalanine (dopa) is formed and then oxidized further to an *o*-quinone. *o*-Quinones are generally red or orange and are spontaneously transformed into melanins.

Ellenby (1946) found evidence that polyphenoloxidase activity is involved in the tanning of *Heterodera* cysts. Bird (1958) detected polyphenoloxidase activity and phenolic compounds in the cuticle of *M. hapla* and *M. javanica*. Bird and Rogers (1965) found this enzyme in the gelatinous matrix with which females of *M. javanica* envelop their eggs. Ellenby and Smith (1967) studied the tanning process of the cyst walls of *H. rostochiensis, H. schachtii,* and *A. avenae* and found evidence that polyphenoloxidase activity is involved in this process.

Polyphenoloxidases appear to be involved in the browning reactions of plant tissues associated with a number of nematodes, but these appear to be of host origin. Rupture of plant cells releases phenols, and host polyphenoloxidases in the presence of air and melanins are eventually formed. Grasses do not form brown lesions, even when cortical cells are broken by lesion nematodes, presumably because polyphenoloxidases are not present in the host (Troll and Rohde, 1966).

Alcohol dehydrogenase and glucose dehydrogenase were not detected in *D. triformis* (Krusberg, 1960).

### B. Transferases

Transferase enzymes catalyze the transfer of an active group from one compound to another. One of the first reactions of the Embden-Meyerhof pathway, the conversion of glucose to glucose-6-phosphate, is an example of transferase enzymic action, the enzyme involved being hexokinase. Hexokinase activity was detected in *D. triformis* and *D. dipsaci* by Krusberg (1960).

Phosphoglucomutase catalyzes the conversion of glucose-1-phosphate to glucose-6-phosphate. An analogous reaction, the conversion of 3-phosphoglyceric acid to 2-phosphoglyceric acid, is brought about by an intermolecular phosphate transfer, the enzyme involved being phosphoglyceric mutase. Activity of both these enzymes was demonstrated in *D. triformis* and *D. dipsaci* also by Krusberg (1960).

Transaminases, enzymes which catalyze the intermolecular transfer of amino groups, were detected with asparatic acid and glutamic acid substrates in *Aphelenchoides ritzemabosi* by Miller and Roberts (1964). Smith and Ellenby (1967) found glutamate-oxaloacetate transaminase (L-glutamine aspartic transaminase) and glutamate-pyruvate transaminase (L-glutamine alanine transaminase) in the cyst walls of *H. rostochiensis.*

### C. Hydrolases

Linford (1937) described the outward flow of "saliva" from the stylets of two *Aphelenchoides* spp. and a *Meloidogyne* sp. followed by liquifaction of tissues. It was assumed that plant parasitic nematodes excreted enzymes to assist in extracorporeal digestion, but it was not until 1957 that identification was attempted. Myuge (1957a,b) and Zinoviev (1957) reported protease, invertase, and amylase activity in exudates of *Meloidogyne* sp., *Ditylenchus destructor*, and *D. dipsaci* incubated in water. Although these findings are difficult to evaluate because of the probable presence of microbial contamination, they mark the beginning of a series of studies on hydrolytic enzymes in plant parasitic nematodes.

Tracey (1958) demonstrated the presence of chitinase and cellulase in three species of *Ditylenchus* in the absence of microorganisms. Since chitin is produced in nematodes only in the eggshell, Tracey speculated that the presence of this enzyme in plant parasites may indicate the ability to occasionally feed on fungal mycelium or may be an ancestral capability no longer used. The bacterial feeder, *Turbatrix aceti*, did not contain chitinase. The detection of chitinase implies digestion of fungal cell walls although the presence of this enzyme in nematode homogenates does not necessarily mean that the enzyme is secreted.

Cellulases have been found in a large number of plant parasitic and fungal feeding nematodes but usually not in bacterial feeders (Table I). For the most part, observations have been made on the production of reducing sugars from water-soluble, partially degraded forms of cellulose ($C_x$ activity). Cellulose in plant cell walls is much less susceptible to enzyme attack. Dropkin's (1963) is the only study thus far to measure swelling and loss of strength in cotton fibers.

TABLE I
CELLULASES IN NEMATODES

| | | | |
|---|---|---|---|
| | $C_x$ in tissue homogenates[a] | | |
| *Aphelenchus avenae* | 1,4 | *Pratylenchus penetrans* | 6,9 |
| *Aphelenchoides sacchari* | 9 | *P. zeae* | 5 |
| *Ditylenchus destructor* | 3,10 | *Heterodera trifolii* | 6 |
| *D. dipsaci* | 4,5,6,10 | *H. schachtii* | 4 |
| *D. myceliophagus* | 4,10 | *Meloidogyne arenaria* | 4,9 |
| *D. triformis* | 4,5,7 | *M. hapla* | 9 |
| *Helicotylenchus nannus* | 4 | *Tylenchulus semipenetrans* | 4 |
| *Radopholus similis* | 9 | *Rhabditis* sp. | 9 |
| | $C_x$ in bathing medium[a] | | |
| *D. triformis* | 8 | *M. incognita* | 2 |
| | S factor in tissue homogenates[a] | | |
| *D. myceliophagus* | 4 | | |
| | No cellulase activity | | |
| *Neodiplogaster* sp. | 4 | *Panagrellus silusiae* | 4 |
| *Turbatrix aceti* | 6,10 | *P. redivivus* | 8,9 |

[a] References indicated by following numbers: 1. Barker (1966); 2. Bird (1966a); 3. Bumbu (1968); 4. Dropkin (1963); 5. Krusberg (1960); 6. Morgan and McAllan (1962); 7. Muse and Williams (1969); 8. Myers (1963); 9. Myers (1965); and 10. Tracey (1958).

Quantities and activities of cellulases in different nematode species vary considerably. *Pratylenchus penetrans* was found to contain 7 times more $C_x$ activity as *H. trifolii.* Tracey (1958) found 4.7 times more $C_x$ activity in *D. dipsaci* than in *D. myceliophagus,* and Dropkin (1963) found 28 times more $C_x$ activity per larva of *M. incognita* when compared to individuals of *D. myceliophagus.*

In none of the above studies was cellulase activity correlated with host cell wall breakdown. Active plant parasites have large amounts of $C_x$, whereas bacterial feeders generally have little although Myers (1965) found high activity in a *Rhabditis* sp. Observations of feeding and intercellular movement of nematodes indicate that mechanical thrusting of the stylet is sufficient to break open cell walls and activity of secreted enzymes is not apparent (cf. Doncaster, Chapter 19). It is generally assumed that nematodes secrete enzymes, but this has never been demonstrated in plant parasitic forms (Bird, 1968). Bird (1968) did have some evidence that $C_x$ was released by *M. incognita* larvae when stimulated by germinating seeds, but he did not consider this conclusive.

Pectic substances are important constituents of the primary cell wall and middle lamella. Breakdown of pectic substances has been shown to be an important aspect of plant diseases caused by bacteria and fungi, and a vast amount of information is now available to show that a number of different enzymes are involved in the breakdown of a complex variety of substances (Bateman and Millar, 1966). The term *pectinase,* used in experiments of only a few years ago, could mean one of perhaps six different enzymes by today's terminology.

Pectic acids are primarily long chain polymers of $\alpha$-1,4-linked D-galacturonic acid but containing other sugars within the chain or bound to it in a number of ways. Carboxyl groups may be methylated, in which case the polymers are called pectinic acids, and pectinic acids sufficiently methylated (about 75%) to form gels with sugar are called pectins. Adjacent chains are usually bridged with $Ca^{2+}$ or $Mg^{2+}$, which increase structural integrity and insolubility.

Pectins can be converted to pectinic acids and pectic acids through the action of the enzyme pectin methylesterase (PME) which removes methyl groups by hydrolysis.

Pectin and pectic acid chains may be cleaved at the glycosidic linkages either by hydrolysis (glycosidase) or a trans-eliminative mechanism (lyase). Endoenzymes break the chain at random into short fragments while exoenzymes remove uronide residues from the ends of the chains (Table II). Measurement of loss of viscosity of citrus pectin, the "standard" method for studying pectinases until recently, is not sufficient to

TABLE II
CLASSIFICATION OF PECTIC GLYCOSIDASES AND LYASES[a]

A. Hydrolytic cleavage of the 1,4-glycosidic bonds of pectic substances
1. Random mechanism of hydrolysis
   a. Pectin attacked in preference to pectic acid . . . endo-polymethylgalacturonase (endo-PMG)
   b. Pectic acid attacked in preference to pectin . . . endo-polygalacturonase (endo-PG)
2. Terminal mechanism of hydrolysis
   a. Pectin attacked in preference to pectic acid . . . exo-polymethylgalacturonase (exo-PMG)
   b. Pectic acid attacked in preference to pectin . . . exo-polygalacturonase (exo-PG)

B. Trans-eliminative cleavage of the 1,4-glycosidic bonds of pectic substances
1. Random mechanism of trans-eliminative degradation
   a. Pectin attacked in preference to pectic acid . . . endo-pectin methyl-*trans*-eliminase (endo-PMTE)
   b. Pectic acid attacked in preference to pectin (endo-PGTE)
2. Terminal mechanism of trans-eliminative attack
   a. Pectin attacked in preference to pectic acid . . . exo-pectin methyl-*trans*-eliminase (exo-PMTE)
   b. Pectic acid attacked in preference to pectin . . . exo-polygalacturonate-*trans*-eliminase (exo-PGTE)

[a] From Bateman and Millar (1966).

distinguish between an endo-PMG and an endo-PTE, and in homogenates more than one pectic enzyme is very likely to be present. Under these conditions pectinase activity was found in *D. dipsaci* (Tracey, 1958; Krusberg, 1963), *D. destructor* (Goffart and Heiling, 1962), *R. similis* (Myers, 1965), *P. penetrans* (Morgan and McAllan, 1962), *H. trifolii* (Morgan and McAllan, 1962), *M. arenaria arenaria* (Myers, 1965), and *M. hapla* (Goffart and Heiling, 1962; Myers, 1965). Negative results were reported from *P. redivivus* (Myers, 1965), *A. sacchari* (Myers, 1965), *D. myceliophagus* (Dropkin, 1963), *D. triformis* (Krusberg, 1963; Myers, 1965), and *M. incognita acrita* (Myers, 1965).

Nematodes found to contain polygalacturonase were *A. avenae* (Barker, 1966), *D. dipsaci* (Krusberg, 1964, 1967; Muse and Williams, 1969; Riedel, 1970), *D. destructor* (Bumbu, 1968), *P. zeae* (Krusberg, 1960), and *D. myceliophagus* (Dropkin, 1963).

Pectin methylesterase was found in *D. triformis* (Krusberg, 1960) and *D. myceliophagus* (Dropkin, 1963). Barker (1966) did not find PME in *A. avenae*. Pectin methylesterase occurs in an alfalfa race of *D. dipsaci* (Krusberg, 1960) but not in an onion race (Riedel, 1970). Production of PME by host tissue may be stimulated by fungus infection, and a similar mechanism could operate with nematodes.

Pectin *trans*-eliminase first found in *D. dipsaci* by Krusberg (1967) has recently been shown to be an endo-PMTE in strains of this species from alfalfa (Muse and Williams, 1969) and onion (Riedel, 1970).

As in the case of cellulases, the activity of pectolytic enzymes can vary in different species. *Radopholus similis* exhibited four times as much pectinase activity as *M. arenaria* (Myers, 1965), and *P. penetrans* had 50% more activity than *H. trifolii* (Myers, 1965).

Maceration of host tissue, presumably from dissolution of the middle lamella, has long been suggested as a mechanism of pathogenesis of *D. dipsaci* (Seinhorst, 1956). As noted above, several workers have shown that this nematode contains pectic enzymes, and recent work has attempted to correlate *D. dipsaci* injury with pectin removal from infected tissue. Krusberg (1963) was not able to demonstrate pectin removal or the accumulation of pectin breakdown products in infected alfalfa tissue and concluded that pectic enzymes were not important in pathogenesis. More recently, Riedel (1970) showed that tissue homogenates from an onion race of *D. dipsaci* would macerate onion and potato tissue. Staining these tissues with hydroxylamine ferric chloride reagent showed that pectic substances had been removed. As Riedel points out, alfalfa tissue becomes necrotic rather than macerated, and perhaps this is why Krusberg was not able to demonstrate histochemically the removal of pectic substances.

Riedel (1970) found both endo-PG and endo-PMTE in tissue extracts but concluded that only endo-PG is involved in tissue maceration. His data indicated that the type of substrate present and pH of tissue would not be favorable for endo-PMTE but would be optimal for endo-PG. As Riedel pointed out, strength of pectic materials is diminished long before pectin removal can be demonstrated histochemically.

Starch occurs as granules in plant cells, and for animals it indirectly serves as a food supply since it is easily hydrolyzed by digestive enzymes. Starch is composed of two polysaccharides, amylose and amylopectin, each of which yield maltose when hydrolyzed by two enzymes: $\alpha$-amylase and $\beta$-amylase. $\alpha$-Amylase attacks any $(1 \rightarrow 4)$ linkage of the starch molecule, whereas $\beta$-amylase removes maltose units from the nonreducing end of a chain of glucose units (amylose). No distinction was made between the two enzymes when their activity in nematodes was studied.

Amylases were present in *R. terricola, P. teres, C. dolichura,* and *A. sacchari* (Myers, 1965), *D. triformis* (Krusberg, 1960; Myers, 1965), *D. dipsaci* (Krusberg, 1960; Goffart and Heiling, 1962), *D. destructor* (Goffart and Heiling, 1962), *T. semipenetrans* (Cohn, 1965), *H. rostochiensis* and *H. schachtii* (Goffart and Heiling, 1962), *M. javanica* (Bird, 1966a), and a *Meloidogyne* sp. (Zinoviev, 1957).

Amylases were not detected in *D. lheritieri* (Myuge, 1959), *R. similis, P. penetrans, M. arenaria arenaria, M. hapla, M. incognita acrita* (Myers, 1965), and *M. incognita* (Myers, 1963).

Amylase activity appears to be more closely related to nutrition than pathogenicity since bacterial feeders and plant parasites both contain these enzymes. Starch disappears in cells adjacent to *R. similis* lesions whereas starch accumulates in syncytia induced by *Heterodera* and *Meloidogyne;* thus, there does not appear to be a correlation between enzymic activity and starch in the host.

Invertase, also known as saccharase, sucrase, or β-fructofuranosidase, catalyzes the hydrolysis of sucrose to glucose and fructose. According to Myers (1965) homogenates of *A. sacchari, M. arenaria arenaria,* and *R. similis* showed invertase activity. Small amounts were detected in *P. penetrans* and *P. redivivus* (Myers, 1965). Unfortunately, data on invertase activity in *D. destructor, D. dipsaci, H. rostochiensis, H. schachtii,* and *Meloidogyne* sp. must be considered only conditionally valid because there is no mention of aseptic conditions under which the studies were performed.

Observations by Linford and Oliveira (1937) implied the release of proteolytic enzymes by an *Aphelenchoides* sp.; however, Dieter (1956) was the first to demonstrate proteolytic enzymes in plant parasitic nematodes, specifically *H. rostochiensis, H. schachtii,* and *Meloidogyne* sp.

Myuge (1956) showed that exudates of *Meloidogyne* sp. had proteolytic properties, and Myuge (1957a,b) and Zinoviev (1957) detected proteolytic enzymes in *D. allii* and *D. destructor*. Later Myuge (1959) demonstrated enzymes with properties of cathepsin in a *Diplogaster* sp. and *Hexatylus viviparus.* Krusberg (1960) detected proteases in *D. triformis* and *D. dipsaci,* whereas Bird (1966a) did not find protease activity in *Meloidogyne* larvae. Gysels (1968) found one proteolytic enzyme in *P. silusiae* and *Caenorhabditis dolichura,* and four in *P. teres* and *R. terricola.* One of the proteases found in *P. teres* was identical with the protease detected in *P. silusiae.* Gysels (1968) concluded that the proteases he detected electrophoretically seemed to be species specific.

Miller and Jenkins (1964) demonstrated proteolytic enzymes in *A. ritzemabosi, D. dipsaci, P. redivivus,* and *T. aceti.* These authors reported trypsin and pepsin in all four species but noted that *A. ritzemabosi* and *D. dipsaci* apparently released no pepsin or trypsin into solution. They found di- and tripeptidases in *P. redivivus* and *A. ritzemabosi.* Prolidase, a dipeptidase, was present in *P. redivivus* and *A. ritzemabosi.*

Histochemical studies on the multivesicular lamellar bodies, sinus canal, and the gelatinous matrix surrounding the eggs of *M. javanica* (Bird and Rogers, 1965; Bird, 1968) did not detect leucine aminopeptidase activity.

Phenylalanine deaminase was reported in *T. agricola* and *P. penetrans* by Zuckerman *et al.* (1966).

Alkaline and acid phosphatases, also known as phosphomonoesterases, catalyze the hydrolysis of phosphoric acid esters. Benton and Myers (1966) found four acid phophatases in *D. triformis* and five in *P. redivivus*, but they found no alkaline phosphatases. In histochemical studies on a *Meloidogyne* sp., Bird and Rogers (1965) detected both alkaline and acid phosphatases. Van Gundy *et al.* (1967) detected acid phosphatase activity in *M. javanica* and *T. semipenetrans*.

Urease, or urea aminohydrolase, was not detectable in *D. triformis* (Myers and Krusberg, 1965).

β-Glucosidases hydrolyze sugars from glycosides and, to some extent, β-glycosidic linkages in a number of polysaccharides. The enzyme has been demonstrated in *Pratylenchus penetrans* (Mountain and Patrick, 1959), *Aphelenchoides fragariae* (de Maeseneer, 1964), and *H. rostochiensis* (Wilski and Giebel, 1966). Mountain and Patrick (1959) demonstrated the release of HCN following hydrolysis of the glycoside amygdalin in *P. penetrans*–infected peach roots. They also noted that some *P. penetrans* individuals left tracks in agar, possibly through enzymic cleavage of glycosidic linkages. Other glycosidases have been demonstrated using glycosides that on hydrolysis release colored products.

Tissue browning, resulting from polymerization of oxidized phenols, is a common response of plants to infection by many nematodes. Browning is particularly evident in lesions caused by migratory endoparasites such as *Pratylenchus* and *Aphelenchoides* and in responses of plants resistant to genera such as *Ditylenchus* and *Heterodera*. In many instances, phenols are presumed to be released from glycosides.

Phenol accumulation and cell browning after infection by lesion nematodes is usually localized in the endodermis and may explain why these nematodes confine their activity to the cortex (Mountain and Patrick, 1959; Rohde, 1963). Oxidized phenols extracted from browned endodermal cells are repellant to specimens of *P. penetrans* and inhibit their respiration rate (Chang and Rohde, 1969). Nemared tomato is resistant to *M. incognita* because larvae are not able to penetrate browned endodermal cells and establish themselves in vascular tissue (Pi and Rohde, 1967). Potato varieties with *Solanum andigenum* type of resistance respond in a similar manner to biotype A of *H. rostochiensis*. Biotype A larvae have considerably more β-glucosidase activity than larvae of biotype B that do not evoke this response (Wilski and Giebel, 1966). When browning is not produced through the production of this enzyme, larvae penetrate the endodermis and develop successfully.

Esterases, or carboxylic acid esterases, catalyze hydrolysis of short chain fatty acids. They are important in the catabolism of fats and in nerve impulse transport through the hydrolysis of acetylcholine. Classification of these enzymes may be based on their susceptibility to specific inhibitors. Cholinesterases are completely inhibited by $10^{-6}\,M$ eserine, whereas nonspecific esterases are not. Of the nonspecific esterases, A-esterases are not inhibited by $10^{-3}\,M$ diethyl *p*-nitrophenyl phosphate but B-esterases are inhibited by concentrations as low as $10^{-8}\,M$.

An esterase that hydrolyzed acetothiocholine was demonstrated histochemically in *T. christiei, P. penetrans, Xiphinema americanum, Helicotylenchus nannus,* and *Dorylaimus* sp. by Rohde (1960). The enzyme was designated as acetylcholinesterase because most activity was localized in the nerve ring and sensory structures and because the reaction was blocked by phosphate insecticides known to be inhibitors of this enzyme. It was, however, pointed out that the histochemical test used was not specific enough to be conclusive. Similar reactions in *M. javanica* and *M. hapla* were insensitive to $10^{-4}\,M$ eserine, indicating that the enzyme was probably not acetylcholinesterase (Bird, 1966b). Lee (1964) detected cholinesterase activity in the nerve ring and around the spear of *Actinolaimus hintoni* and *Dorylaimus keilini.* The cholinesterase of *A. hintoni* was associated with a B-esterase. It was suggested that esterases were associated with feeding of these nematodes.

Benton and Myers (1966) detected electrophoretically four esterases in *D. triformis* and eight in *P. redivivus.* Two enzymes that hydrolyzed choline esters plus a third esterase were separated from *P. redivivus* homogenates (Spurr and Chancey, 1967). The cholinesterases were inhibited by phosphate and carbamate nematicides supporting the hypothesis that these nematicides act through esterase inhibition.

### D. Lyases

Lyases remove groups from substrates nonhydrolytically and leave double bonds. They may also add groups to existing double bonds. Pectinmethyl *trans*-eliminase and polygalacturonate *trans*-eliminase are lyases and were discussed above in the section on hydrolytic pectic enzymes.

Rothstein and Mayoh (1964, 1965) discussed the presence of isocitrate lyase in *Caenorhabditis briggsae, Panagrellus redivivus, Rhabditis anomala,* and *Turbatrix aceti.* This enzyme reversibly cleaves L-isocitric acid to glyoxylic acid and succinic acid and is essential to the function of the glyoxylic acid pathway.

Aconitase, which catalyzes the reversible interconversion of citric acid to isocitric acid in the Krebs cycle, has been demonstrated in *Panagrellus* (Read, 1960) and *Turbatrix aceti* (Ells and Read, 1961).

Another Krebs cycle enzyme, fumarate hydratase (fumarase), was found by Krusberg (1960) in *D. triformis*. This enzyme catalyzes the hydration of fumarate to L-malate.

Phosphopyruvate hydratase (enolase), an enzyme of the Embden-Meyerhof pathway, catalyzes the conversion of 2-phosphoglyceric acid to 2-phosphoenolpyruvic acid. The enzyme was detected in *D. triformis* and *D. dipsaci* by Krusberg (1960).

### E. Ligases

Enzymes of this class are generally known as synthetases. They catalyze the formation of C–O, C–S, C–N, or C–C bonds. Evidence for the activity of only two enzymes of this class was found in nematodes. The presence of citrate synthetase in nematodes was shown by Krusberg (1960) and Ells and Read (1961). This enzyme is also known as citrogenase and condensing enzyme. It catalyzes the first reaction of the Krebs cycle, the condensation of acetyl-CoA with oxalacetic acid resulting in the formation of citric acid. Krusberg (1960) detected activity of this enzyme in *D. triformis, D. dipsaci,* and *Pratylenchus zeae;* and Ells and Read (1961) found it in *Turbatrix aceti*.

Malate synthetase activity was detected in *C. briggsae, P. redivivus, R. anomala,* and *T. aceti* (Rothstein and Mayoh, 1966). This enzyme catalyzes the formation of malic acid by combination of glyoxylic acid with acetyl-CoA. Malate synthetase is another enzyme of the glyoxylic acid pathway.

## IV. SUMMARY

Relatively few species of plant parasitic nematodes have been cultured in a quantity sufficient to provide homogenates for conventional enzyme analysis. Histochemical localization of enzymes in nematodes has been successful in some cases although these techniques have been mostly used to study nematode-infected plant tissues. The cuticle of intact nematodes is impervious to many reagents, whereas the small size and soft tissues of these animals makes sectioning difficult.

The metabolic enzymes that have been found differ only slightly from corresponding enzymes in other organisms. Cell wall degrading enzymes,

$\beta$-glycosidases, and other enzymes generally associated with plant pathogens are present in plant parasites in large quantity.

In only a few cases have enzymes been demonstrated in the environment of aseptic nematodes. Presumably, enzymes in secretions would be in very small amounts and hard to detect. Perhaps enzymes are secreted only in response to a stimulus from the host.

Animal parasitic nematodes secrete enzymes that influence their hosts, but similar enzymes have not been detected so far in stylet exudates of plant parasitic nematodes (Bird, 1968). An interesting possibility, suggested by Bird's work is that histonelike proteins in saliva could de-repress DNA sites and regulate enzyme production by the host.

## REFERENCES


Acedo, J. (1968). M. S. Thesis, Univ. of Massachusetts, Amherst, Massachusetts.

Barker, K. R. (1966). *Proc. Helminthol. Soc. Wash.* **33,** 134–138.

Bateman, D. F., and Millar, R. L. (1966). *Annu. Rev. Phytopathol.* **4,** 119–146.

Benton, A. W., and Myers, R. F. (1966). *Nematologica* **12,** 495–500.

Bird, A. F. (1958). *Nematologica* **3,** 205–212.

Bird, A. F. (1966a). *Nematologica* **12,** 471–482.

Bird, A. F. (1966b). *Nematologica* **12,** 359–361.

Bird, A. F. (1968). *J. Parasitol.* **54,** 879–890.

Bird, A. F., and Rogers, G. E. (1965). *Nematologica* **11,** 231–238.

Bryant, C., Nicholas, W. L., and Jantunen, R. (1967). *Nematologica* **13,** 197–209.

Bumbu, I. V. (1968). *In* "Plant Nematodes of Crops in Moldavia" (A. A. Spasski, ed.), pp. 16–37. Akad. Nauk Mold.

Chang, L. M., and Rohde, R. A. (1969). *Phytopathology* **59,** 398. (Abstr.)

Cohn, E. (1965). *Nematologica* **11,** 47–54.

de Maeseneer, J. (1964). *Nematologica* **10,** 403–408.

Deubert, K. H., and Zuckerman, B. M. (1968). *Nematologica* **14,** 453–455.

Dieter, A. (1956). *Wiss. Z. Univ. Halle, Math.-Naturwiss. Reihe* **5,** 157–184.

Dropkin, V. H. (1963). *Nematologica* **9,** 444–454.

Ellenby, C. (1946). *Nature* (*London*) **157,** 302–303.

Ellenby, C., and Smith, L. (1967). *Comp. Biochem. Physiol.* **21,** 51–57.

Ells, H. A., and Read, C. P. (1961). *Biol. Bull.* **120,** 326–336.

Goffart, H., and Heiling A. (1962). *Nematologica* **7,** 173–176.

Gysels, H. (1968). *Nematologica* **14,** 489–496.

Krusberg, L. R. (1960). *Phytopathology* **50,** 9–22.

Krusberg, L. R. (1963). *Nematologica* **9,** 341–346.

Krusberg, L. R. (1964). *Nematologica* **10,** 72. (Abstr.)

Krusberg, L. R. (1967). *Nematologica* **13,** 443–451.

Lee, D. L. (1964). *Proc. Helminthol. Soc. Wash.* **31,** 285–288.

Linford, M. B. (1937). *Proc. Helminthol. Soc. Wash.* **4,** 41–46.

Linford, M. B., and Oliveira, J. M. (1937). *Science* **85,** 295–297.

Miller, C. W., and Jenkins, W. R. (1964). *Nematologica* **10,** 480–488.

Miller, C. W., and Roberts, R. N. (1964). *Phytopathology* **54,** 1177.

Morgan, G. T., and McAllan, J. W. (1962). *Nematologica* **8**, 209–215.
Morikawa, O. (1965). *Nippon Oyo Dobutsu Konchu Gakkai-Shi* **9**, 187–190.
Mountain, W. B., and Patrick, Z. A. (1959). *Can. J. Bot.* **37**, 459–470.
Muse, B. D., and Williams, A. S. (1969). *J. Nematol.* **1**, 19–20.
Myers, R. F. (1963). *Phytopathology* **53**, 884. (Abstr.)
Myers, R. F. (1965). *Nematologica* **11**, 441–448.
Myers, R. F., and Krusberg, L. R. (1965). *Phytopathology* **55**, 429–437.
Myuge, S. G. (1956). *Zh. Obshch. Biol.* **17**, 396–399.
Myuge, S. G. (1957a). *Zool. Zh.* **36**, 620–622.
Myuge, S. G. (1957b). *Akad. Nauk SSSR* **3**, 357–359.
Myuge, S. G. (1959). *Helm. Bratislava* **1**, 43–50.
Pennoit-de Cooman, E. (1950). *Ann. Soc. Roy. Belg.* **81**, 5–13.
Pi, C.-l., and Rohde, R. A. (1967). *Phytopathology* **57**, 344. (Abstr.)
Read, C. P. (1960). *In* "Comparative Physiology of Carbohydrate Metabolism in Heterothermic Animals" (West. Soc. Natur., eds.), pp. 3–34. Univ. of Washington Press, Seattle, Washington.
Riedel, R. M. (1970). Ph.D. Thesis, Cornell Univ., Ithaca, New York.
Rohde, R. A. (1960). *Proc. Helminthol. Soc. Wash.* **27**, 121–123.
Rohde, R. A. (1963). *Phytopathology* **53**, 886–887. (Abstr.)
Rothstein, M., and Mayoh, H. (1964). *Arch. Biochem. Biophys.* **108**, 134–142.
Rothstein, M., and Mayoh, H. (1965). *Comp. Biochem. Physiol.* **16**, 361–365.
Rothstein, M., and Mayoh, H. (1966). *Comp. Biochem. Physiol.* **17**, 1181–1188.
Seinhorst, J. W. (1956). *Tijdschr. Plantenziekten* **62**, 179–188.
Smith, L., and Ellenby, C. (1967). *Nematologica* **13**, 395–405.
Spurr, H. W., and Chancey, E. L. (1967). *Phytopathology* **57**, 832. (Abstr.)
Tracey, M. V. (1958). *Nematologica* **3**, 179–183.
Troll, J., and Rohde, R. A. (1966). *Phytopathology* **56**, 995–998.
Van Gundy, S. D., Bird, A. F., and Wallace, H. R. (1967). *Phytopathology* **57**, 559–571.
Wilski, A., and Giebel, J. (1966). *Nematologica* **12**, 219–224.
Zinoviev, V. G. (1957). *Zool. Zh.* **36**, 617–620.
Zuckerman, B. M., Miller, C. W., and Deubert, K. (1966). *Nematologica* **12**, 428–430.

## CHAPTER 17

# Nematode-Induced Syncytia (Giant Cells). Host–Parasite Relationships of Heteroderidae

BURTON Y. ENDO

*Crops Research Division, Agricultural Research Service, U. S. Department of Agriculture, Beltsville, Maryland*

I. Introduction . . . . . . . . . . . . . . . . 91
II. Nematode Penetration and Migration . . . . . . . . . 92
III. Mechanism of Feeding in the Heteroderidae . . . . . . . 93
IV. Stimulation of Galls . . . . . . . . . . . . . . 94
V. Formation of Syncytia . . . . . . . . . . . . . 97
A. Nuclear and Nucleolar Condition of Syncytia . . . . . . 101
B. Syncytial Cytoplasm . . . . . . . . . . . . . 108
VI. Nature of Resistance . . . . . . . . . . . . . 111
VII. Conclusions . . . . . . . . . . . . . . . . 114
References . . . . . . . . . . . . . . . . 115

## I. INTRODUCTION*

The formation of syncytia is an integral part of the host–parasite relationships of nematodes belonging to the family Heteroderidae and, to a limited extent, the Tylenchidae. The family Heteroderidae consists of nematodes that have evolved to a large number of species. All members of the family are sessile endoparasites that induce infected plants to form specialized, local tumors—the syncytia. These are sites of high metabolic activity upon which the organism depends. The activity includes protein synthesis and enzyme function. In 1887, Treub first called attention to the enlarged multinucleate cells adjacent to the lip region of nematodes in plant roots (Treub, 1887). Since that time, considerable

* The following abbreviations are used: DNA, deoxyribonucleic acid; IAA, indoleacetic acid; RNA, ribonucleic acid.

interest and research have been devoted to the morphology, function, and mechanisms involved in the formation of syncytia and the associated hyperplastic and hypertrophied tissues.

Syncytia induced by the root-knot nematode, *Meloidogyne* spp., usually occur in the form of a cluster of multinuclear cells near the lip region of the nematode. Hyperplasia and hypertrophy often surround the region of infection and usually cause galls to be formed terminally or subterminally on the infected root. Syncytia induced by the cyst nematode, *Heterodera,* usually are elongate with ends merging with normal tissue and each syncytium is generally associated with only one larva. However, when multiple infections occur within a small area of root tissue, the syncytium may become elongate as a result of the several syncytia coalescing. Syncytia are also associated with the genus *Meloidodera* (Ruehle, 1962) of the same family grouping as the root-knot and cyst nematodes and in *Nacobbus* (Schuster *et al.,* 1965) and *Rotylenchulus* (Rebois *et al.,* 1970) of the family Tylenchidae.

Unless noted otherwise, the terms *syncytium* and *giant cell* will be used interchangeably with the understanding that both terms refer to a multinuclear unit induced in the host by the nematode. Based on the process of cell wall dissolution in giant cell formation, Crittenden (1962) introduced the term *lysigenoma* which denotes a morbid condition formed by the breakdown of adjoining cells. Reviews are available on various aspects of the host–parasitic interactions of plant parasitic nematodes (Franklin, 1951; Chitwood and Oteifa, 1952; Dropkin, 1955, 1969a; Fielding, 1959; Seinhorst, 1961; Krusberg, 1963; Flegg, 1965; Mountain, 1965; Cadman, 1963; Powell, 1963; Miller, 1965; Pitcher, 1965; Dalmasso, 1967).

## II. NEMATODE PENETRATION AND MIGRATION

Root-knot nematode larvae readily penetrate plant roots near the apical meristem (Nemec, 1910); however, other regions of the root are not immune to attack (Christie, 1936). Nemec (1910) found that root-knot nematode larvae tended to penetrate at the root-cap with subsequent penetration into the meristematic zone of the root apex. Once in the root, migration occurred both inter- and intracellularly. When larvae entered from the cortex, the lip region was often located at the edge of the vascular system with the tail region extending into the cortex. Root-knot nematode larvae usually move intercellularly through the meristem without inducing necrosis of host cells surrounding the nematode body.

Christie (1936) also reported that larvae tended to enter the apical meristem but could enter more mature tissue, depending upon the point of inoculation. They found that larvae usually entered the roots intercellularly but also noted some intracellular migration. Krusberg and Nielsen (1958) reported that root-knot nematodes penetrated sweet potato roots at the terminals as well as farther up the root. In addition to intracellular migration, they found that larvae entered and migrated through ruptured tissues caused by secondary root emergence and at abnormal openings such as cracks in the surface of enlarged roots.

The rapidity of penetration of cowpea and pineapple roots by root-knot nematode larvae was reported by Godfrey and Oliveira (1932); in less than 6 hr, single and multiple larval penetrations were readily detected. Furthermore, they found that larvae migrated within the root tissues for as long as 3 days before becoming sedentary. Mankau and Linford (1960) observed that *Heterodera trifolii* penetrated clover roots within 15 min after coming into contact with the root surface. The larvae generally entered the young piliferous zone of the roots, but penetration was also common along the length of the root up to the mature tissue. Mankau's observation of larvae at the base of lateral roots indicated that larvae could also enter the root where secondary roots emerged.

## III. MECHANISM OF FEEDING IN THE HETERODERIDAE

Two major concepts concerning the mechanism of nematode feeding have been proposed. Kostoff and Kendall (1930) proposed that the root-knot nematode secured its nutrition by injecting a secretion of its salivary glands into plant tissues. The secretion was thought to increase permeability of the affected cells, thus allowing the nematode to utilize host cell fluids that accumulated in intercellular spaces. This was a reasonable proposal considering that, at that time, there was no evidence of stylet penetration of the syncytial walls. Huang and Maggenti (1969b) also suggested that root-knot nematodes infecting *Vicia faba* might feed in the intercellular spaces between giant cells. In contrast to this mode of feeding, Linford (1937) reported that "the root-knot nematode obtains its food by penetrating cells with its slender stylet and feeding directly from their substance." He found that the esophageal bulb of the root-knot nematode pulsated while the stylet was extended into a living giant cell of *Pisum sativum*. In *in vitro* experiments, salivary extrusions were observed emanating from the extended stylet tip. The extrusion apparently arose from the dorsal gland which was seen to give rise to a flow of material. Actual observations of the salivary extrusions into live giant cells were not observable because of the dense cytoplasm of the giant cells.

The functions of the stylet and glandular secretions were studied by Bird (1967, 1968). He found that the subventral esophageal glands were primarily concerned with the activities related to emergence of the larva from the egg and with the penetration of the larva into the plant tissue. However, the feeding or parasitic stage was correlated with increased feeding activity and flow of materials apparently originating in the dorsal gland. Both the subventral and dorsal glands increased in size during the first 1–3 days after larval penetration. The subsequent disappearance of the granules and change in chemical composition led to a postulate that these structures are associated with the production of enzymes which aid both penetration of the egg shell and of the plant root (Bird and Saurer, 1967).

## IV. STIMULATION OF GALLS

One of the earliest host responses to root-knot nematode stimulation is galling of plant roots. Schuster and Sullivan (1960) reported that galls were induced in tomato roots by larvae of *Meloidogyne incognita* without actual entry of larvae into the root. They concluded that the stylet penetrated the root surface and secreted materials that stimulated host tissues to form galls. Chen and Mai (1965) also observed stylet penetration and secretion of substances from an ectoparasitic nematode, *Trichodorus christiei,* into host cells. *Trichodorus christiei* was observed to thrust its stylet into a cell and release a viscous substance which accumulated near the stylet tip. As a result, cyclosis stopped and an ingestion of a portion of the host cytoplasm by the nematode occurred. *Trichodorus christiei* caused moderate hypertrophy and a reduction in terminal growth of roots. The stunting of root terminals was accompanied by secondary root formation; the terminals of the secondary roots in turn also became stunted.

It is apparent that the root-knot nematode, once within the root, secretes a material which causes hypertrophy and to some extent hyperplasia. The comprehensive study of root-knot nematodes on tomato roots by Christie (1936) and the study of galling responses reported by the early workers Treub (1887) and Nemec (1910, 1911), have laid ground work for more recent studies on the agents of gall formation.

Owens and Specht (1964) suggested that hypertrophy of host tissues could be a response to mechanical pressure exerted by the enlarging nematode or to lateral movement of growth substances around the cavity created by the growth of the nematode. The mechanical pressure concept of Owens and Specht (1964) does not seem adequate in explaining Schuster and Sullivan's (1960) findings that gall formation can occur without the entry of the root-knot nematode. The alternate proposal by Owens and Specht (1964) suggested that the role of growth substances seems plausible when one considers the generalized response of tissues to the nematode.

Usually *Heterodera* infections do not result in galling of root tissue. However, the reports of Feldmesser (1952) and Sembdner (1963) point out that the gall reaction does occur in tomato roots infected with the potato cyst nematode, *H. rostochiensis*. Sembdner (1963) noted that hyperplastic reactions were most prevalent in tissues of the vascular system. Cellular proliferation occurred after giant cells were formed and the hyperplastic tissue was well in advance of the syncytial zone, often surrounding it and located at the leading edge of the syncytium. The newly formed cell masses often displaced the vascular tissues and resulted in an abnormal morphology of the root.

Owens and Specht (1966) examined the chemical composition of normal and root-knot galled tomato. They found that in galled tissues, carbohydrates and pectins decreased 36% and cellulose and lignin decreased 31%. Other components in the gall increased as follows: hemicelluloses 36%, organic acids 67%, free amino acids 304%, proteins 80%, nucleotides 29%, RNA 87%, DNA 70%, lipids 154%, and minerals 4%.

Myuge (1958) found that the buffering capacity of gall tissue was 1.7–3.2 times higher than that of normal root tissue. These differences were attributed to the accumulation of free amino acids resulting from protein hydrolysis by nematode secretions and to the high free phosphoric acid content.

A search for indole compounds was conducted on root extracts of *Abelmoschus esculentus* infected with the root-knot nematode, *M. javanica*. When chromatographed with known samples of IAA and indolebutyric acid, extracts of galled tissues were found to contain indole compounds which were thought to play a part in gall formation (Balasubramanian and Rangaswami, 1962).

Bird (1962) found that extracts of tomato roots infected with *M. javanica* yielded a 3-indolyl derivative. Extracts of galled tissue gave a positive reaction but surrounding plant tissue did not. The growth promoting substance in the gall material had an $R_f$ value slightly higher than that recorded for IAA. He noted that this substance did not necessarily emanate from the host tissue but could come from the nematode.

Based on previous studies of growth regulator distribution in nematode-infected tissues, Viglierchio and Yu (1968) found that the kind of auxin attributed to the presence of a particular species of nematode appeared to be superimposed on the normal auxin content of the host. Setty and Wheeler (1968) reported that the roots of tomato plants with galls caused by larvae of *Meloidogyne* spp. contained auxin concentrations similar to noninfected roots but in a larger total amount because infected roots were heavier.

Nematode secretions ejected into plant tissues may induce gall formation in a number of ways including changes in cell wall permeability and the production of toxic or growth-promoting substances in the vicinity of the nematode. Kostoff and Kendall (1930) stated that nematode secretions are ejected into the parenchyma tissue of *Nicotiana* near the vascular region causing an increase in permeability of surrounding plant tissue. This presence of accumulated nutrients is thought to account for the accelerated growth of galls. In extracted tissues of sugar beet and rape infected with *Heterodera schachtii,* Nolte and Köhler (1952) found a material which caused curling, rolling, discolorations, and folding of beet leaves. In the same study, leaf extracts of *Ditylenchus dipsaci*-infected rye plants contained a growth-promoting substance. A similar component was extracted from crushed nematodes separated from the tissues. It was concluded that nematodes secrete materials which act as growth substances or are incorporated into the growth regulator system of the plant which may influence the elongation of the root.

Ustinov (1951) observed the effect of the growth regulator $\alpha$-naphthalene acetic acid on root-knot nematode-infected plants. When the material was sprayed on infected roots, galling was intensified. Myuge (1956a) found a low level of auxins in extracted tissues of root-knot nematode-infected plants. He suggested that amino acids secreted by the nematode caused gall formation by disturbance of the meristematic region of the root. When alcohol extracts of galled tissue were mixed in gelatin and applied to root surfaces, root swelling occurred (Myuge, 1956b). This gall reaction was attributed to toxins which could arise from the interaction of the nematode and plant.

Sayre discussed the possibility of interactions of nematode secretions with plant hormones (Mountain, 1960). He postulated that in the process of feeding, the nematode secretes a proteolytic enzyme which releases IAA or tryptophan, an IAA precursor, from protein chains in the tissue. The released IAA would then contribute to the galling of tissues surrounding the nematode.

## V. FORMATION OF SYNCYTIA

The formation of giant cells or syncytia through the dissolution of cell walls and the coalescing of their contents was reported for root-knot infections of *Nicotiana* hybrids (Kostoff and Kendall, 1930). Christie (1936) gave further support to this process of syncytial formation in his study of root-knot infections of tomato. He observed that undifferentiated cells near the head of the nematode enlarged slightly and, following a slight swelling of nuclei, the cell walls adjacent to the initially stimulated cell disintegrated. There was progressive dissolution of cell walls followed by the coalescence of protoplasm during the expansion of the giant cell. Similar evidence of cell wall dissolution has been reported for a number of hosts infected by species of *Meloidogyne* (Krusberg and Nielsen, 1958; Dropkin and Nelson, 1960; Owens and Specht, 1964; Littrell, 1966). In contrast to the above reports, Huang and Maggenti (1969b) found no evidence of cell wall dissolution or breakdown in the formation of syncytia in root-knot infections of *Vicia faba* when syncytial boundaries were observed through the electron microscope. They concluded that cell wall breakdown played no part in giant cell formation and the multinucleate condition arose primarily from mitosis without cytokinesis.

As a syncytium increases in size and its walls become thicker (Krusberg and Nielson, 1958; Roman, 1961), knoblike projections characterized as nonpectic, isotropic, and fast green positive have been found extending from thickened giant cell walls (Krusberg and Nielsen, 1958). Similar wall materials were observed in root-knot infection of soybeans by Dropkin and Nelson (1960). The cell wall thickenings tested positive for cellulose and pectin but negative for lignin, suberin, starch, and amino acids (ninhydrin). Bird (1961) found that in giant cells of tomato roots, all normal soluble polysaccharides of the cell wall were present as were pectic substances, hemicellulose, insoluble polysaccharides, and cellulose.

Further insight into the structure of the syncytial wall was provided by an electron microscope study of root-knot infections (Huang and Maggenti, 1969b). Among the interesting features described were the boundary formations which apparently arose from the primary wall and formed into plasmalemma invaginations (Fig. 4). Numerous tubules and vesicles formed between the primary wall and the plasmalemma before the secondary walls formed (Figs. 1–6). The secondary wall materials were at first rodlike, later became branched and eventually formed

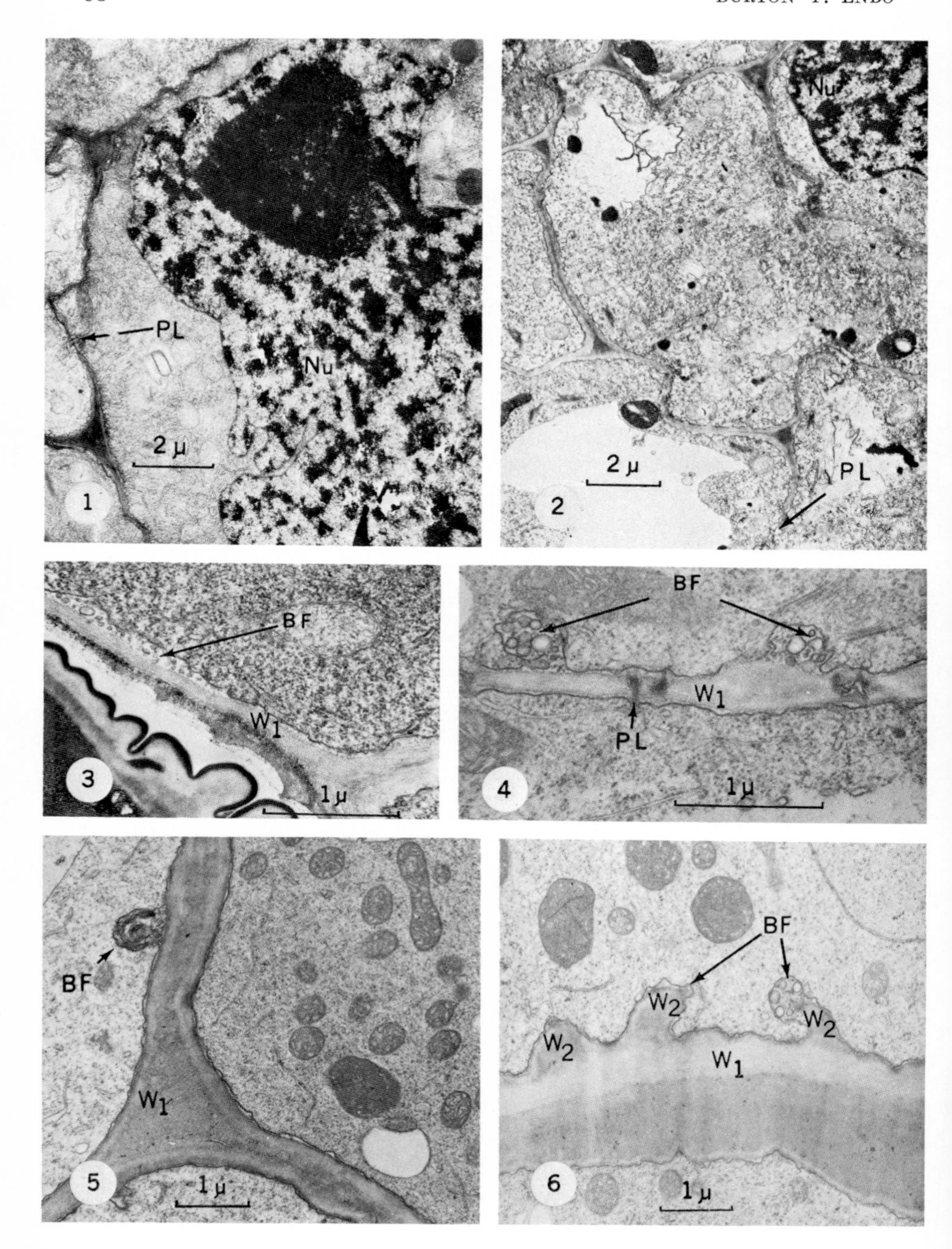
PL
Nu
2 μ
1
Nu
2 μ
PL
2
BF
W1
1μ
3
BF
W1
PL
1μ
4
BF
W1
1μ
5
BF
W2
W2
W2
W1
1μ
6

anastomosing trabeculae. During this period, the plasmalemma remained intact (Figs. 7–10).

In *Heterodera* infections, cell wall dissolution in the process of syncytial formation has been reported on a large number of plant species (Nemec, 1932; Mankau and Linford, 1960; Sembdner, 1963; Endo, 1964, 1965). Nemec (1932) described syncytial formation in sugar beet roots infected with *H. schachtii.* He found nuclei associated with openings in the cell wall and postulated that nuclei may be involved in cell wall dissolution.

Syncytial walls of *Heterodera* infections were especially thickened near the lip region of the nematode and along the syncytial walls adjacent to the xylem (Triffitt, 1931; Nemec, 1932; Bergman, 1958; Mankau and Linford, 1960; Sembdner, 1963; Endo, 1964). Cell wall thickening was minimal where syncytia appear to merge with normal tissues. In an electron microscopic study of the early stages of *H. glycines* infection of soybean, Gipson *et al.* (1969) stated that cell wall disintegration appeared to be the initial phenomenon leading to syncytial development. Within 42 hr after inoculation, abnormal perforations of cell walls were visible. Progressive deterioration of cell walls occurred such that by 196 hr after inoculation portions of the cell walls were no longer present.

It should be stressed that although this review emphasizes nematode-induced giant cells in plant tissues, other agents have been implicated in causing the polynuclear and/or the polyploid stage of modified plant cells. Through a determination of DNA content in nuclei of pea root nodules induced by the nitrogen fixing bacterium *Rhizobium leguminosarum,* Mitchell (1965) determined the ploidy condition of nodules which ranged from $2N$ and $4N$ for uninfected cortex to $8N$ and $16N$ comple-

---

FIGS. 1–6. Fig. 1. Electron micrograph of a giant cell from the root of *Vicia faba* 2 days after penetration by *Meloidogyne javanica.* It has only the primary wall, which is not uniform in thickness. Plasmodesmata (PL) are frequently located at the thinner portion of the wall. The lobed nucleus (Nu) is characteristic for giant cells. Fig. 2. Cortex parenchyma cells in the zone of differentiation from an uninfected root of *V. faba.* Fig. 3. Portion of giant cell from *V. faba* showing the boundary formation (BF) which consists of invaginated plasmalemma and microtubules. The boundary formation is formed in the boundary between the primary wall ($W_1$) and protoplast. Fig. 4. A giant cell wall at an early stage of secondary thickening. Note that the boundary formations contain both tubules and vesicles. Plasmodesmata on the primary wall are clearly visible. Fig. 5. Giant cell wall showing a longitudinal view of tubules in the boundary formation. Fig. 6. A giant cell wall in the process of secondary thickening. Note that the secondary wall material ($W_2$) is deposited between the primary wall ($W_1$) and the boundary formation. (Courtesy of C. S. Huang and A. R. Maggenti, University of California, Davis. Reprinted from *Phytopathology.*)

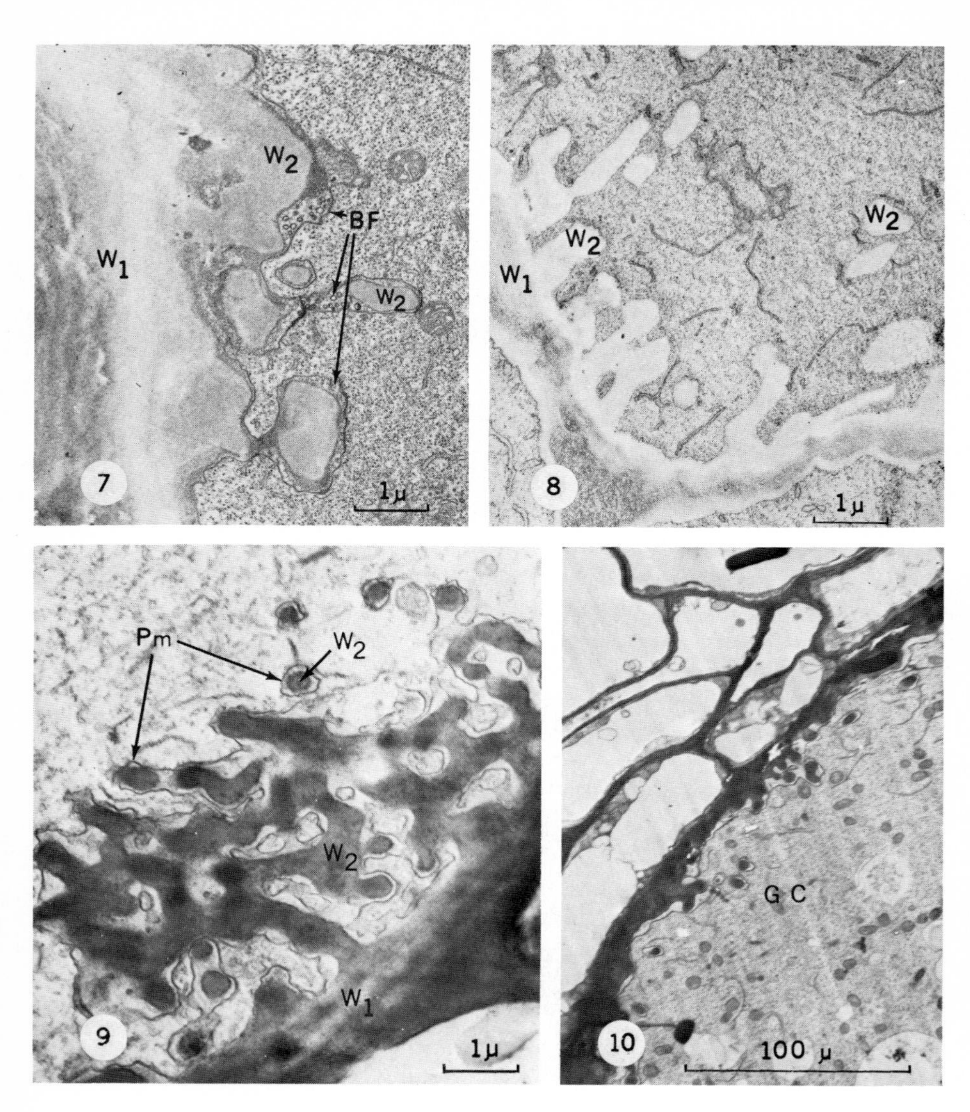


FIGS. 7–10. Fig. 7. Electron micrograph of a giant cell wall from *Vicia faba* 14 days after penetration by *Meloidogyne javanica* in an advanced stage of secondary thickening. Note that the secondary wall depositions ($W_2$) are still closely associated with boundary formations (BF). Fig. 8. A giant cell wall from *V. faba* at the same stage of secondary thickening as the one in Fig. 7. The specimen was fixed with glutaraldehyde, followed by $KMnO_4$. Note that the boundary formations are not resolved by $KMnO_4$ fixation. Fig. 9. A portion of a giant cell wall from *Cucumis sativus* 19 days after root-knot nematode penetration assumes a form of anastomosing trabeculae and is delineated from the protoplast by a continuous plasmalemma (Pm). Fig. 10. A portion of a giant cell (GC) from *C. sativus* 19 days after nematode penetration. Note uneven thickening of the wall. (Courtesy of C. S. Huang and A. R. Maggenti, Univ. of California, Davis. Reprinted from *Phytopathology*.)

ment for a centrally infected zone. A number of herbicides induced abnormal mitosis which showed chromosome bridges resulting from stickiness, a retardation of chromosome movement in anaphase, a binucleate cell, a multinucleate cell with multisepta, the formation of an incomplete cell wall, and an impediment in the differentiation of meristematic tissues (Sawamura, 1964). Multinucleate cells and ploidy ($2N$–$8N$) in onion root tips were induced by artificial treatment with a drug, caffeine, which inhibits the process of cytokinesis (Giménez-Martín *et al.*, 1968). Syncytial formation is not restricted to abnormal or diseased tissue of plants but is a feature of normal development in both plants and animals. For example, the multinucleate state occurs in endosperm development in plants and in osteoclast formation (Fullmer, 1965) of animal tissue, the latter being involved in bone resorption during and after skeletal formation.

## A. Nuclear and Nucleolar Condition of Syncytia

Syncytia induced by nematodes have been reported to attain the polynuclear state by (1) mitoses in the absence of cytokinesis, (2) amitosis or nuclear fragmentation, and (3) the merging of nuclei from several cells that coalesce as a result of cell wall dissolution. Within the polynuclear syncytium, nuclear hypertrophy has been attributed to swelling, nuclear coalescence, and fusion or incorporation of chromosomal sets. Tischler (1902) reported that an increase in the number of nuclei in young giant cells initially occurs through normal mitosis. Later, further multinucleation involved amitosis or nuclear fragmentation through budding, in which nuclei gave rise to a whole series of new nuclei. In a report of giant cells in galled glorybower (*Clerodendron*) roots, Nemec (1910) observed that when the head region of the nematode reached the plerome, the host cells surrounding the oral opening of the larva enlarged. The plasma content of these cells increased, and their nuclei divided without the formation of cell walls; thus, multinucleation of cells occurred.

Nemec (1910) agreed with Tischler (1902) that in the early stages of giant cell formation the first nuclear divisions were mitotic. Nemec also noted that in small cells with one or two irregularly formed nuclei, these nuclei could be interpreted as arising from amitosis or fragmentation. In addition to these small nuclei, numerous large nuclei were found in the syncytium. Nemec (1910) concluded that these nuclei arose from nuclear fusion and speculated that nuclei apparently undergoing fragmentation, as described by Tischler, may in fact be nuclei in the process of fusion.

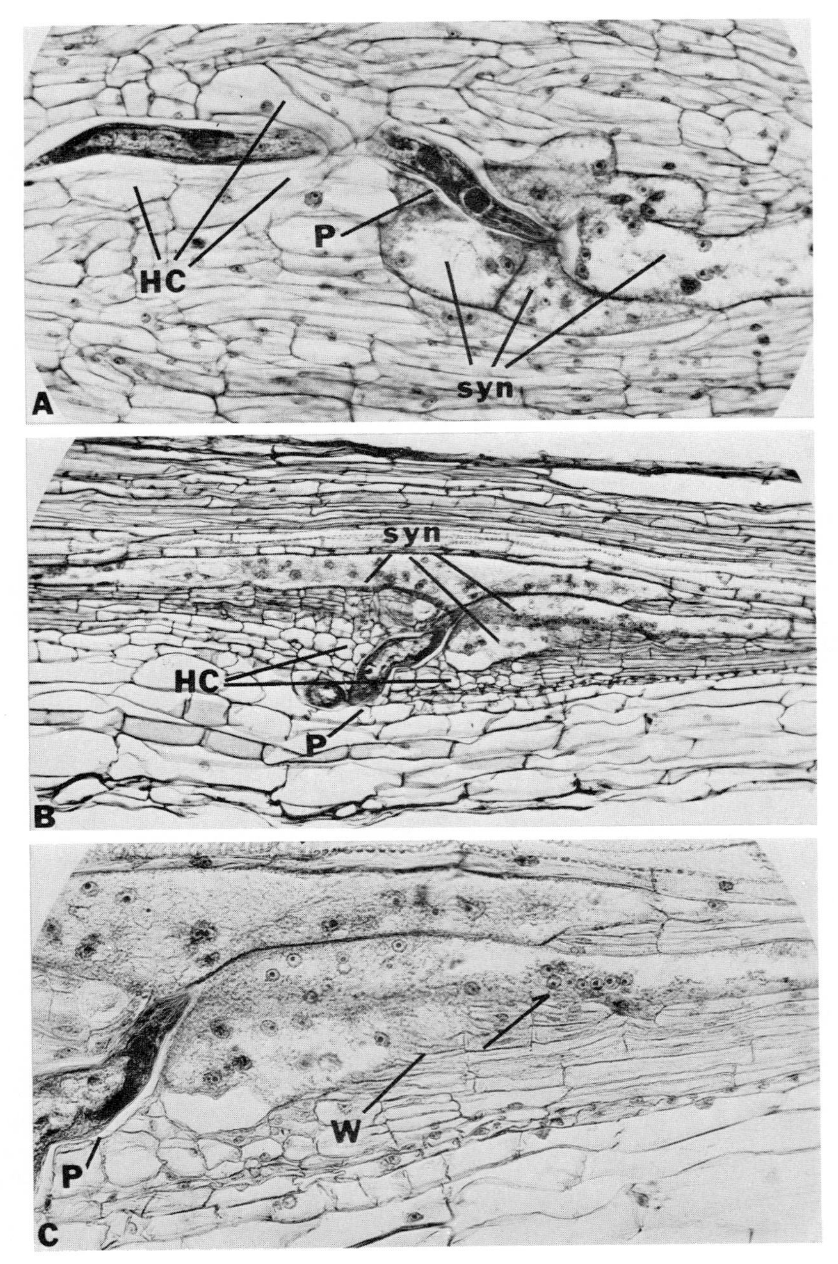


FIG. 11. (A) Section of a tumor at the beginning of the second molt of the parasite (P) showing well developed syncytia (syn) and incipient hypertrophy of

Owens and Specht (1964) found that the number of nuclei within the developing syncytium could be correlated with the number of host cells that would normally occupy the volume of the syncytium. Thus, multinucleation in the early stages of syncytial development was thought to result solely from the dissolution of host cell walls and the coalescence of their protoplasm (Fig. 11). Nuclear changes ranged from a nucleus near the stylet of a nematode, which showed a hypertrophied nucleolus with the apparent absence of a nuclear membrane, to nuclei with various stages of membrane deterioration and a lobulated periphery (Fig. 12). Other nuclear aberrations included nucleolar fragmentation so that small granules, which stained like nucleoli, were scattered throughout the nucleus. Irregularly shaped dumbbell- or sickle-shaped nuclei were observed. Similar observations of sickle-shaped nuclei were reported on root-knot infections of sweet potato (Krusberg and Nielsen, 1958).

In the absence of observable cell wall breakdown in giant cells in *Vicia faba* and *Cucumis sativus* induced by root-knot nematodes, Huang and Maggenti (1969a) attributed the multinucleation of giant cells to repeated mitoses without cytokinesis within the stimulated cell.

Nuclear enlargement in syncytia of root-knot infected tomato, cucumber, and hawk's-beard (*Crepis capillaris*) (Owens and Specht, 1964) appeared to result from swelling and in some cases from nuclear fusion. Such nuclear fusions in syncytia of tomato account for the extreme enlargement of some nuclei which had diameters of 35 $\mu$ as compared to normal cell nuclei of 6 $\mu$. These differences represented a 200-fold increase in volume. However, other data (Rubinstein and Owens, 1964) indicate that a 10–12-fold increase in nuclear volume is more usual in tomato roots. Hairy vetch plants inoculated with *M. incognita* had syncytial nuclei with over 100 chromosomes, whereas normal cells have a $2N$ number of 14 (Dropkin, 1965).

Huang and Maggenti (1969a) studied the nuclear behavior in *Vicia faba* infected with *M. javanica.* They were able to show the prevalence of nuclear enlargement, synchronous division of nuclei, and ploidy of these giant cell nuclei. Determinations of ploidy were facilitated by the relatively low chromosomal complement of *V. faba* ($2N = 12$) and the

---

cells (HC) around the mid and posterior sections of the parasite (×230). (B) Section of young tomato root showing the third-stage parasite (P), arrangement of forming syncytia along the vascular elements, and arrangement of hypertrophic stelar and cortical cells (HC) around the body of the parasite (×100). (C) Enlargement of a sector of Fig. 11B showing fragmentation of cell walls (W) at the lower periphery of the lower syncytium, the variable size of syncytial nuclei, and the tendency of some nuclei to agglomerate (→) (×240). (Courtesy of R. G. Owens and H. N. Specht, Boyce Thompson Institute.)

presence of one pair of marker chromosomes. Chromosome ploidy of 4*N*, 8*N*, 16*N*, 32*N*, and 64*N* were established by chromosome counts (Figs. 13 and 14). In a given giant cell mitosis was generally synchronized, but exceptions were found where prophase and metaphase chromosomes existed concurrently in the same cell. The nuclear hypertrophy referred to in

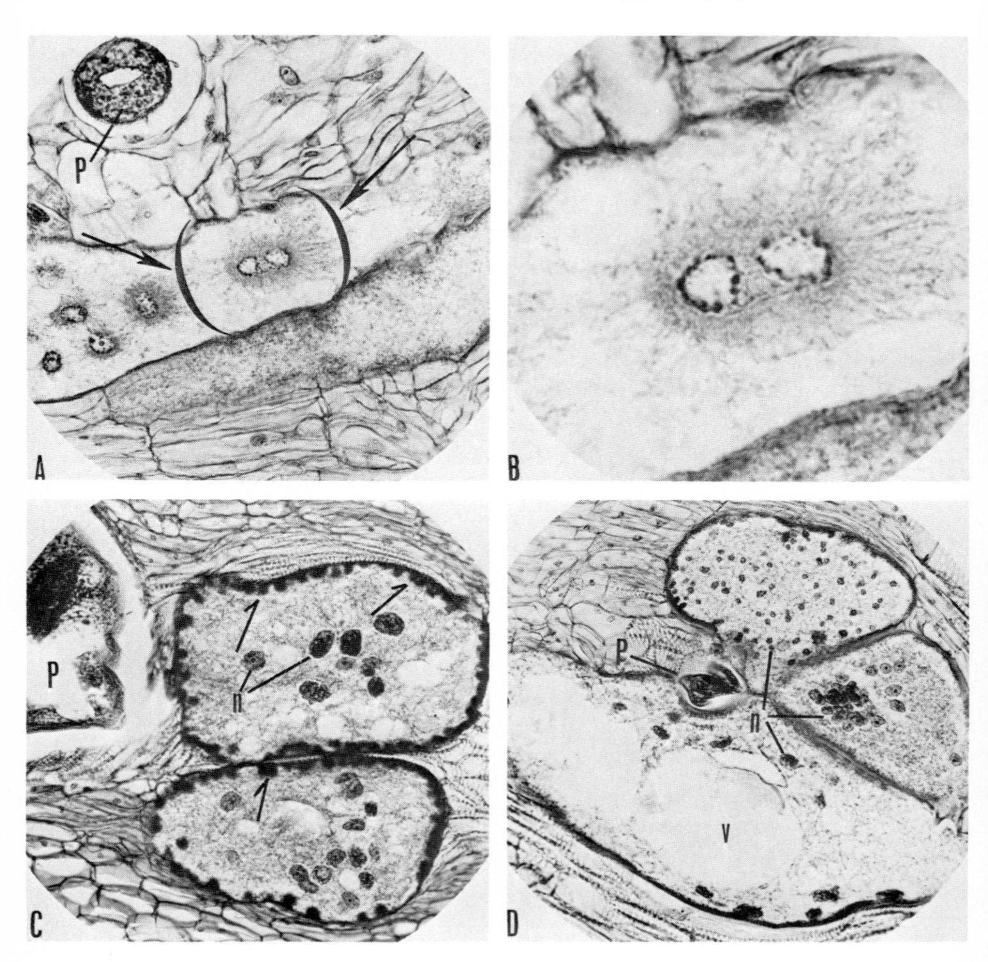


FIG. 12. (A) Nuclear and cytoplasmic aberrations during the fourth or fifth stage of parasite development; P, parasite. (B) Enlargement of the circled sector of Fig. 12A showing the peripheral arrangement of chromatin in the nuclei. (C) Protuberances (→) on syncytial walls during the fifth stage of the parasite; n, nuclei. (D) Three syncytia in a cluster showing various degrees of degeneration of the cytoplasm and nuclei during the adult stage of the parasite; v, vacuole. (Courtesy of R. G. Owens and H. N. Specht, Boyce Thompson Institute.)

root-knot infections could be explained in part by the increased ploidy of syncytial nuclei. The authors attribute the increased size of nuclei to the accumulation of chromosome sets. Three mechanisms for these accumulations were postulated, operating singly or in combination: (1) fusion between homologous chromosome sets within the same mitotic apparatus, (2) incorporation of many chromosome sets into one metaphase plate with the daughter chromosome sets subsequently conducting their anaphase migrations together, and (3) fusion between two chromosome sets of two adjacent mitotic apparatuses at anaphase.

In contrast to the reports on *Meloidogyne,* Mankau and Linford (1960) did not observe nuclear division or fusion in the syncytia induced by *Heterodera trifolii* in clover roots. Disintegration of some nuclei occurred continuously in syncytia. Therefore, the number of nuclei within a developing syncytium was less than the number estimated from cells that would have merged through cell wall dissolution in the formation of the syncytium. In the study of *H. rostochiensis* infection of tomato roots (Sembdner, 1963) and *H. glycines* infection of soybean root (Endo, 1964), mitosis was not evident in syncytia.

Irregular nuclear membranes and enlarged nucleoli appear to be common phenomena in syncytia (Treub, 1887; Tischler, 1902; Nemec, 1910; Kostoff and Kendall, 1930; Christie, 1936; Krusberg and Nielsen, 1958; Davis and Jenkins, 1960; Dropkin and Nelson, 1960; Paulson and Webster, 1969).

In observations of the ultrastructure of giant cell nuclei, Bird (1961) found that an irregularly shaped nucleus with a large nucleolus contained Feulgen-positive bodies scattered along the nuclear membrane which at times connected to the nucleolus. The granular nucleoplasm was bounded by a double-membraned nuclear envelope which was often disrupted in the region of the Feulgen-stained bodies. These areas were suggested to be sites of transport of material from the nucleus into the cytoplasm. Bird also reported clumps of chromatin near the nuclear membrane.

Nuclear morphology of syncytia of clover roots infected by *Heterodera trifolii* was described by Mankau and Linford (1960). Upon stimulation by the nematode, the cell nucleus enlarged and the cytoplasm increased in density and became hyperchromatic. Cells adjacent to syncytia showed nuclear enlargement and cellular hypertrophy, which was followed by a dissolution of the cell wall and a merging of cytoplasm.

Studies of *Heterodera* infections emphasize the wide range of size, shape, and staining qualities of syncytial nuclei (Mankau and Linford, 1960; Sembdner, 1963; Endo, 1964). However, a clear developmental sequence of nuclear changes could not be applied to specific stages of syncytial development (Sembdner, 1963).

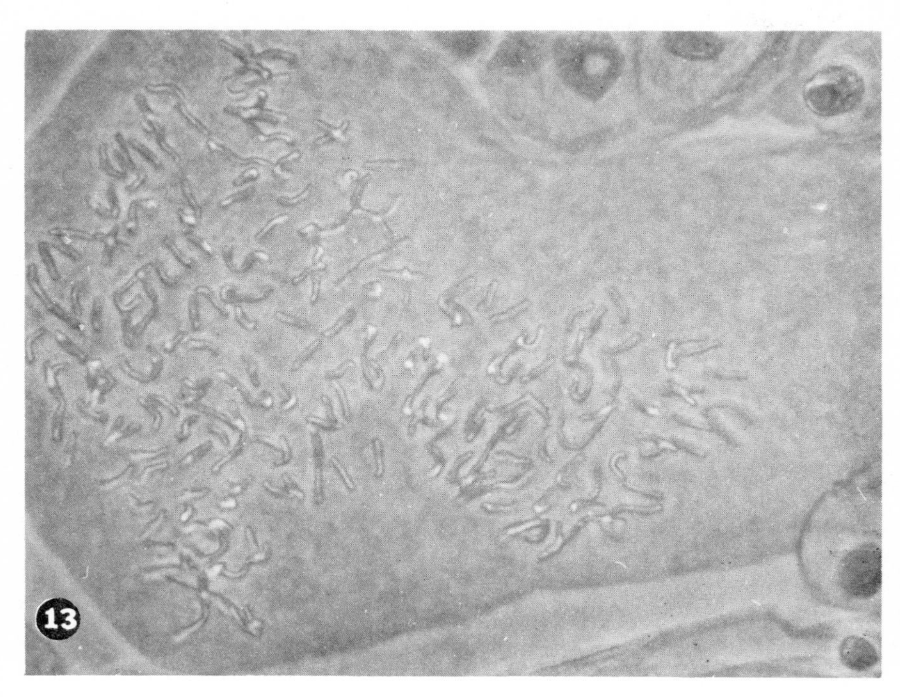

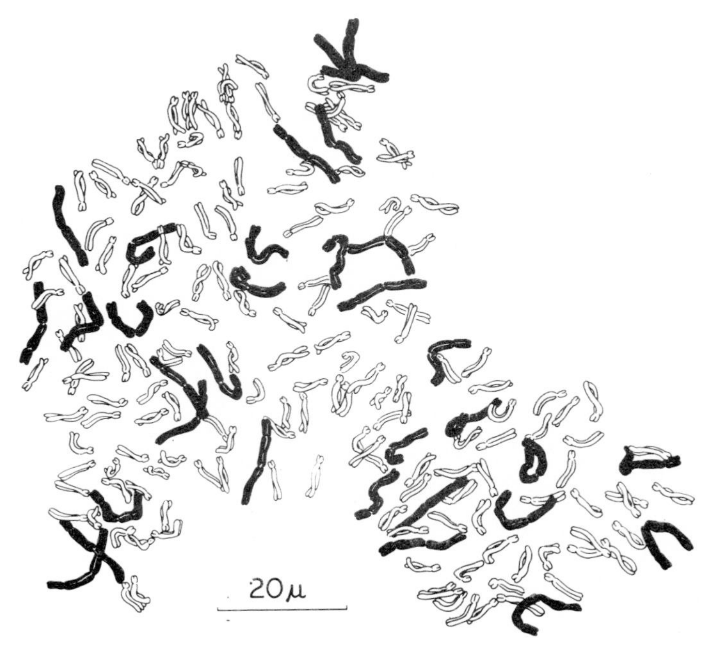
20μ

The diversity of nuclear changes found in syncytia of nematode-infected plants shows some similarities to changes that occur in normal and abnormal tissues of animals. Among these features are hypertrophied nuclei with the nuclear membrane having irregular contours and deep fissures which give rise to the lobulated nuclear outline. The concept that nuclear volume increases in cells with active metabolism (Oberling and Bernhard, 1961) for animal tissue may apply to nuclei of nematode-induced syncytia. Recent histochemical studies have revealed increased rates of enzymic activity in syncytia stimulated by root-knot nematodes (Endo and Veech, 1969; Veech and Endo, 1969).

Cytochemical and tracer techniques have contributed to our knowledge of syncytial changes induced by root-knot nematodes. Relative amounts of DNA in syncytial and normal nuclei were measured microspectrophotometrically (Rubinstein and Owens, 1964). Syncytial nuclei had 2–11 times more DNA and 4–11 times larger volume than normal cortical cell nuclei. Using tritiated thymidine and uridine (Fig. 15). Rubinstein and Owens determined that DNA synthesis within a syncytium was dependent upon the close association of a feeding nematode, whereas RNA synthesis, once initiated, was apparently independent of the nematode. Labeling experiments illustrated in Fig. 15 give added support to data indicating active DNA synthesis and the features of synchronous mitoses which were shown in this report (Rubinstein and Owens, 1964) and in others (Bird, 1961; Dropkin, 1965).

The chemical nature of root-knot nematode secretions that cause increased nucleic acid and protein synthesis is unknown. However, the action of these nematode secretions that influence host metabolism was shown to be influenced by an antimetabolite, 5-bromo-2-deoxycytidine (Bird and McGuire, 1966), which is presumed to influence thymidine incorporation into DNA of giant cells (Rubinstein and Owens, 1964). Nematode development was inhibited by the application of this antimetabolite of DNA synthesis.

In observing nuclei of syncytia in clover roots induced by the clover cyst nematode, Mankau and Linford (1960) reported a gradient of Feulgen stain for DNA. Normal and hypertrophied nuclei along the inside edge of the syncytium stained heavily, whereas other hypertrophied nuclei within the syncytium stained very lightly. This difference was

---

FIGS. 13 and 14. Fig. 13. Phase contrast micrograph of a $32N$ giant cell at metaphase with 192 chromosomes induced by *Meloidogyne javanica* on *Vicia faba* (same scale as Fig. 14). Fig. 14. A camera lucida drawing of the 192 chromosomes in Fig. 13 (shaded, M chromosomes; unshaded, S chromosomes). (Courtesy of C. S. Huang and A. R. Maggenti, University of California, Davis.)

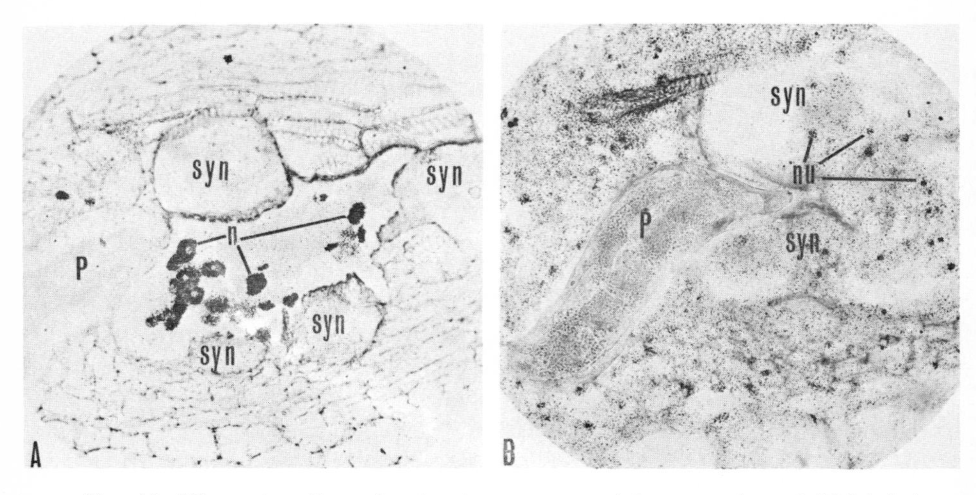


Fig. 15. Microautoradiographs showing patterns of incorporation of $^3$H-labeled nucleosides in nuclei (n), nucleoli (nu), and cytoplasm of root-knot syncytia (syn) at the second-stage larva of the nematode (P). (A) Thymidine incorporation identified by black silver deposits in emulsion above labeled nuclei. (B) Uridine incorporation in cytoplasm and nucleoli of syncytium. (Courtesy of J. H. Rubinstein and R. G. Owens, Boyce Thompson Institute.)

attributed to an apparent destruction of DNA in the large, older nuclei of the syncytium.

## B. Syncytial Cytoplasm

The granular texture of cytoplasm of giant cells in root-knot infections has been reported for numerous host–parasite interactions. Christie (1936) noted that newly formed giant cells varied in appearance depending on the tissues from which they were derived. When first formed, syncytia usually had large clear areas which were separated from one another by strands of cytoplasm. The protoplasmic density of the syncytia progressively increased and eventually gave rise to a homogeneous internal structure. There was a tendency for the syncytial cytoplasm near the head region of the nematode to be more dense and hyperchromatic than the remainder of the syncytial contents. Observations of the syncytial cytoplasm by Owens and Specht (1964) showed that newly formed syncytia contained large vacuoles. In older syncytia, vacuoles were smaller and more numerous.

Nemec (1910) observed elongated structures which he at first termed mitochondria but later referred to as protein crystals (Nemec, 1926).

Bird (1961) showed that cytoplasm of syncytia contained an abundance of mitochondria and the usual cellular components such as Golgi bodies, proplastids, and dense endoplasmic reticulum. Cytoplasm of syncytia closely resembled the contents of meristematic cells at interphase. In terms of composition, giant cell cytoplasm had up to 10 times more protein than normal cell cytoplasm. Traces of carbohydrates and fats were demonstrated as were positive reactions for RNA using gallocyanin as a stain.

In a study of the ultrastructure of various stages of root-knot giant cells in tomato roots, Paulson and Webster (1969) found little change in the morphology of plastids, whereas mitochondria appeared to form vesicles. The increased number of dictyosomes appeared to produce smooth and coated vesicles. Older giant cells contained both smooth and rough endoplasmic reticulum in the form of vesicles and cisternae.

Several enzymes were histochemically localized in fresh tissue sections of soybeans infected with root-knot nematodes (Endo and Veech, 1969). Seven oxidoreductive enzymes, dehydrogenases, and diaphorases were found in greater concentration in syncytia than in surrounding host tissue. The intensity of reaction was comparable to meristematic tissue of the root apex or secondary root primordia. In a later study (Veech and Endo, 1969) the sites of activity of several hydrolytic and oxidative enzymes were demonstrated in fresh tissue sections of root-knot nematode-infected soybean roots. Increases in enzymic activity were found to parallel development of syncytia and nematodes. These observations indicated that, while there was little increase in enzymic activity in the apparently nonaffected root tissues surrounding the syncytia, the syncytia themselves supported greater than normal levels of enzymic activity.

The condition of the cytoplasm of syncytia in *Heterodera* infections shows similarities to syncytial cytoplasm of root-knot nematode infections. Bergman (1958) noted the fine granular texture of syncytial cytoplasm in sugar beets during early stages of *H. schachtii* infection. At later stages, the cytoplasm became turbid and stained heavily with fuchsin.

In addition to a granular appearance of the cytoplasm, Sembdner (1963) noted the presence of fine, elongated osmiophilic structures, which he identified as mitochondria, in syncytia of tomato roots infected with *H. rostochiensis*.

The persistance and texture of syncytial cytoplasm in soybean roots infected by *H. glycines* varied with the feeding habits of the male and female nematode (Endo, 1964). Initially both sexes stimulated a reticulate cytoplasmic network which later became granular. Syncytia induced by males were ephemeral, and their deterioration began within a few

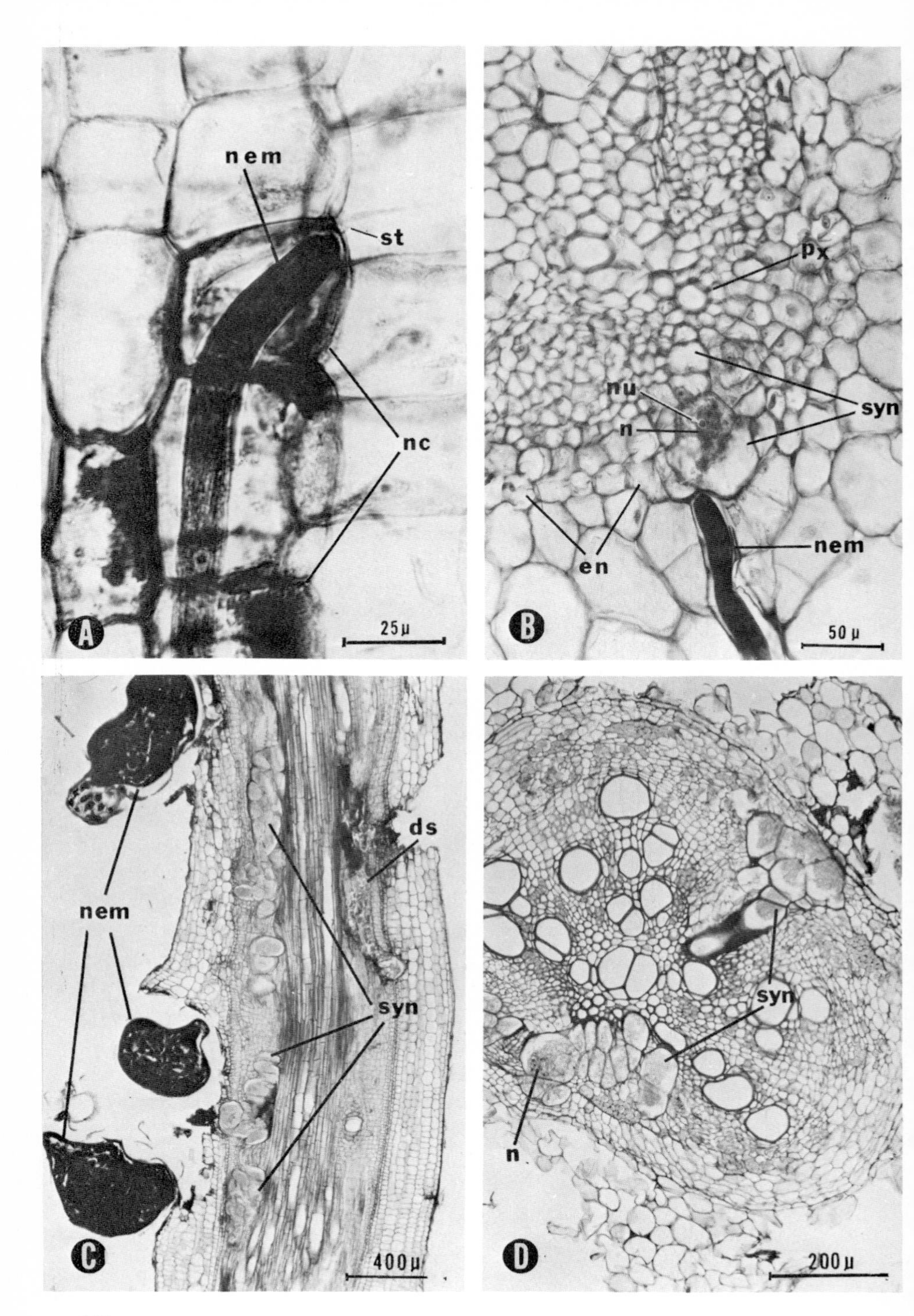
nem
st
nc
A
25μ
px
nu
syn
n
en
nem
B
50 μ
ds
nem
syn
C
400μ
syn
n
D
200μ

days after inoculation. This is indicative of the relatively short feeding time of developing male larvae. The extensive syncytial development and dense granular cytoplasm, characteristic of syncytia induced by female nematodes, can be attributed to the prolonged feeding time required for their growth and development (Fig. 16).

Gipson *et al.* (1969) reported on the ultrastructure of syncytia induced by *H. glycines*. They found that in the first 6 days of development, the central vacuoles of the cells being incorporated into the syncytium appeared to be replaced with cytoplasm containing large numbers of microtubules and resembling smooth endoplasmic reticulum. The remaining vacuoles progressively decreased in size as the syncytium developed.

Mankau and Linford (1960) reported the presence of hyaline, tubular structures in the syncytial cytoplasm of host cells stimulated by *H. trifolii*. Since structures ($2\ \mu \times 50$–$70\ \mu$) were often seen near the lip region of the nematode, the authors suggested that these materials could be either secretions injected by the stylet or products of a reaction between such secretions and the surrounding host cytoplasm. Similar elongated bodies were observed (Endo, 1964) in syncytia near the larval lip region of the soybean cyst nematode feeding on soybean roots. These structures were stained with safranin. In root-knot infections of soybeans, Dropkin and Nelson (1960) reported that long narrow rods measuring up to 70 $\mu$ long and 1 $\mu$ wide were prominent features of some giant cells.

## VI. NATURE OF RESISTANCE

Currently, considerable interest is focused on plant resistance to the Heteroderidae. This part of the review brings together some of the information concerning the mechanisms by which plants resist infection.

Rohde (1960) noted that resistance is any characteristic of a plant

---

FIG. 16. Penetration and infection of *Heterodera glycines* in soybean roots, a susceptible reaction. (A) Intracellular migration of larva (nem) through cortex causing necrosis of cells (nc); note stylet (st) penetration of cortical cell (c) wall (1 day after inoculation). (B) Nematode (nem) in contact with early stage of syncytium (syn) development extending from the endodermis (en) into the protoxylem region (px). Enlarged nucleus (n) surrounds hypertrophied nucleolus (nu). Other nuclei are present but not readily distinguishable (2 days after inoculation). (C) Adult female nematodes (nem) adjacent to syncytia (syn) in vascular region. After the nematode matures and is no longer feeding, deteriorated syncytium (ds) occurs (24 days after inoculation). (D) Vascular damage from syncytial development; syncytia inhibited wide regions of secondary xylem and phloem tissue. Note the cluster of nuclei in a syncytium (27 days after inoculation). From Endo (1964).

or any interaction between host and parasite which retards or prevents the occurrence of a parasitic relationship between the plant parasitic nematode and its host.

One aspect of resistance may be related to the entry of the nematode into the plant. It is apparent among species of *Meloidogyne* and *Heterodera* that larvae of both genera readily enter resistant plants (de Guiran, 1960; Ichinohe and Asai, 1956; Riggs and Winstead, 1959; Williams, 1956). However, Dropkin and Webb (1967) reported that the number of *M. hapla* larvae entering resistant tomato plants was less than in susceptible plants. Once within the plant roots, larvae of Heteroderidae migrate for a period of time and, depending on the suitability of the host, eventually stimulate the formation of multinucleate giant cells or syncytia. However, Christie (1936) found that root-knot nematodes develop to maturity while feeding on parenchyma cells of *Pelargonium* cuttings. Linford (1937) stated that root-knot nematodes feed on many kinds of nonvacuolated, thin-walled cells. Recent histochemical studies (Endo and Veech, 1969) of root-knot infected soybeans showed enzymic activation along the path of the migrating nematode. This might indicate some degree of probing and possibly feeding by the nematode prior to syncytial initiation.

Research on resistance in the Heteroderidae has, for the most part, been directed toward the nature of giant cell or syncytial formation in plant tissue. Barrons (1939) suggested that resistance may result from chemicals within the resistant plants that counteract or neutralize the giant-cell-inducing effect of the salivary secretions of nematodes. Christie (1936) pointed out that if cells of resistant plants failed to respond, the parasite was surrounded by cortical and vascular tissues that became vacuolate and failed as a source of food for the organism.

Environmental conditions such as temperature have been reported to affect the levels of resistance and susceptibility. The soybean cultivar Chief changes from a poor host for root-knot nematode at low temperatures to a better one at high temperatures (Dropkin, 1963). Similarly, Holtzmann (1965) found that several tomato cultivars which were considered resistant to root knot at 20°–25°C became less resistant at 30°–34.5°C. In a survey of 18 known resistant plant species and varieties, Dropkin (1969b) showed that resistance to *Meloidogyne* is dependent on temperature in some plants and not in others and that resistant plants do not always develop cell necrosis in response to nematodes.

The induction of host necrosis near the nematode is considered a form of resistance to nematode infections. Necrosis can result in death of the nematode either through direct toxicity or by the destruction of the plant tissues, thus causing starvation. The term *hypersensitivity* was applied to the response of the resistant soybean, Peking, to infection by *H. glycines* (Ross, 1958). In a study of susceptible and resistant plant responses to potato cyst nematode, Kühn (1958) found a necrogenic defense reaction among certain resistant *Solanum* progeny. Wilski and Giebel (1966) hypothesized that in a resistant plant, hydrolytic enzymes are secreted by the nematode which act on a nontoxic phenolic glucoside in the host cells. The enzymic hydrolysis of the glucoside yields a toxic aglycone which results in host necrosis. In terms of resistance-breaking biotypes, some nematodes with low levels of $\beta$-glucosidase may induce the accumulation of lower quantities of the toxic phenol, thus allowing these biotypes to develop on varieties of plants previously considered to be resistant.

Addition of exogenous cytokinins to resistant tomato seedlings blocked the necrotic response (hypersensitive reaction) and allowed root-knot nematodes to grow and induce gall formation. Necrosis was not blocked with indoleacetic acid, naphthalene acetic acid, and gibberellic acid. In contrast, when kinetin, 6-benzylaminopurine, zeatin, and 6-($\gamma$,$\gamma$-dimethyl-allylamino) purine were added to the agar medium, a large proportion of larvae enlarged and gall formation was increased (Dropkin *et al.*, 1969). Histochemical studies revealed that the root-knot nematode, soon after penetration in soybean roots, caused a slight increase in the activity of enzymes (oxidoreductive, hydrolytic, and oxidative) at the feeding site in both susceptible and resistant plants. However, by the time syncytial induction occurred in the susceptible host, the most common response of the resistant host was cell necrosis (Veech and Endo, 1970).

In *H. glycines* infections of soybeans (Endo, 1965) resistance was in the form of cortical tissue damage caused by migrating larvae and the deterioration of affected cells. Usually nematodes failed to develop beyond the second larval stage. Occasionally nematodes stimulated small syncytia which allowed limited growth of the nematode. Development of larvae into males was occasionally found, whereas female development was rare. In general, the extent of syncytial development coincided closely with the degree of nematode development. At the termination of feeding, necrosis and collapse of the syncytium occurred and often parenchyma tissue invaded the site.

Resistance to the potato cyst nematode, *H. rostochiensis*, has been incorporated into progeny of the resistant *Solanum tuberosum* subspecies *andigena* (Jones and Parrott, 1968). Environmentally controlled sex determination is an integral feature of resistance and has been correlated with the relative size of the giant cells induced by invading larvae. Larvae inducing small syncytia developed into males, whereas larvae inducing large syncytia developed into females. In the roots of the resistant potatoes, it was found that nearly all larvae of some field populations developed into males and only a few females developed. Thus, resistance was reflected in the low rate of reproduction and lower populations of nematodes.

Myuge (1964) postulated that enzyme neutralizers in plant roots respond to enzymic secretions of sedentary nematodes. He felt that complete neutralization would prevent giant cell formation whereas partial neutralization would allow limited giant cell formation and restricted nematode growth and reproduction. However, when a balance between the plant neutralizer and the nematode secretions is reached, giant cell and nematode development occurs optimally. In the absence of the neutralizer, necrosis of the plant cell occurs and the nematode dies.

## VII. CONCLUSIONS

Interest in the nature of giant cells is reflected in the numerous reports on various facets of their anatomy and physiology. Recently reported ultrastructural observations of host responses to infection should add greater dimensions to the understanding of the mechanisms of syncytial formation. Some clarification is still needed however on the rather fundamental question of cell wall dissolution. This process, though accepted by several workers, has been challenged in a recent electron microscopic study in which cell wall dissolution was not observed (Huang and Maggenti, 1969b).

The fundamental question of gall formation and its relation to syncytia and nematode activity will require further investigation. It has been postulated that stimuli for syncytial formation differ from those which initiate gall reaction (Owens and Specht, 1964). The separation and identification of such specific stimulants is a challenge. Excretory products may be a factor in gall formation, especially where nematodes are embedded in the root tissues (Myers and Krusberg, 1965; Endo and Veech, 1969). The concept of inhibitors (Myuge, 1964) and the role of growth regulators (Balasubramanian and Rangaswami, 1962; Yu and Viglierchio, 1964; Dropkin *et al.*, 1969) should be of continuing interest and provide fruitful avenues of research.

The interdependence of nematode maturation and syncytial development is closely related to breeding programs for developing nematode-resistant plants. The extent of syncytial formation has been correlated with the development of male or female nematodes (Trudgill, 1967). Resistance of potato roots to *H. rostochiensis* is reflected in small syncytia and predominance of males (Jones and Parrott, 1968). The recent decline in resistance to *H. glycines* of certain soybean varieties reemphasizes the importance of the host–parasite interaction and the development of new biological races or biotypes.

The need for more information concerning the host–parasite interaction as it relates to giant cell formation should stimulate further research in this area of the biological relations of the organism, its host, and the environment.

## REFERENCES


Balasubramanian, M., and Rangaswami, G. (1962). *Nature (London)* **194,** 774–775.

Barrons, K. C. (1939). *J. Agr. Res.* **58,** 263–271.

Bergman, B. H. H. (1958). *Meded. Inst. Rationff Suikerprod.* **28,** 151–168.

Bird, A. F. (1961). *J Biophys. Biochem. Cytol.* **11,** 701–715.

Bird, A. F. (1962). *Nematologica* **8,** 1–10.

Bird, A. F. (1967). *J. Parasitol.* **53,** 768–776.

Bird, A. F. (1968). *J. Parasitol.* **54,** 475–489.

Bird, A. F., and McGuire, R. J. (1966). *Nematologica* **12,** 637–640.

Bird, A. F., and Saurer, W. (1967). *J. Parasitol.* **53,** 1262–1269.

Cadman, C. H. (1963). *Annu. Rev. Phytopathol.* **1,** 143–172.

Chen, T. A., and Mai, W. F. (1965). *Phytopathology* **55,** 128.

Chitwood, B. G., and Oteifa, B. A. (1952). *Annu. Rev. Microbiol.* **6,** 151–184.

Christie, J. R. (1936). *Phytopathology* **26,** 1–22.

Crittenden, H. W. (1962). *Phytopathology* **52,** 163.

Dalmasso, A. (1967). *Ann. Epiphyt.* **18,** 249–272.

Davis, R. A., and Jenkins, W. R. (1960). *Nematologica* **5,** 228–230.

de Guiran, G. (1960). *Meded. Landbouwhogesch. Opzoekingssta. Staat Gent* **25,** 1047–1056.

Dropkin, V. H. (1955). *Exp. Parasitol.* **4,** 282–322.

Dropkin, V. H. (1963). *Phytopathology* **53,** 663–666.

Dropkin, V. H. (1965). *Nematologica* **11,** 36.

Dropkin, V. H. (1969a). *Annu. Rev. Phytopathol.* **7,** 101–122.

Dropkin, V. H. (1969b). *Phytopathology* **59,** 1632–1637.

Dropkin, V. H., and Nelson, P. E. (1960). *Phytopathology* **50,** 442–447.

Dropkin, V. H., and Webb, R. E. (1967). *Phytopathology* **57,** 584–587.

Dropkin, V. H., Helgeson, J. P., and Upper, C. D. (1969). *J. Nematol.* **1,** 55–61.

Endo, B. Y. (1964). *Phytopathology* **54,** 79–88.

Endo, B. Y. (1965). *Phytopathology* **55,** 375–381.

Endo, B. Y., and Veech, J. A. (1969). *Phytopathology* **59,** 418–425.

Feldmesser, J. (1952). *Phytopathology* **42,** 466.

Fielding, M. J. (1959). *Annu. Rev. Microbiol.* **13,** 239–254.

Flegg, J. J. M. (1965). *Annu. Rep., East Malling Res. Sta., Kent* 1964, pp. 62–70.

Franklin, M. T. (1951). "The Cyst-Forming Species of *Heterodera*." Commonw. Agr. Bur., Farnham Royal, Bucks, England.

Fullmer, H. M. (1965). *Int. Rev. Connect. Tissue Res.* **3,** 1–76.

Giménez-Martín, G., Lopez-Saez J. F., Moreno, P., and Gonzales-Fernandez, A. (1968). *Chromosoma* **25**(3), 282–296.
Gipson, I., Kim, K. S., and Riggs, R. D. (1969). *Phytopathology* **59**, 1027–1028.
Godfrey, G. H., and Oliveira, J. (1932). *Phytopathology* **22**, 325–348.
Holtzmann, O. (1965). *Phytopathology* **55**, 990–992.
Huang, C. S., and Maggenti, A. R. (1969a). *Phytopathology* **59**, 447–455.
Huang, C. S., and Maggenti, A. R. (1969b). *Phytopathology* **59**, 931–937.
Ichinohe, M., and Asai, K. (1956). *Hokkaido Nogyo Shikenjo Iho* **71.**
Jones, F. G. W., and Parrott, D. M. (1968). *Outlook Agr.* **5**, 215–222.
Kostoff, D., and Kendall, J. (1930). *Zentralbl. Bakteriol., Parasitenk, Infektionskr. Hyg., Abt. 2* **81**, 86–91.
Krusberg, L. R. (1963). *Annu. Rev. Phytopathol.* **1**, 219–240.
Krusberg, L. R., and Nielsen, L. W. (1958). *Phytopathology* **48**, 30–39.
Kühn, H. (1958). *Z. Pflanzenkr.* (*Pflanzenpathol.*) *Pflanzenschutz* **65**, 465–472.
Linford, M. B. (1937). *Phytopathology* **27**, 824–835.
Littrell, R. H. (1966). *Phytopathology* **56**, 540–544.
Mankau, R., and Linford, M. B. (1960). *Ill. Agr. Exp. Sta., Bull.* **667**, 50 pp.
Miller, H. N. (1965). *Soil Crop Sci. Soc. Fla., Proc.* **24**, 310–325.
Mitchell, J. P. (1965). *Ann. Bot.* (*London*) **29**, 371–376.
Mountain, W. B. (1960). *In* "Nematology: Fundamentals and Recent Advances with Emphasis on Plant Parasitic and Soil Forms" (J. N. Sasser and W. R. Jenkins, eds.), pp. 426–431. Univ. of North Carolina Press, Chapel Hill, North Carolina.
Mountain, W. B. (1965). *In* "Ecology of Soil-borne Plant Pathogens" (K. F. Baker and W. C. Synder, eds.), pp. 285–301. Univ. of California Press, Berkeley, California.
Myers, R. F., and Krusberg, L. R. (1965). *Phytopathology* **55**, 429–437.
Myuge, S. G. (1956a). *Dokl. Akad. Nauk SSSR* **108**, 164–165.
Myuge, S. G. (1956b). *Zh. Obshch. Biol.* **17**, 396–399.
Myuge, S. G. (1958). *Akad. Nauk SSSR Moskva* **30**, 91–92.
Myuge, S. G. (1964). Plant Parasitic Nematodes. Feeding of Phytohelminths and their Relationships with Plants. 47 pp. Izd. Kolos, Moskva. (Available from U. S. Dept. of Commerce Clearinghouse for Federal Sci. and Tech. Inform., Springfield, Virginia 22151.)
Nemec, B. (1910). "Das Problem der Befruchtungsvorgänge und andere zytologische Fragen. VI. Vielkernige Riesenzellen in *Heterodera* Gallen," pp. 151–173. Gebrüder Bornträger, Berlin.
Nemec, B. (1911). *Z. Pflanzenkr.* **21**, 1–16.
Nemec, B. (1926). *Biol. Gen.* **2**, 96–103.
Nemec, B. (1932). *Stud. Plant Physiol. Lab., Charles Univ., Prague* **4**, 1–14.
Nolte, H. W., and Köhler, H. (1952). *Nachrichtenbl. Deut. Pflanzenschutzdienst* **32**, 24–28
Oberling, C. and Bernhard, W. (1961). *Cell* **5**, 405–496.
Owens, R. G., and Specht, H. N. (1964). *Contrib. Boyce Thompson Inst.* **22**, 471–489.
Owens, R. G., and Specht, H. N. (1966). *Contrib. Boyce Thompson Inst.* **23**, 181–198.
Paulson, R. E., and Webster, J. M. (1969). *J. Nematol.* **1**, 22–23. (Abstr.)
Pitcher, R. S. (1965). *Helminthol. Abstr.* **34**, 1–17.
Powell, N. T. (1963). *Phytopathology* **53**, 28–35.

Rebois, R. V., Epps, J. M., and Hartwig, E. E. (1970). *Phytopathology* **60**, 695–700.
Riggs, R. D., and Winstead, N. N. (1959). *Phytopathology* **49**, 716–724.
Rohde, R. A. (1960). *In* "Nematology: Fundamentals and Recent Advances with Emphasis on Plant Parasitic and Soil Forms" (J. N. Sasser and W. R. Jenkins, eds.), pp. 447–453. Univ. of North Carolina Press, Chapel Hill, North Carolina.
Roman, J. (1961). *J. Agr. Univ. P. R.* **45**, 55–84.
Ross, J. P. (1958). *Phytopathology* **48**, 578–579.
Rubinstein, J. H., and Owens, R. G. (1964). *Contrib. Boyce Thompson Inst.* **22**, 491–502
Ruehle, J. L. (1962). *Phytopathology* **52**, 68–71.
Sawamura, S. (1964). *Cytologia* **29**, 86–102.
Schuster, M. L., and Sullivan, T. (1960). *Phytopathology* **50**, 874–876.
Schuster, M. L., Sandstedt, R., and Estes, L. W. (1965). *J. Amer. Soc. Sugar Beet Technol.* **13**, 523–537.
Seinhorst, J. W. (1961). *Annu. Rev. Microbiol.* **15**, 177–196.
Sembdner, G. (1963). *Nematologica* **9**, 55–64.
Setty, K. G. H., and Wheeler, A. W. (1968). *Ann. Appl. Biol.* **61**, 495–501.
Tischler, G. (1902). *Ber. Deut. Bot. Ges.* **19**, *Gen-Versamml.*, pp. 95–107.
Treub, M. (1887). *Ann. Jard. Bot. Buitenzorg* **6**, 93–96.
Triffitt, M. J. (1931). *J. Helminthol.* **9**, 1–16.
Trudgill, D. L. (1967). *Nematologica* **13**, 263–272.
Ustinov, A. A. (1951). *Trans. Inst. Zool. Acad. Sci. SSSR* **9**, 405–459.
Veech, J. A., and Endo, B. Y. (1969). *J. Nematol.* **1**, 265–276.
Veech, J. A., and Endo, B. Y. (1970). *Phytopathology* **60**, 896–902.
Viglierchio, D. R., and Yu, P. K. (1968). *Exp. Parasitol.* **23**, 88–95.
Williams, T. D. (1956). *Nematologica* **1**, 88–93.
Wilski, A., and Giebel, J. (1966). *Nematologica* **12**, 219–224.
Yu, P. K., and Viglierchio, D. R. (1964). *Exp. Parasitol.* **15**, 242–248.

# CHAPTER 18

# Interaction of Plant Parasitic Nematodes with Other Disease-Causing Agents

N. T. Powell

*Department of Plant Pathology, North Carolina State University, Raleigh, North Carolina*

I. Introduction . . . . . . . . . . . . . . . 119
II. Nematode–Fungus Complexes . . . . . . . . . . . 120
A. Nematode–Fungus Wilt Disease Interactions . . . . . . 120
B. Nematode–Fungus Root Rot Disease Interactions . . . . 123
C. Nematode–Fungus Seedling Disease Interactions . . . . 126
III. Nematode–Bacteria Interactions . . . . . . . . . . 127
A. Nematode–Bacteria Wilt Disease Interactions . . . . . . 127
B. Nematode–Bacteria Root Disease Interactions . . . . . . 128
C. Nematode–Bacteria Foliage Disease Interactions . . . . 129
IV. Nematode–Virus Relationships . . . . . . . . . . 129
V. Some Effects of Complexes on Nematode Populations . . . 131
VI. The Nature of Complexes Involving Nematodes . . . . . . 132
VII. Conclusions . . . . . . . . . . . . . . . 134
References . . . . . . . . . . . . . . . 135

## I. INTRODUCTION

Interactions of plant parasitic nematodes with other pathogens should receive serious and sophisticated study. That nematodes can cause important plant diseases quite independent of other microorganisms is well established, and successful control of these pathogens has contributed greatly to the reduction of disease losses. It is also well-known that certain fungi, bacteria, and viruses are important pathogens and often cause damage without influences from other biotic agents.

However, plants are rarely, if indeed ever, subject to association with only one potential pathogen. This applies whether or not the plants are under cultivation. Rather, they are constantly exposed to numerous other organisms, many of which are common components of the soil biosphere. It thus seems appropriate to consider interrelationships among such organisms and the ultimate effects of these complexes upon host plants. It is not realistic to assume that a plant although infected with one pathogen will not be affected by another. In fact, it seems much more appropriate to assume that because a host is infected by one pathogen its response to additional invaders will be altered. These alterations may have significant influences upon disease development within a particular host, epidemiology of all pathogens involved, and, ultimately, on disease control. In plant pathology, an understanding of disease complexes may be at least as important as the understanding of "monopathogenic" situations. In diseases caused by soil-borne pathogens, the latter must be a rarity rather than the rule. Plant parasitic nematodes are usually a part of the soil microfauna, and should be considered when disease complexes are studied.

Evidence that injury from even established pathogens is often increased by plant parasitic nematodes is convincing, and evidence of these damaging combinations is accumulating rapidly. In addition, knowledge that certain microorganisms are destructive only when they occur in combination with other biological agents may be of equal or even greater importance.

This section will be devoted to the examination of certain known interactions involving nematodes and other disease-causing agents. Literature on this subject has been reviewed recently by several authors (Pitcher, 1963, 1965; Powell, 1963; Miller, 1965; Weischer, 1968); therefore, only representative examples of various types of complexes and possible mechanisms involved in such interactions will be presented.

## II. NEMATODE–FUNGUS COMPLEXES

### A. Nematode–Fungus Wilt Disease Interactions

By far the greatest number of complexes reported involving nematodes has included a fungus as the other component, and many reports of nematode–fungus interrelationships have involved a nematode pathogen along with one of the wilt-inducing fungi. Atkinson (1892) was among the first to notice this association when he observed that infection by root-knot nematodes (*Meloidogyne* spp.) increased the severity of *Fusarium* wilt in cotton. Since then, the interaction of these two pathogens has received much attention on a number of different host plants such as tomato and tobacco. Early work with each of these hosts noted the deleterious effects of root-knot nematode infestations on the expression of *Fusarium* wilt resistance in breeding programs. These reports often depicted close correlations between nematode populations and wilt severity as measured by increased death rates, along with accelerated and more severe symptom development. Indirect control of *Fusarium* wilt in cotton and tomato, as well as other crops, by soil fumigation with nematicides and rotations, underscored further the relationship between these two pathogens. Most of this earlier work has been summarized previously (Powell, 1963).

Following early leads, investigators began to work with the root-knot–*Fusarium* interaction on various hosts and in controlled inoculation experiments, which generally supported previous observations and indirectly suggested some possible mechanisms involved. Jenkins and Coursen (1957) induced wilting in *Fusarium* wilt-resistant tomato variety Chesapeake only when root-knot nematodes were present along with fungal inoculum. Furthermore, when *M. hapla* was combined with the fungus, only 60% of the plants wilted, whereas *M. incognita acrita* promoted wilt in 100% of the plants. This might suggest that these two nematodes affect the host in somewhat different physiological manners, and that these differences are meaningful in predisposition to *Fusarium* wilt.

Later research expanded earlier reports of the occurrence of the root-knot–*Fusarium* wilt complex in various crops and dealt with more analytical approaches. Minton and Minton (1963) studied the histopathology of sites infected with *M. incognita acrita* and *F. oxysporum* f. *vasinfectum* in cotton. The fungus grew well in sloughing epidermal cells and decaying cortex but poorly in healthy cortical tissue. Abundant growth occurred in the xylem and nematode-induced giant cells, but the fungus was not found in healthy phloem or cambium. They noted that in nematode-infected roots, the xylem is exposed to fungus attack. The portals of fungus entry to the xylem were through decaying tissues.

The root-knot–*Fusarium* wilt interaction in tobacco has received similar analysis (Melendéz and Powell, 1967). Giant cells and adjacent xylem elements in both *Fusarium*-susceptible and -resistant plants were extensively colonized. Giant cells, however, were sensitive to fungus invasion and became devoid of protoplasm soon after invasion. Fungal hyphae present in such cells gradually became debilitated. Occasionally, the female nematode and gelatinous matrix of the egg mass were invaded.

In some hosts, the nematodes apparently are capable of maximum predisposition only after 3–4 weeks (Porter and Powell, 1967). This time interval would be sufficient to permit the formation of galls, along with the accompanying morphological and physiological changes. In fact, this relationship appears so important in tobacco that the appearance of *Fusarium* wilt symptoms is more closely related to the time of nematode inoculation than to time of fungus inoculation. However, the same may not hold true for all crops (Johnson and Littrell, 1969). Goode and McGuire (1967) have pointed to another possibly important aspect of the *Fusarium*–root-knot interrelationship. They found that nematode infection enables certain races of *F. oxysporum* f. *lycopersici* to attack tomato varieties ordinarily resistant to those races and suggested that the fungus mutates within the host. If such observations are true for a number of nematode–fungus interactions, it could force reexamination of the established "host range" concept.

*Fusarium* wilt diseases appear to be greatly influenced by host predisposition from several nematode pathogens, although most of the complexes studied involve the root-knot group with the *Fusarium* wilt fungi. Certain varieties of soybean, inoculated with the *Fusarium* fungus and *Heterodera glycines,* wilt to a greater extent than those predisposed by *M. incognita* (Ross, 1965). This difference may be more nearly related to differences in nematode penetration and migration than to physiological differences. It is also true that soybeans are generally more tolerant to root-knot nematodes than to cyst nematodes.

Nematodes other than sedentary endoparasites are also capable of forming disease complexes with *Fusarium* wilt fungi (Powell, 1963). The sting nematode, *Belonolaimus gracilis,* is as efficient as root-knot nematodes in promoting wilt in cotton (Cooper and Brodie, 1963). Although cotton seedling emergence is not affected by either sting or root-knot nematodes, damage becomes apparent later among older plants exposed to either nematode and the fungus (Minton and Minton, 1966). Wilt symptoms do not appear unless one of the nematode pathogens is present. A nematode–fungus relationship occurs in peas. The chief symptom of the "early yellowing" disease, root rot, is dependent upon the presence of both *Hoplolaimus uniformis* and *F. oxysporum* f. *pisi* race 3 (Labruyére *et al.,* 1959).

*Verticillium* wilt is a typical vascular wilt disease and is similar to *Fusarium* wilt in a number of ways. It, too, is often increased by the presence of plant parasitic nematodes along with *Verticillium* species causing the disease. Early work with *Verticillium* wilt complexes on eggplant, pepper, tomatoes, and cotton has been bolstered by research with other crops. Root-lesion nematodes (*Pratylenchus* spp.) are evidently the most important nematodes involved in interactions with *Verticillium* wilt fungi. An important interrelationship exists between these two pathogens in peppermint (Bergeson, 1963; Faulkner and Skotland, 1965) and potatoes (Morsink and Rich, 1968). In both situations, either pathogen alone may cause disease but damage is greater when the nematodes and fungus occur together.

In peppermint, reduction in incubation period in combined infections is often more noticeable than differences in ultimate disease development (Bergeson, 1963). When *V. dahliae* and *P. minyus* co-infect peppermint, the optimum temperature for wilt symptom expression is shifted (Faulkner and Bolander, 1969). Although the nematode promotes an increase in wilt at all temperatures that have been tested, the optimum soil temperature for disease development when the fungus is present alone is 24°C, as compared to 27°C in plants exposed to both pathogens. The optimum temperature for nematode reproduction is also apparently changed by the presence of the fungus.

Black shank of tobacco, caused by *Phytophthora parasitica* var. *nicotianae,* is a wilt disease, although root decay is also an important part of the disease syndrome. The fungus is a very aggressive pathogen in this host and causes rapid death in susceptible individuals. Nonetheless, an important complex involves the black shank fungus and *M. incognita* (Powell and Nusbaum, 1960). A similar complex in tobacco with *P. parasitica* var. *nicotianae* and *M. javanica* has been reported (Miller, 1968).

Root-lesion nematodes interact with the black shank fungus in tobacco varieties susceptible to this fungus disease (Inagaki and Powell, 1969). They have not, however, been shown capable of altering the reaction of black shank-resistant varieties to the disease. Significantly, black shank symptoms are more severe if both pathogens are added to plants simultaneously or 1 week apart than if the nematode precedes the fungus by 4 weeks. This is contrary to certain complexes involving *Meloidogyne* spp.

### B. Nematode–Fungus Root Rot Disease Interactions

Nematodes representing several genera are important and, at times, vital to development of root rots ostensibly caused by fungi. In fact, associations of this type may have greater overall importance than complexes involving nematodes with wilt-inducing fungi. This should be expected, however, when the normal habitat and behavior of the pathogens are considered. Recent reviews have outlined instances in which nematodes and fungi combine in a root rot disease (Powell, 1963; Pitcher, 1965).

Migratory nematode pathogens are involved in several root rot disease complexes, and a significant number include species of *Pratylenchus*. Root rot of winter wheat in Canada, in which *P. minyus* combines with *Rhizoctonia solani,* significantly reduces wheat growth (Benedict and Mountain, 1956; Mountain, 1954). Although no definite interaction has been described, nematodes of the genus *Pratylenchus* along with a *Pythium* fungus have been associated with a large percentage of peach trees suffering from peach decline. *Criconemoides* spp. and *Tylenchorhynchus* spp. have also been associated with this condition (Hendrix *et al.,* 1965). *Pratylenchus penetrans* and *Trichoderma viride* caused more reduction in root and shoot growth in both alfalfa and celery than either organism alone. This deserves special attention since *Trichoderma* spp. are not generally regarded as important pathogens (Edmunds and Mai, 1966a).

At certain temperatures, *P. scribneri* and *F. moniliforme* cause greater reduction of corn fresh weight than either pathogen alone (Palmer *et al.,* 1967). Although *P. penetrans* did not alter resistance of burley tobacco to black root rot, caused by *Thielaviopsis basicola,* the fungus seems to exercise some effect on root penetration by the nematodes (Olthof, 1968).

Migratory ectoparasitic nematodes may enhance the severity of certain fungus root rots. Chrysanthemum roots inoculated with *Belonolaimus longicaudatus* and *Pythium aphanidermatum* develop symptoms of *Pythium* root rot earlier and more extensively than those inoculated with the fungus alone (Littrell and Johnson, 1969).

*Tylenchulus semipenetrans* increases root decay by *Fusarium* spp. in lemon. There are differences, however, depending upon whether the fungus present is *F. oxysporum* or *F. solani.* Temperature also exerts an influence on the interaction (O'Bannon *et al.,* 1967). *Tylenchus agricola* contributes to an increase in root rot of corn caused by *F. roseum,* but it fails to influence disease development by *P. ultimum* (Kisiel *et al.,* 1969).

The burrowing nematode, *Radopholus similis,* has been implicated in root rot complexes on banana. Recent studies show that *F. solani* is more commonly found than the *Rhizoctonia* fungus in the deep root lesions formed by this nematode, whereas the opposite is true of shallow lesions caused by *Helicotylenchus* spp. (Stover, 1966).

The root-knot nematode group is also important in root decay complexes. Much of the research confirming this has been done quite recently, although their importance in wilt diseases has been recognized for some time.

The frequency with which *Aspergillus flavus* is isolated from Argentine Spanish peanut shells is increased if *M. arenaria* is present (Minton and Jackson, 1967). This could have significance beyond actual disease loss since this fungus is capable of producing aflatoxin, a potent hepatocarcinogen. Decay in alfalfa roots by either *F. roseum* or *F. oxysporum* f. *batatas* is increased when *M. incognita acrita* occurs with the fungal pathogen. The addition of KCl influences root breakdown by this complex (Kushner and Crittenden, 1967).

Research on tobacco has presented additional evidence for the importance of root-knot nematodes in root decay interactions and has shown that the status of normally unimportant fungal pathogens may be elevated to major significance by nematode infection. Both *Rhizoctonia solani* and *Pythium* spp. can cause damping-off of tobacco seedlings, but usually damage is insignificant. Also, the *Rhizoctonia* fungus occasionally causes a root and stalk disease known as *soreshin.* However, the occurrence of this condition is sporadic and unimportant. Thus, neither *Pythium* nor *Rhizoctonia* is regarded as important pathogens after tobacco plants have progressed beyond juvenile stages. However, if either of these fungi is added to plants which have been previously infected by *M. incognita* (3–4 weeks), they are capable of invading the roots and causing extensive decay. On the other hand, little damage results if the fungus and nematode pathogen are applied simultaneously and virtually no decay occurs resulting from the fungus alone (Powell and Batten, 1967; Melendéz and Powell, 1969).

The same effect is evident with fungi which, although pathogenic on other crops, have not been reported on tobacco. Species of *Curvularia, Botrytis, Aspergillus,* and *Penicillium* will invade and cause decay of tobacco roots if roots have been predisposed by prior root-knot nematode infection (Powell, unpublished data). Without this predisposition, these fungi appear incapable of establishing a parasitic relationship in tobacco.

The *Trichoderma* fungus has a very poor "pathogenic reputation" on any host, although various species of this genus commonly inhabit most cultivated soils. If, however, tobacco roots are infected with *M. incognita, Trichoderma* is able to move into such roots and induce decay. Under these conditions, the syndrome is comparable to necrosis resulting from the combination of *Pythium* with *M. incognita* (Melendéz and Powell, 1969).

The realization that disease complexes occur, which involve root-knot nematodes and other microorganisms considered to be soil saprophytes, should elevate the importance of biopredisposition by nematodes. It means that host ranges may be altered, depending upon nematode populations present in the soil where the plant is growing. It might further indicate that much of the damage such as root decay, heretofore ascribed to nematodes alone, is in reality, damage resulting from an interaction between the nematode and another, ostensibly innocuous, soil inhabitant. All root-knot nematode–fungus combinations do not interact in all hosts to result in increased root decay. An example of this may be found in the reaction of oats to *M. incognita acrita* and *Helminthosporium victoriae* (Stavely and Crittenden, 1967).

Cyst-forming nematodes, belonging to the genus *Heterodera,* are capable of promoting increases in certain root rots caused by fungi. *Heterodera schachtii* and *R. solani* combine to promote root and seedling damage in sugar beet (Polychronopoulas *et al.,* 1969). Seedling infection by both pathogens produces a rapid softening and decay of tissue. Penetration and establishment by the fungus appears to be facilitated by nematode-induced wounds and root proliferation. Giant cells resulting from nematode activity are a very suitable substrate for fungus growth, and mycelium readily invades adjacent areas as well.

*Heterodera rostochiensis* has been reported to interact with both *R. solani* and *Colletotrichum atramentarium* (the causal fungus of brown root rot) in tomato (Dunn, 1968). This complex appears to be complicated by possible interrelationships between the two fungi. James (1968) did not observe an increase in susceptibility to *C. atramentarium* resulting from nematode activity but did notice a fungal effect on nematode behavior.

It is evident that root-rot complexes involving nematodes and fungi are among the most widespread and important interactions. Only recently have investigators begun to give deserved attention to problems of this type.

### C. Nematode–Fungus Seedling Disease Interactions

Fungi and nematodes combine to increase damage from disorders which may be classified as seedling diseases. Under greenhouse conditions, the citrus nematode apparently interacts with *F. solani* to reduce citrus seedling growth, and the effect of the pathogen combination is noticeably greater than that of either pathogen alone. This may be important in subsequent root rot in citrus (Van Gundy and Tsao, 1963).

Cotton seedlings grown in soil infested with *M. arenaria, M. hapla, M. incognita, Rotylenchulus reniformis,* or *Hoplolaimus tylenchiformis* are susceptible to *R. solani* longer than those grown in the absence of nematodes. Plants are susceptible longer to *P. debaryanum* only when either *M. incognita* or *M. hapla* is present. On the other hand, *Tylenchorhynchus claytoni* and *Trichodorus christiei* do not lengthen the period of susceptibility to either fungus (Brodie and Cooper, 1964). The differential abilities of various nematodes to predispose seedlings to infection by different fungi certainly point to variations in host responses to nematodes.

Most research in plant pathology has dealt with the effect of one or, at most, two pathogens on the host. In nature, however, plants may be expected to be infected by more than two pathogens simultaneously or in various sequences. Because of their exposure to whatever is present in the environment, it would be naive to assume that plants somehow escape the influence of several pathogens at once. Needless to say, the consequences of any one of these invaders could be detrimental, but the combined action of several could be devastating. Such is the case in tobacco in which *M. incognita,* F. *oxysporum* f. *nicotianae,* and *Alternaria tenuis* interact to form a destructive complex. The root-knot nematodes predispose plants to *Fusarium* wilt, and in turn, plants infected by these two pathogens are noticeably predisposed to *A. tenuis,* a foliage pathogen (Powell and Batten, 1969). No one can estimate how many truly multiple interrelationships such as this exist in nature or how much loss they cause.

## III. NEMATODE–BACTERIA INTERACTIONS

There are relatively few described complexes involving nematodes and bacteria when compared to the number including nematodes and fungi. Of course, the number of bacteria recognized as plant pathogens are few compared to the number of fungus pathogens. There are valid cases, however, of interactions between bacteria and nematodes which have real economic consequences. The nematode–bacteria complexes have also been the subject of recent reviews (Pitcher, 1963, 1965; Miller, 1965; Weischer, 1968).

### A. Nematode–Bacteria Wilt Disease Interactions

As with fungi, some of the most striking complexes between nematodes and bacteria include those bacteria which induce wilting in the host as one of the primary symptoms. Bacterial wilt (or Granville wilt) of several host plants, caused by *Pseudomonas solanacearum,* is one of the world's best-known diseases of this type. It has been known for a long time that nematodes, present along with the bacterial pathogen, greatly increase the incidence of this disease (Pitcher, 1963). Most early work with this interaction was concerned with inoculation experiments or observations showing an increase in disease when both pathogens were present. More recent research has been designed to determine differential abilities among various nematodes to promote disease incidence. In tomato, both the root-knot nematode, *M. hapla,* and the spiral nematode, *Helicotylenchus nannus,* contribute to an increase in wilt development.

*Helicotylenchus nannus* is not quite as effective as *M. hapla* in promoting wilt but still exercises a significant effect. In the same experiments, another spiral nematode *Rotylenchus* sp. (nonparasitic on tomato) did not influence the rate or severity of wilt symptoms (Libman *et al.*, 1964).

*Meloidogyne incognita* greatly influences bacterial wilt development in tobacco. Plants exposed to nematodes 3 or 4 weeks before exposure to *P. solanacearum* developed more severe wilt symptoms earlier than when plants were exposed to only bacteria or to both pathogens simultaneously (Johnson and Powell, 1969). Giant cells in jointly infected roots harbored bacteria-like inclusions and degenerated rapidly. Root decay soon followed.

A number of ectoparasitic nematodes also increase disease development in plants infected by *P. solanacearum* (Pitcher, 1965).

Bacterial wilt of alfalfa, caused by *Corynebacterium insidiosum*, is increased by the stem nematode *Ditylenchus dipsaci*. The effects have been demonstrated in field as well as in greenhouse tests (Hawn, 1963, 1965). The basis for this interaction may lie in transmission of the bacteria by *D. dipsaci*. Even varieties which have high resistance to wilt become diseased in the presence of *D. dipsaci*. On the other hand, varieties with nematode resistance remain relatively free from wilt when exposed to both pathogens (Hawn and Hanna, 1967).

### B. Nematode–Bacteria Root Disease Interactions

A number of bacterial pathogens normally inhabit soil and may become pathogenic on roots of susceptible plants. Symptoms of such infections may be manifest as galling or as root decay. For example, hairy root of roses, caused by *Agrobacterium rhizogenes*, is usually of minor importance, but serious losses have been recorded recently in southern California. This condition, apparently caused by a new strain of *A. rhizogenes*, is frequently correlated with infestations of the lesion nematode, *P. vulnus* (Munnecke *et al.*, 1963).

Crown gall of peach, caused by *A. tumefaciens*, is increased by high populations of the Javanese root-knot nematode, *M. javanica*. At low nematode population levels, however, crown gall symptoms are no more severe than those occurring in plants inoculated with bacteria alone after wounding (Nigh, 1966).

Some bacteria, usually regarded as secondary invaders, appear to be implicated in necrosis of tomato roots infected by *M. incognita* (Mayol and Bergeson, 1969). Reductions in foliage growth occur only when plants are inoculated with nematodes and maintained septically. Galls on such plants become necrotic and, upon isolation, yield many bacterial colonies when compared to similar plants grown under aseptic conditions. Aseptically grown roots, although heavily galled, remain firm. Most of the bacteria isolated from necrotic roots under septic conditions belong to the nutritional group associated with root rots. Fungi are also found occasionally in such roots. Knowledge such as this, as is true with fungal root rots, aids significantly in understanding damage resulting from complexes and should place renewed emphasis on their importance.

### C. Nematode–Bacteria Foliage Disease Interactions

The foliage and meristem disorder in strawberry known as *cauliflower* involving *Aphelenchoides ritzema-bosi* and the bacterium, *Corynebacterium fascians,* provides another striking example of a nematode–bacterium complex (Crosse and Pitcher, 1952; Pitcher and Crosse, 1958; Pitcher, 1963). Evidently, the cauliflower disease is a complex requiring both the nematode and the bacterium for expression of the complete disease syndrome. Different bacterial strains, most of which are saprophytic unless nematodes are present, apparently are responsible for the variety of cauliflower symptoms observed.

## IV. NEMATODE–VIRUS RELATIONSHIPS

Interest has increased recently in possible interrelationships between nematode and virus pathogens of plants. Most of this interest has been expressed primarily in studies of virus–nematode vector relationships and is a significant aspect of plant pathology. This topic is discussed separately elsewhere in this treatise. These two types of pathogens may interact in complexes that are manifest in other ways. For example, soybeans infected by both root-knot nematodes and the tobacco ring spot virus, a strain of which causes a bud blight disease in this host, suffer from extensive galling and greatly reduced root systems. Roots of plants infected by the nematodes alone are galled but not reduced in size and the virus alone has no obvious effects on the roots (Ryder and Crittenden, 1962). Accompanying changes in histopathology occur in infected plants. These phenomena occur in this complex although infection by each pathogen takes place singly without a vector relationship.

The tobacco ring spot virus and the tobacco mosaic virus influence the relationship established between *M. javanica* and root tissue (Bird, 1969). More nematode larvae enter the roots of tobacco ring spot virus-infected bean plants than virus-free plants, but there is no difference in rate of nematode development; nor is there evidence of synergism between the two pathogens on root development. These root-knot nematodes grow more rapidly in roots of tomato plants infected with tobacco mosaic virus than in virus-free roots. In the latter case, the virus has no observable effect on the number of nematodes entering the root. No vector relationship exists in either instance.

Populations of both *A. ritzema-bosi* and *D. dipsaci* are greatly decreased in tobacco plants which are also infected with the tobacco mosaic virus (Weischer, 1969). Also, growth retardation of plants under combined attack is "unexpectedly high." This implies that damage by a combination of the virus and nematode pathogen is greater than would be evident when either pathogen is present alone.

Preliminary research on tobacco in North Carolina suggests that certain systemic viruses and nematodes do interact within the host, although no vector relationship is involved (Osores-Duran, 1967). The expression of the interaction appears to be dependent upon the combination of nematode, virus, and tobacco cultivar involved. Plants of the $F_2C_1$ tobacco cultivar are stunted less when inoculated with both *M. incognita* and the tobacco ring spot virus than when the nematodes are present alone. Also, plants inoculated with both pathogens have more galling than plants receiving the nematode alone.

A somewhat similar effect occurs with *M. incognita* and the cucumber mosaic virus in the $F_2C_1$ cultivar. The pronounced stunting effect noted when the virus is applied alone is reduced when nematodes are present. Plants receiving both pathogens are generally taller and have more lateral roots than those inoculated with only the virus.

The typical reduction in top weight of plants of the tobacco cultivar, Xanthi, by root-knot nematodes is largely overcome when plants are also infected by the potato virus X. Conversely, the reduced top weight in $F_2C_1$ tobacco resulting when plants are infected with the potato virus X is substantially overcome by the presence of *M. incognita*.

An additive effect on plant height occurs in the association of root-knot nematodes and the potato virus Y in Xanthi tobacco plants. Stunting in plants inoculated with both pathogens is approximately equal to the sum of the stunting effects in plants harboring each pathogen separately. Although the disease syndromes with virus–nematode interactions are generally similar to those occurring as a result of the individual pathogens, there may be occasional deviations. For example, plants of the Xanthi tobacco cultivar jointly infected with *M. incognita* and the cucumber mosaic virus often exhibit a curvature of the stem with the production of new shoots where the stem is bent. This symptom is not evident in plants infected with either pathogen alone (Osores-Duran, 1967).

There are often changes in root-knot nematode reproduction rates when plants are jointly infected with a virus and this nematode. These changes are usually reflected as increases in the number of nematode eggs in roots of plants inoculated with both types of pathogens in combination.

It should not be surprising to find striking interactions between nematode and virus pathogens, even in the absence of a vector relationship. Actually, it is inconceivable that pathogens such as these would not have noticeable effects on host physiology. Accordingly, both pathogens, being obligate parasites, must be influenced by any physiological change in the host.

## V. SOME EFFECTS OF COMPLEXES ON NEMATODE POPULATIONS

Research with complexes involving plant parasitic nematodes has consistently and increasingly pointed to obvious effects on nematode populations by other components of the complex. Although these effects are frequently noted, uncertainty exists as to why these effects occur.

*Verticillium* wilt, for example, seems to render plants more suitable for colonization by lesion nematodes, *Pratylenchus* spp. When the fungus is added to soil already infested with the nematodes, reproduction of the latter is increased in roots of both eggplant and tomato. Reproduction of *Tylenchorhynchus capitatus* also increases in tomato roots when the *Verticillium* fungus is present. Physiological changes evidently occur in fungus-infected plants, making them more susceptible or attractive to the nematode pathogen (Mountain and McKeen, 1962).

Reproduction by *P. penetrans* is also enhanced by the presence of *V. dahliae* in roots of American elm and sugar maple. Similar effects are noted when certain nutrients, particularly nitrogen, are applied (Dwinell and Sinclair, 1967). Potatoes infected by *V. albo-atrum* also support greater reproduction of *P. penetrans* than fungus-free plants (Morsink and Rich, 1968). Although most evidence indicates strongly that *Verticillium* wilt generally promotes increases in reproduction of *Pratylenchus* spp., the effect is not apparent in all cases. Roots of *Verticillium*-infected pepper plants do not support higher nematode populations than roots of *Verticillium*-free plants (Mountain and McKeen, 1962; Olthof and Reyes, 1969).

The *Fusarium* wilt fungi seem to enhance reproduction of migratory nematodes. *Tylenchorhynchus claytoni*, although not contributing to wilt development, increases more rapidly on pea roots infected with *F. oxysporum* f. *pisi* race 1 than in roots free from the fungus (Davis and Jenkins, 1963). More *P. penetrans* larvae enter alfalfa roots infected with

*F. oxysporum* than noninfected roots. The same is true if roots are treated with *Trichoderma viride* (Edmunds and Mai, 1966a; Hirano and Kawamura, 1965). Also, more larvae of this species migrate toward alfalfa roots inoculated with the *Fusarium* wilt fungus than toward noninfected roots. This attraction may result from an affinity of nematodes for $CO_2$, which is released from fungus-infected roots in greater quantity than from healthy roots (Edmunds and Mai, 1967).

Although most examples of fungal effects on nematode populations involve one of the migratory nematodes, sedentary parasites are influenced in some instances. In soybean, higher populations of *Heterodera glycines* develop in *Fusarium*-infested soil than in soil free from the fungus (Ross, 1965).

Effects on nematode population such as those exercised by *Fusarium* fungi obviously are real and important. These interrelationships could have profound influences on population dynamics in soil (Seinhorst, 1968). Of course, not all fungal pathogens would result in comparable effects on nematode populations (Kisiel *et al.*, 1969).

Several situations occur wherein fungi have depressing effects on nematode activity. Most of these examples involve a sedentary endoparasite. The brown root rot fungus of potatoes significantly reduces egg hatching in *Heterodera rostochiensis*. This seems to result from the influence of exudates produced by the fungus (James, 1966). The male sex ratio of *H. rostochiensis* on tomato is increased when either *Rhizoctonia solani*, *Verticillium albo-atrum*, or the gray sterile fungus is present, and this increase is closely correlated with an increase in fungus inoculum. *Verticillium* has more influence on sex ratios than the other fungi. These effects may be related to giant cell destruction by the fungi (Ketudat, 1969).

Reproduction of *M. incognita* is depressed in chrysanthemum roots which are also infected with *Pythium aphanidermatum* (Littrell and Johnson, 1969). *Plasmodiophora brassicae*, which causes clubroot in cabbage, has a marked effect on development of *M. incognita acrita* in roots. In plants jointly infected by both pathogens, lysigenomata (giant cells) around the nematode head are often occupied by the fungus and the nematodes seldom develop to maturity, thus reducing the nematode population (Ryder and Crittenden, 1965).

## VI. THE NATURE OF COMPLEXES INVOLVING NEMATODES

In few, if any, interactions can the true basis or nature of the complex be stated definitely and categorically. Indeed, this may never be possible with many interrelationships between nematodes and other disease-causing agents. However, it does seem that a general pattern may be emerging from the rather diverse and intensive investigation that these problems have been receiving recently.

It is fairly evident that mechanical wounding induced by nematodes as they penetrate plant tissue is important with certain disease complexes. Such a factor may be especially relevant in those cases involving a bacterial pathogen and a nematode. It is also evident that mechanical effects of nematodes cannot account for all the predisposition of plants to subsequent invasion by other microorganisms. Evidence, some of which is indirect, is accumulating that physiological changes in the host, wrought by nematode activity, are responsible for changes in host susceptibility to pathogens of other types.

Histopathological analyses of multiple infection sites provide a portion of this evidence. In a number of cases, the locus of nematode activity in the plant is more extensively colonized by a fungus than comparable tissue which has not been influenced by nematodes (Powell and Nusbaum, 1960; Meléndez and Powell, 1967; Minton and Minton, 1963; Polychronopoulos *et al.*, 1969). This indicates that such tissue has been physiologically changed, and that the changes influence fungal growth and development.

Alterations in host physiology owing to nematode parasitism are not restricted to the site of nematode activity, however. The predisposition to *Fusarium* wilt of the side of the root system opposite from the location of nematode inoculation shows that these changes extend within the plant (Bowman and Bloom, 1966). Research with *Verticillium* wilt and *Pratylenchus minyus* in peppermint points to similar conclusions (Faulkner and Skotland, 1965), as do interrelationships between certain viruses and nematodes (Osores-Duran, 1967).

Increasing evidence of involvement of nematodes, particularly members of the root-knot group, in root decay complexes also suggest physiological changes in nematode-infected roots. Interactions of *Meloidogyne* spp. with fungi or bacteria, which appear harmless *unless the nematode is present,* emphasize the fact that the root systems have been substantially modified (Powell, unpublished data; Mayol and Bergeson, 1969; Edmunds and Mai, 1966b). Modification obviously permits invasion by microorganisms which are unable to colonize nematode-free roots. The fact that predisposition to certain diseases by root-knot nematodes reaches its maximum only after the nematodes have been within the host for several weeks further emphasizes the importance of physiological preconditioning.

There is ample evidence that biochemical constituents of plant roots are substantially altered by nematode infection, and these changes include a large range of both organic and inorganic materials. However, which of these are directly involved in predisposition to subsequent infections is not clear. Changes of the magnitude that have been reported must have sizeable impact on this overall problem.

## VII. CONCLUSIONS

It must now be concluded that nematodes are of tremendous importance as components of disease complexes along with other disease-causing agents. In fact, interactions may be the major economic hazard posed by these pathogens. Some of the damage normally ascribed to nematodes alone may result from a complex of which nematodes are only a part—but a vital part. This may be the case in several root decay syndromes involving root-knot nematodes. At any rate, it appears that as components of interactions, nematodes reach their ultimate in causing destruction. This places renewed emphasis on the importance of their control in crop culture.

Another logical conclusion is that nematodes contribute to disease complexes mostly by modifying the physiology of the host plant and less by mechanical effects they exert on the host. As part of an interaction, nematodes are one of at least three different biological systems operative within the host; namely, that of the host, the nematode, and the other disease-causing agent which forms the complex. The metabolic activities of any one of these must influence those of the other components.

Although a sizable number of interactions have been documented, especially during recent years, the surface of this vast area of plant pathology has hardly been touched. No one is capable of estimating how many more complexes exist. These may be subtle cases which are not readily evident, but which are, nevertheless, of profound agricultural importance. In reality, the subtle or incipient interrelationship may, in the final analysis, be the one of greatest significance. Thus, research should be directed toward further determination of the roles played by disease complexes in plant disease losses. It is to be suspected that they are far more important than has been previously realized. This is especially true if the interaction includes a nematode along with a normally nonaggressive pathogen, or one which does not attack a particular host plant unless it has been predisposed by nematode infection. The latter instance provokes serious questions regarding established host ranges, especially when soil-borne pathogens are involved.

Epidemiology must also be considered in disease complex investigations. Effects of interactions on this important facet of plant pathology could be profound. For example, changes in nematode reproduction in the presence of certain other organisms appear to be great.

Information on nematode-induced physiological changes in plants should be applied to answer questions related to the nature of interactions. Without doubt, some of these alterations render plants susceptible to attack by other types of pathogens. Development of satisfactory control of certain diseases may be realized only through an understanding of these basic principles of plant pathology.

## REFERENCES

Atkinson, G. F. (1892). *Ala. Polytech. Inst. Agr. Exp. Sta. Bull.* **41,** 61–65.
Benedict, W. G., and Mountain, W. B. (1956). *Can. J. Bot.* **34,** 159–174.
Bergeson, G. B. (1963). *Phytopathology* **53,** 1164–1166.
Bird, A. F. (1969). *Nematologica* **15,** 201–209.
Bowman, P., and Bloom, J. R. (1966). *Phytopathology* **56,** 871. (Abstr.)
Brodie, B. B., and Cooper, W. E. (1964). *Phytopathology* **54,** 1023–1027.
Cooper, W. E., and Brodie, B. B. (1963). *Phytopathology* **53,** 1077–1080.
Crosse, J. E., and Pitcher, R. S. (1952). *Ann. Appl. Biol.* **39,** 475–486.
Davis, R. A., and Jenkins, W. R. (1963). *Phytopathology* **53,** 745. (Abstr.)
Dunn, E. (1968). *8th Int. Symp. Nematol., Antibes, 1965* p. 115.
Dwinell, L. D., and Sinclair, W. A. (1967). *Phytopathology* **57,** 810. (Abstr.)
Edmunds, J. E., and Mai, W. F. (1966a). *Phytopathology* **56,** 1320–1321.
Edmunds, J. E., and Mai, W. F. (1966b). *Phytopathology* **56,** 1132–1135.
Edmunds, J. E., and Mai, W. F. (1967). *Phytopathology* **57,** 468–471.
Faulkner, L. R., and Bolander, W. J. (1969). *Phytopathology* **59,** 868–870.
Faulkner, L. R., and Skotland, C. B. (1965). *Phytopathology* **55,** 583–586.
Goode, M. J., and McGuire, J. M. (1967). *Phytopathology* **57,** 812. (Abstr.)
Hawn, E. J. (1963). *Nematologica* **9,** 65–68.
Hawn, E. J. (1965). *Nematologica* **11,** 39. (Abstr.)
Hawn, E. J., and Hanna, M. R. (1967). *Can. J. Plant Sci.* **47,** 203–208.
Hendrix, F. F., Powell, W. M., Owen, J. H., and Campbell, W. A. (1965). *Phytopathology* **55,** 1061. (Abstr.)
Hirano, K., and Kawamura, T. (1965). *Nipp'on Shokubutsu Byori Gakkaiho* **30,** 24–30.
Inagaki, H., and Powell, N. T. (1969). *Phytopathology* **59,** 1350–1355.
James, G. L. (1966). *Nature (London)* **212,** 1466.
James, G. L. (1968). *Ann. Appl. Biol.* **61,** 503–510.
Jenkins, W. R., and Coursen, B. W. (1957). *Plant Dis. Rep.* **41,** 182–186.
Johnson, A. W., and Littrell, R. H. (1969). *J. Nematol.* **1,** 122–125.
Johnson, H. A., and Powell, N. T. (1969). *Phytopathology* **59,** 486–491.
Ketudat, U. (1969). *Nematologica* **15,** 229–233.
Kisiel, M., Deubert, K., and Zuckerman, B. M. (1969). *Phytopathology* **59,** 1387–1390.
Kushner, V. D., and Crittenden, H. W. (1967). *Phytopathology* **57,** 646. (Abstr.)

Labruyére, R. E., Den Ouden, H., and Seinhorst, J. W. (1959). *Nematologica* **4,** 336–343.
Libman, G., Leach, J. G., and Adams, R. E. (1964). *Phytopathology* **54,** 151–153.
Littrell, R. H., and Johnson, A. W. (1969). *Phytopathology* **59,** 115–116 (Abstr.)
Mayol, P. S., and Bergeson, G. B. (1969). *J. Nematol.* **1,** 17. (Abstr.)
Melendéz, P. L., and Powell, N. T. (1967). *Phytopathology* **57,** 286–292.
Melendéz, P. L., and Powell, N. T. (1969). *Phytopathology* **59,** 1348. (Abstr.)
Miller, C. R. (1968). *Phytopathology* **58,** 553. (Abstr.)
Miller, H. N. (1965). *Soil Crop Sci. Soc. Fla., Proc.* **24,** 310–325.
Minton, N. A., and Jackson, C. R. (1967). *Nematologica* **13,** 146. (Abstr.)
Minton, N. A., and Minton, E. B. (1963). *Phytopathology* **53,** 624. (Abstr.)
Minton, N. A., and Minton, E. B. (1966). *Phytopathology* **56,** 319–322.
Morsink, F., and Rich, A. E. (1968). *Phytopathology* **58,** 401. (Abstr.)
Mountain, W. B. (1954). *Can. J. Bot.* **32,** 737–759.
Mountain, W. B., and McKeen, C. D. (1962). *Nematologica* **7,** 261–266.
Munnecke, D. E., Chandler, P. A., and Starr, M. P. (1963). *Phytopathology* **53,** 788–799.
Nigh, E. L., Jr. (1966). *Phytopathology* **56,** 150. (Abstr.)
O'Bannon, J. H., Leathers, C. R., and Reynolds, H. W. (1967). *Phytopathology* **57,** 414–417.
Olthof, T. H. A. (1968). *Nematologica* **14,** 482–488.
Olthof, T. H. A., and Reyes, A. A. (1969). *J. Nematol.* **1,** 21–22. (Abstr.)
Osores-Duran, A. (1967). M.S. Thesis, 146 pp. North Carolina State Univ., Raleigh, North Carolina.
Palmer, L. T., MacDonald, D., and Kommedahl, T. (1967). *Phytopathology* **57,** 825. (Abstr.)
Pitcher, R. S. (1963). *Phytopathology* **53,** 35–39.
Pitcher, R. S. (1965). *Helminthol. Abstr.* **34,** 1–17.
Pitcher, R. S., and Crosse, J. E. (1958). *Nematologica* **3,** 244–256.
Polychronopoulos, A. G., Houston, B. R., and Lownsbery, B. F. (1969). *Phytopathology* **59,** 482–485.
Porter, D. M., and Powell, N. T. (1967). *Phytopathology* **57,** 282–285.
Powell, N. T. (1963). *Phytopathology* **53,** 28–35.
Powell, N. T., and Batten, C. K. (1967). *Phytopathology* **57,** 826. (Abstr.)
Powell, N. T., and Batten, C. K. (1969). *Phytopathology* **59,** 1044. (Abstr.)
Powell, N. T., and Nusbaum, C. J. (1960). *Phytopathology* **50,** 899–906.
Ross, J. P. (1965). *Phytopathology* **55,** 361–364.
Ryder, H. W., and Crittenden, H. W. (1962). *Phytopathology* **52,** 165–166. (Abstr.)
Ryder, H. W., and Crittenden, H. W. (1965). *Phytopathology* **55,** 506. (Abstr.)
Seinhorst, J. W. (1968). *Nematologica* **14,** 549–553.
Stavely, J. R., and Crittenden, H. W. (1967). *Plant Dis. Rep.* **51,** 470–473.
Stover, R. H. (1966). *Can. J. Bot.* **44,** 1703–1710.
Van Gundy, S. D., and Tsao, P. H. (1963). *Phytopathology* **53,** 488–489.
Weischer, B. (1968). *8th Int. Symp. Nematol., Antibes, 1965* p. 103. (Engl. sum.)
Weischer, B. (1969). *Nematologica* **15,** 334–336.

CHAPTER 19

# Feeding in Plant Parasitic Nematodes: Mechanisms and Behavior

C. C. DONCASTER

*Rothamsted Experimental Station, Harpenden, Herts., England*

I. Introduction . . . . . 137
II. Behavior Leading to Feeding . . . . . 139
A. Exploration . . . . . 139
B. Penetration . . . . . 140
III. Pharyngeal Gland Secretions . . . . . 141
IV. Ingestion . . . . . 143
A. Host–Parasite Relationships . . . . . 143
B. Ingestion Mechanisms . . . . . 144
References . . . . . 156

## I. INTRODUCTION

This review of feeding by plant parasitic nematodes refers only to those members of the Tylenchida and Dorylaimida that attack higher plants or fungi. Besides outlining our present knowledge, this chapter is intended to show the features of feeding on which information is needed to clarify the principles of the process.

Feeding has been studied most often in agar preparations which are convenient for microscopy and for maintaining sterility but which nevertheless constitute an unnatural environment that affects penetration and other behavior. In Linford's (1942a) methods of observation the physical conditions were nearly natural and the optical conditions good enough for large microscopic magnifications, but only Pitcher (1967) showed the behavior of nematodes directly in relation to their hosts in an almost unrestricted, if disturbed, soil environment. In Table I (feeding behavior) (p. 148), observational techniques are not mentioned when they were standard agar culture methods.

In this account, Goodey's (1963) classification of nematodes is followed. In referring to some morphological structures that seem inappropriately named, an attempt is made to use unambiguous alternatives that describe the structures' probable functions. The author regards the *feeding apparatus* as consisting not only of the spear, but also all parts of the alimentary system concerned with ingestion, together with the glands; the *head* is regarded as extending back to a short distance behind the stylet knobs. The term *valve* will be used only when it refers to a specific part of a tubular system, e.g., the pharynx that by opening and closing helps to maintain unidirectional flow of a fluid passing through it. A region that seems able to impel its fluid contents, with the aid of valves will be called *pump*, e.g., the metacorpal *valvular apparatus*. *Pharynx* is preferable to *esophagus* for the reason given in Hyman (1951). Hirumi *et al.* (1968) and Raski *et al.* (1969) used *pharynx* and *esophagus* for anterior and posterior parts, respectively, of the alimentary canal between the stoma and the intestine of *Trichodorus*. Their terminology will be followed in this genus, although homologies with other nematodes are uncertain. In Table I, *probing* is used to denote gentle, exploratory stylet thrusts, whereas *thrusting* implies a powerful, piercing stroke.

Compared with other invertebrates, the class Nematodea is remarkably uniform morphologically but outstandingly successful and well adapted to a wide range of environments. Nematodes' adaptability seems to lie more in their physiology and behavior than in their morphology. Nevertheless, the mouth spear is a morphological adaptation which enables many Tylenchida and Dorylaimida to lead parasitic or predatory modes of life. However, not even the major roles of the spear were appreciated until Christie and Arndt (1936) saw the feeding of *Aphelenchus avenae* and *Aphelenchoides parietinus*. Until 1960, feeding observations were on a few economically important species (reviewed by Mankau and Linford, 1960; Seinhorst, 1961), but since then three times as many phytoparasitic nematodes have been studied.

Determining the fine structure of the feeding apparatus of some nematodes has contributed much to an understanding of mechanical principles, but more analyses of movements of the structures in living specimens are needed. Furthermore, relationships of the internal pressures of hosts and parasites, though difficult to study directly, are important and not yet understood.

## II. BEHAVIOR LEADING TO FEEDING

### A. Exploration

Jones (1960), Klingler (1965), and C. D. Green (Chapter 24) reviewed the physical and chemical nature of stimuli that attract nematodes to their hosts; therefore, this section is restricted to the behavior of nematodes on reaching their hosts.

Nematodes differ in their specificity for hosts and for parts of hosts and this sometimes changes with age, e.g., *Xiphinema index* (Fisher and Raski, 1967), see Table I.

On first reaching their host, most nematodes seem to explore it. They move over it, frequently turn their head toward it, and probe it with the mouth stylet. Probing tends to increase, but locomotion decreases and finally stops. Usually, the body lies more or less bent about projections from the substratum so that forward thrust can be applied. The nematode presses its lip region against the host and apparently obtains the tactile or chemical stimuli necessary to start penetration or feeding by a quick side-to-side rubbing with the lips. The stylet is thrust with increasing vigor, sometimes first at two or three selected points, then at one of them until it penetrates. The nematode may thrust its stylet into two or three cells before selecting one for feeding. If trial penetrations are to sample the contents, they seem not to include ingestion.

This general scheme is modified in different species by eliminating, shortening, or prolonging different stages. Predatory nematodes and some parasitic ones, e.g., *Paraphelenchus acontioides* (Taylor and Pillai, 1967), shorten their prefeeding behavior most and may simply penetrate by a few rapid stylet thrusts the first cell they touch. The exploratory behavior of the slow-feeding ectoparasites and of endoparasites is prolonged and sometimes suggests the nature of their sensory perceptions as discussed in Section II, B.

Larvae of *Heterodera marioni* (syn. *Meloidogyne* sp.) (Linford, 1937c, 1942b), *H. trifolii* (Mankau and Linford, 1960), and *H. cruciferae* (Doncaster *et al.*, 1968) moved less while exploring and thrust their stylets more when cell contents escaped from damaged areas and when their lips and stylets made contact with the root. Peacock (1959) noted that more *Meloidogyne incognita* larvae were attracted to tomato roots where a penetration hole existed than to unattacked roots.

### B. Penetration

Doncaster and Shepherd (1967) and Doncaster *et al.* (1968) showed that *H. rostochiensis* and *H. cruciferae* larvae thrust their stylets more strongly and more often at semirigid membranes than at softened or flexible ones. In a living root *H. cruciferae* perforated first the periphery of the cross wall of a cortical cell where the wall was held most rigidly by being joined to the lateral walls. Subsequent perforations merged with the preceding ones and weakened the cell wall, which eventually broke under continuous pressure from the nematode. By similar methods, second-stage *H. rostochiensis* larvae cut slits in their eggshells for hatching and they accurately controlled the strength and placement of their stylet thrusts. When a thrust failed to perforate the eggshell, the nematode thrust harder at the same point or repositioned its thrust nearer the last to keep the slit continuous. By probing with its stylet, a larva located the end of a short slit made previously and continued it. Such behavior suggests that the stylet combines sensory functions with mechanical ones. The labial papillae seem to be tactile and, at each thrust of the stylet against a solid object, may feel an opposing reaction by diminished pressure on the lips. Such a mechanism would provide the nematode with only an indirect assessment of the strength and effects of stylet thrusts, which might be modified by factors such as the texture of the object thrust against and the force with which the body pushes forward, because this force must oppose back thrust of the lip region. However, proprioreception associated with the stylet protractor muscles, could provide a direct assessment of the effect of each thrust and would best explain the fine degree of control that these nematodes appear to possess.

Several observers (Dickinson, 1959; Seinhorst, 1961; Wallace, 1963; Fisher and Evans, 1967) considered how nematodes resist the back thrust produced by exserting the stylet against a resistant surface. Most think that frictional forces between the body and the substratum are more important than adhesion by suction between the lips and the host. McElroy and Van Gundy (1968) described how *Hemicycliophora arenaria* thrust its long stylet deep into roots and how, toward the end of penetration, the lips broke contact with the host's epidermis and then the nematode drove the stylet in by thrusting forward with its whole body. Frictional forces similar to those operating in locomotion were therefore essential for penetration, although secretions might have softened the cell walls and lessened resistance to the stylet.

Few publications refer to mechanisms of spear retraction. Wright

(1965) first demonstrated stylet retractor muscles in *Xiphinema,* but none are known in the Tylenchida. Raski *et al.* (1969) concluded that in *Trichodorus,* pharyngeal wall turgor increases by contraction of the spear protractor muscles and retracts the stylet when the muscles relax. Perhaps hydrostatic pressure somehow causes stylet retraction in the Tylenchida, also.

## III. PHARYNGEAL GLAND SECRETIONS

When secretions include granules, they can be seen in their ducts by phase contrast or normal, high-power microscopy, but nongranular secretions may not show without histochemical staining. Granular secretions include a fluid phase which often forms large clear zones close to the duct's outlet, apparently when emission is rapid (Anderson, 1964; Doncaster, 1966). No phytoparasitic nematode is known to emit secretions in granular form; thus, outside it in a living, unstained preparation, secretions might only show as a zone, differing in refractivity from the cytoplasm of the food cell, or their presence might be inferred from the host's response (see also Bird, 1969).

Mechanisms for controlling secretory movements need further study. Although valves have not been found in the gland ducts of phytoparasitic nematodes, Pillai (1966a) assumed that in *Ditylenchus triformis* a valve is associated with the dorsal duct's outlet. However, filling the ducts and discharging secretions probably depend on pressure gradients in all nematodes. The dorsal duct opens into the fore part of the pharynx which is permanently open to the exterior. The subventral ducts sometimes open into the pumping region, where diminished pressures withdraw their secretions, and sometimes into the valvular region close behind it. Ducts can usually be distended into ampullae or collecting reservoirs near their outlets; in *Trichodorus* the reservoirs have a complex structure (Raski *et al.,* 1969). In some *Ditylenchus* and *Heterodera* spp. and no doubt in many others, reservoirs can also form along most of the dorsal duct's passage through the procorpus and in the metacorpus (Anderson, 1964; Doncaster, 1966; Doncaster *et al.,* 1968).

In the ectoparasites, secretions are sometimes made to surge rapidly forward when they are to be injected into the host. In *Tylenchorhynchus dubius* secretions were expelled by a wriggling contraction of the body (Klinkenberg, 1963). Doncaster (1966) found that the body of *Ditylenchus myceliophagus* contracted when secretory outflow began, and Pillai (1966a) noted similar effects of body contractions in *D.*

*triformis*. Species that produce secretory surges probably always control them by controlling their body turgor, whereas in *Heterodera* and other species, salivary flow is slow and may depend on pressure developed in the gland during synthesis. Often, before pumping, the metacorpus of some Tylenchida twitches because of contractions by individual muscles and in *H. cruciferae* and *D. dipsaci*, the author has seen such contractions as well as those that operated the pump, squeezing the metacorpal reservoirs and ejecting their contents.

On leaving their ducts, secretions are either emitted (e.g., *D. myceliophagus*) or they mix with the food in the pharynx [e.g., *Hexatylus* (see Table I)] or they are passed to the intestine after the feed, as seems likely in *Aphelenchoides blastophthorus* and *A. ritzemabosi*. Because the dorsal duct opens in front of the pumping region, one would expect the dorsal gland secretions to be emitted the most easily, whereas secretions from the subventral glands would tend to pass to the intestine. Nevertheless, the short pump of many Tylenchida seems to be no barrier to the emission of subventral gland secretions, and Bird (1967; 1968a,b) and Bird and Saurer (1967) showed this could happen in *Meloidogyne*. Furthermore, they found that pharyngeal glands of *M. javanica* produced different substances according to developmental stage and environment. Shortly before the larvae hatched and penetrated the host, their subventral glands synthesized granules (6–8 μ diam) each bounded by a membrane. These secretions were mainly protein; the additional production of enzymes needs confirmation. Within 1–3 days of entering the host (i.e., up to the time when the larvae could no longer leave the host and reinvade it), the subventral ducts filled with smaller granules without membranes; these were mainly of carbohydrates. The three glands then enlarged, nongranular neutral mucopolysaccharide secretions were present in the subventral ducts, and giant cells began to form in the host. The dorsal gland synthesized granular, histonelike basic proteins of which most were produced by adult females, whose dorsal glands enlarged and subventrals diminished. Bird (1968b) believed dorsal gland secretions accelerate development of giant cells and thus facilitate feeding.

Though subventral secretions predominated, granular secretions occurred in all three ducts of larvae of *H. rostochiensis* and of *H. cruciferae* shortly before egg hatch; they also occurred in *H. cruciferae* larvae while penetrating the host and in newly emerged males which do not feed (Doncaster and Shepherd, 1967; Doncaster *et al.*, 1968). A week after entering their host, the larvae of *H. cruciferae*, like those of *Meloidogyne*, had the dorsal gland slightly enlarged and the subventrals diminished and without granular secretions.

In some mycophagous Tylenchida, unlike those that fed on higher plants, only dorsal gland secretions were visible (Anderson, 1964; Doncaster, 1966; Hechler, 1962; Pillai, 1966a). *Ditylenchus destructor* attacks both types of host, and when it fed on epidermis of vetch or carrot the subventral ducts quickly filled with granular secretions, but when it fed on fungi these ducts seemed empty. Soon after the stylet penetrated an epidermal cell, a zone of gland secretions became faintly visible in the cell surrounding the stylet tip. The zone did not change in size or shape until the last few minutes of the feed, but it became progressively more refractive, suggesting that an interface formed and the fluids on either side changed their relative densities. The volume of the food cell was not noticeably altered, yet feeds were prolonged; thus, some contents of neighboring cells might have diffused through the intervening cell walls and been ingested. McElroy and Van Gundy (1968) thought the subventral secretions of *H. arenaria* passed to the intestine. They believed the granular dorsal gland secretions were responsible for the host's reactions which seem to have involved increased permeability of cell walls. Apparently the contents of many cells around the food cell diffused into it, and ingestion could therefore proceed uninterrupted from one cell for some days.

## IV. INGESTION

### A. Host–Parasite Relationships

Harris and Crofton (1957) measured the hydrostatic body pressures of live *Ascaris* and obtained a mean value of 0.092 atm above atmospheric pressure (range, 0.021–0.296), probably considerably greater than normal internal pressures of other invertebrates. Internal pressures of the same order of magnitude probably occur in most nematodes and account for the almost universal possession of a muscular, pumping pharynx (Crofton, 1966). Yet, while most phytoparasitic and mycophagous nematodes seem actively to pump the contents of host cells into their intestines, higher plants and fungi have osmotic potentials of 5, 10, or even more than 20 atm. Although few direct measurements have been made of hydrostatic pressures of intact, living plant cells, Green (1968) measured the turgor of cells of the freshwater alga, *Nitella*, by introducing a fine capillary, closed at the distal end and containing an air bubble, the length of which was calibrated in terms of cell pressures. He found that turgor was normally about 4–6 atm, i.e., about 15 times the maximum Harris and Crofton recorded for *Ascaris*.

If, on puncturing plant cells, nematodes can inject gland secretions into them and then need to pump out the contents, they presumably dominate the pressure relationship. Several observers have found that very elastic plant cells (e.g., from parenchyma of pith or storage organs) may increase in volume by up to 40% of their minimal unplasmolized volume but can afterwards contract again. However, many other cells can swell by only 2–5% (see Bennet-Clark, 1959). Assuming such inelastic cells were punctured by phytoparasitic nematodes, their turgor would be at equilibrium with the nematode or the environment after very little of their contents had escaped and the nematode could then ingest the remaining contents only by pumping them out. But fungi, in which many cells are interconnected by perforated septa, might expel larger amounts of cytoplasm. This could, perhaps, be ingested under pressure by nematodes having a large enough stylet lumen to allow the most viscous fungal ingredients to pass (e.g., *Hexatylus* and *Aphelenchoides* spp.).

Nematodes with fine stylets, which inject secretions into the host, may do so to prevent viscous or granular material from blocking the stylet; thus, *D. myceliophagus* seems to gel and perhaps to predigest the semisolid components of the hyphal contents before it ingests the fluid ingredients and products of digestion (Doncaster, 1966).

Intestinal movements, such as those of *Neotylenchus* (Hechler, 1962), *D. destructor* (Anderson, 1964), *D. myceliophagus* (Doncaster, 1966), and *D. triformis* (Pillai, 1966a), if achieved by changing the body length, no doubt reflect changes of body turgor. They may thus significantly influence the fungus–nematode pressure relationship and therefore ingestion.

In several Tylenchida the lips stick to the surface of the host after the stylet has penetrated. In so doing, they might seal in the stylet and allow the plant cell to redevelop turgor and force its contents into the nematode. However, permeability of the cell, on which turgor partly depends, is regulated by the physiological state of the protoplast (Miller, 1938) which is no doubt affected by the nematode's gland secretions.

### B. Ingestion Mechanisms

Despite the different pharyngeal patterns in phytoparasitic nematodes, the principles involved in active ingestion must be similar to those described by Croften (1966). Where the pharyngeal lumen has the primitive triradiate pattern throughout, as in *Trichodorus*, *Hexatylus*, etc., a one-way flow of fluid into the intestine can be produced by dilating the forepart of the lumen to suck in fluid, followed by posterior dilation with simultaneous anterior closure, then finally closing the hind part. Dilation is achieved by contracting the full complement of radial muscles crossing the pharyngeal wall; their action is opposed by body turgor pressure. Dilation and muscular contraction also produce an additional opposing pressure within the wall because the pharynx is bounded by an elastic membrane; thus, when the muscles relax, the excess pressure in the wall collapses the lumen, the contents are expelled into the intestine, and equilibrium turgor is restored.

Many of the Tylenchida have pharynxes modified by shortening the pumping region. *Ditylenchus dipsaci,* described and figured in detail by Yuen (1968), illustrates such specialization. The mouth stylet leads directly into the procorpal lumen that is circular in section, thickly lined, noncollapsible, and about 0.08-$\mu$ radius. In the anterior third of the metacorpal pump the lumen changes from circular to triradiate, and there it is closed except when dilated by muscular contraction. The lumen continues back to the intestine in triradiate form but without dilator muscles. The isthmus lumen is incompletely closed, having a central radius of about 0.07 $\mu$; but when dilated, its radius would be about 0.23 $\mu$ and that of the postcorpus about 0.22 $\mu$. Analyses of cine film of *Ditylenchus dipsaci* and of *Aphelenchoides blastophthorus* feeding show no moving pharyngointestinal valve in either, but in both, the metacorpal lumen just behind the pump is compressed at pump dilation and dilated at pump closure. In one *A. blastophthorus* feeding on *Botrytis cinerea* a globule of food close behind the metacorpus did not move during pump dilation, but during closure it shot several microns down the isthmus. The lumen through the hind part of the metacorpus therefore functioned as an outlet valve from the pump. It was probably closed by pressure in the bulbar wall produced by muscular contraction and pump dilation. Hence, continued use of the term *valve apparatus* for the pump is misleading, as McElroy and Van Gundy (1968) have already said.

The existence of a pump with only an outlet valve is not, at first sight, sufficient to explain unidirectional flow. In the *A. blastophthorus* example described, one-way flow of food might have been caused by the fungus dominating the pressure relationship, and it might be argued that the operation of the pump was to impede the rate of inflow of food under pressure from the host. Nevertheless, by pharyngeal pumping, Tylenchida can actively impel fluids into their bodies against their internal pressures, as shown by several species ingesting fluids at atmospheric pressure (Myers, 1967). Therefore, a means must exist of limiting reverse flow when the metacorpal outlet valve opens. If a tightly closing inlet valve is present in *D. dipsaci,* it cannot be situated farther forward than the anterior third of the pump. However, reference to Poiseuille's law (see Barr, 1931) shows that a pump can operate with only an outlet valve present. This law states that the volume of liquid which flows per second $Q$ under pressure $P$ through a capillary is related to the length $L$ and radius $R$ of the capillary and to the viscosity $\eta$ by the equation $Q = PR^4/8L\eta$. If in a hypothetical nematode the pump lies midway along the pharynx, having a radius of 0.08 $\mu$ in front of the pump and 0.23 $\mu$ behind, then the resistance of the anterior tube to liquid flow will be $0.23^4/0.08^4 = 68.32$ times that of the posterior tube. With the posterior valve closed, the pump will fill via the anterior tube under the action of the strong radial muscles. With the pump full and the radial muscles relaxed, the residual wall pressure will expel the fluid both through the anterior tube and, via the valve, through the posterior tube; but the relative resistance will ensure that 68 times as much flows into the posterior tube and thence into the intestine. Over the whole cycle the net flow through the anterior tube is rearward, and although the tube is always open it acts as a "leaky valve." In practice, the posterior tube may not initially be distended to the full 0.23-$\mu$ radius, but the excess pressure exerted by the pump wall over the body turgor pressure acting on the wall of the posterior tube, will ensure acceptance of more fluid by the hind pharynx than by the anterior tube.

McElroy and Van Gundy (1968) commented on the extreme development of the metacorpus in *Hemicycliophora* and other Criconematidae. In *H. arenaria* they found inlet and outlet valves associated with a large metacorpal pump. Ingestion was assisted by a wave of muscular contractions passing along the isthmus.

The feeding mechanism of *Xiphinema index,* described by Wright (1965) and Roggen *et al.* (1967), differs from the specialized tylenchid pattern in the greater length of both stylet and pharynx and in the possession of at least two muscular systems that perhaps perform or assist functions carried out in the Tylenchida by hydrostatic pressure. Thus, Wright demonstrated a sheath composed of muscles and basement membrane (connective tissue) that surrounds the pharyngeal bulb. This may provide an additional means of increasing pharyngeal wall pressure above pseudocoelomic pressures, enabling the long lumen to be closed completely. Roggen *et al.* showed that a muscular system in the bulbar wall could close the lumen. Muscles that dilate the lumen against the body turgor share a peculiarity with those that close it, in being attached to thickenings of the lumen wall that act as levers. However, the production of unidirectional flow of food in *Xiphinema* is not specifically explained. The anterior pharyngeal lumen is circular in section and wider than the tylenchid counterpart (0.15–0.25 $\mu$ radius), but it can evidently be partly constricted by a further unusual set of muscles in the form of three radiating series (Wright, 1965). An analysis of the sequence of pharyngeal movements during ingestion would be valuable.

Hirumi *et al.* (1968) described in detail the spear and associated structures of *Trichodorus* and Raski *et al.* (1969) the pharynx and esophagus. Extra muscles exist which evert the stoma before spear exsertion and the spear protractor muscles are highly developed. However, the repeated thrusting and mode of operation of the spear in *Trichodorus* is not yet understood (Hirumi *et al.*, 1968). The spear consists of a protrusible outer part into which an inner spear is inserted obliquely, but neither is tubular throughout and food passes alongside them. Nevertheless, the vigor of spear thrusting during feeding suggests it is somehow involved in ingestion.

TABLE I

FEEDING BEHAVIOR OF PLANT PARASITIC AND MYCOPHAGOUS NEMATODES

| Nematodes | Sources of food (these observations) | Behavior leading to feeding | Pharyngeal gland secretions | Ingestion | General feeding characteristics, other data, effects on host | References |
|---|---|---|---|---|---|---|
| **Order Tylenchida, Superfamily Tylenchoidea** | | | | | | |
| **Family Tylenchidae** | | | | | | |
| *Tylenchus emarginatus* | Root epidermis *Picea mariana*, *P. rubens*, other *Picea* and *Pinus* spp. (No feeding on 10 fungal spp. from rhizospheres). | Preferences shown for host sp., not for region of root. 2 in 3 trial penetrations led to feeding; took 10 sec at 2 thrusts/sec. | Not seen. ≈ 20 sec quiescence before ingestion ?suggests injection.[a] | Regular pump pulsation lasting up to few minutes. Granular parts of food cell collected around stylet but? only sap ingested. | Ectoparasite. Feeds lasted 21–210 sec (mean, 102); 6–132 sec between feeds (mean, 39). Slight damage to host; other pathogens entered. | Sutherland (1967) |
| *Tetylenchus joctus* | (i)[c] Root epidermis *Vaccinium macrocarpon;* (ii) *Brassica oleracea* L.v. *gongylodes*, *Lycopersicon esculentum*, *Lactuca sativa*. | Region of differentiation, or about tip of root preferred. Several trial penetrations before feeding. 3–6 thrusts/sec. Stylet penetrated cell 2–4 μ. | No data | Rapid pump pulsation followed stylet penetration. Cell sap ingested first; cytoplasm later. | Ectoparasite. Feeds lasted ½–7 min. Some stylet thrusting at completion of feed. Food cells emptied. | Zuckerman (1960) (i), Khera and Zuckerman (1963) (ii) |
| *Tylenchorhynchus claytoni* | Root epidermis of *Nicotiana tabacum* (a);[c] *V. macrocarpon* (b); *Pinus resinosa* (c). | On (a), region of cell elongation and between root hairs preferred; on (c), also any parts not previously fed on. Quicker thrusting, pressing lips against food cell for penetration. | No data | Rapid pump pulsation followed stylet penetration. Continued up to 30 min (a); 2–4½ min (b). | Ectoparasite, but Steiner (1937) found *T. claytoni* in root parenchyma (a). Gregariousness noted (c), where groups fed 7–10 days then moved en masse to another root. Many nematodes stunted hosts; other pathogens entered. | (a) Steiner (1937), Krusberg (1959); (b) Zuckerman (1961); (c) Sutherland and Adams (1964) |
| *T. dubius* | Root epidermis *Poa annua*, *Trifolium repens*, *T. pratense*, and root hair *Lolium ?perenne* in polyethylene-bag agar cultures. | Region of cell elongation preferred. 20–50 rapid stylet thrusts penetrated food cell. | Sudden contraction of body-wall muscles, wriggling movement assisted injection; zone[b] formed. | Rapid pump pulsation followed ½ min quiescence. Zone ingested. | Ectoparasitic. Feeds lasted up to 10 min; less in crowded conditions. Often gregarious. | Klinkenberg (1963) |
| *Ditylenchus destructor* (1); *D. intermedius* (2); *D. myceliophagus* (3).[c] | (1) Hyphae of *Chaetomium indicum;* (1,3) *Botrytis* | Stylet probing; lips rubbed and pressed | Dorsal secretions surged forward after penetra- | After injection, regular postpharyngeal pulsa- | Ectoparasitic. (1) Feeds lasted up to 30 min, *C.* | (2) Linford (1937a); (1) on *C. indicum*, etc., |

| | | | | | | |
|---|---|---|---|---|---|---|
| | *cinerea* (has perforated septa); (1,2) Other fungal spp. also. | against hypha; quicker stylet thrusting; trial penetrations. | tion. Reservoirs in procorpus, also in metacorpus (1,3). Secretions granular and by duct outlets, also fluid. Injection by increased turgor and spasmodic metacorpal muscular contractions lasted up to one half total feed; stopped cyclosis in host. | tions ?induced slow ingestion (1,3). Occasional pump pulsation for strong suction; became regular in last ½ min of feed; brief streaming in host ended feed. Regular intestinal movements mixed contents. | *indicum*, ≈ 1 hr, *B. cinerea* and others; (2) 1–2 hr; (3) ¾–2½ hr. (1,3) solitary feeding or in pairs (♂, ♀). Feeds ended by breaking adhesion between lips and hypha. 5–10 host's cells died per feed. | Anderson (1964); (1,3) on *B. cinerea*, Doncaster (1966) |
| *D. triformis* | Hyphae (septate) of *Pyrenochaeta terrestris*. | Stylet probing with lips brushing cell wall. Penetration by 4–8 sets of 4–6 thrusts (fewer needed near septa). | Injection of dorsal glandular secretions by increasing turgor by periodic preanal body contractions (2–5/min). Continued through ingestion. | Irregular pump pulsation (1–5, sometimes more, every 2 sec to 4 min). | Ectoparasite. Feeds lasted 2–100 min. Body contractions/relaxations also moved intestinal contents. Host cells became more granular, then translucent and died after feed. | Pillai (1966a) |
| *D. destructor* | Stem epidermis *Vicia sativa;* epidermis tap root *Daucus carota* in water preparations between agar slip and coverslip. | Stylet probing. Penetration by stylet thrusting with lips appressed to cell wall. | Dorsal and subventral. Reservoirs—1 each duct in metacorpus, also dorsal in procorpus. Secretions granular; subventral also fluid in reservoirs. Zone formed within 5–7 min, became increasingly refractive, but size unchanged until near end of feed. | Postpharynx pulsed irregularly after injection, but during penetration also. Irregular pump pulsation latter half of feed became more regular, stronger? in last few min when zone rapidly collapsed. | Endoparasite. Feeds lasted 15–60 min. Solitary feeding or in pairs (♂, ♀). Refractive boundary of zone remained in food cell. | Doncaster (this chapter) |
| *D. dipsaci* | Hyphae of unspecified fungi (a). Leaf epidermis, mesophyll of *Stellaria media* in water preparations between agar slip and coverslip (b). | Same as *D. destructor* on stem epidermis, but more persistent; few trial penetrations. Firm surfaces stimulated thrusting. | On (b), as *D. destructor* on stem epidermis but quicker surge of secretions. Injection? Dorsal secretions moved by anterior metacorpal muscle contractions. Part subventral secretions ingested with food. | On (a), pump pulsation followed stylet entry. In (b), vibratory pump pulsations (4-5/sec) developed from anterior metacorpal contractions. Pumping squeezed subventral reservoirs. Ingestion lasted ≈ ½ min. | Ectoparasite (a). Endoparasite (b). Feeds lasted up to 5 min (all hosts). | (a) Linford (1937b); (b) Doncaster (this chapter) |

TABLE I (*Continued*)

| Nematodes | Sources of food (these observations) | Behavior leading to feeding | Pharyngeal gland secretions | Ingestion | General feeding characteristics, other data, effects on host | References |
|---|---|---|---|---|---|---|
| **Family Heteroderidae** | | | | | | |
| *Heterodera cruciferae* (1); *H. trifolii* (2), second-stage larvae. | (1) Root cortex, *B. oleracea* v. *capitata* and *gemmifera* in glass-ball soil substitute in observation chambers; thick root sections. (2) Root cortex, *T. repens* v. Ladino in black sand in observation boxes. | Explored root surface with ≈ 1.5 stylet probes per sec. Intracellular penetration *in toto* into cortex by cutting with stylet (2.5/sec) and thrusting with body; took 10–20 min. Preferred root-hair region or beside branch roots. Transient feeding then permanent site chosen by (1?,2), usually at endodermis. | Granular secretions in all ducts; reservoir by each duct's outlet. Host reaction suggests injection. In (1), subventral glands active until 5–7 days after entering host. | Periods of pump pulsation lasted few min, often with short rests. | No ectoparasitism seen. Feeds usually brief but (1) at temporary site fed 7 hr. Penetration holes attract other larvae. Cortical tunnels in host attracted other pathogens. Syncytia in stele. | (2) Mankau and Linford (1960); (1) Doncaster *et al.* (1968) |
| *H. cruciferae* (1); *H. trifolii* (2), intermediate larval stages, adult ♀♀. | Same as second-stage larvae. Nematodes observed in thick root sections or after dissecting them out in water. | Irregular stylet thrusting ½–1 min. Larvae direct stylet through widest angle; head loses mobility with age. | Dorsal gland increased activity with age; granular secretions, ?slow, steady outflow cf. *Ditylenchus* but metacorpal muscular contractions assisted. All ducts adult ♂♂ (1) with secretions. | Pump pulsation developed from spasmodic metacorpal contractions; regular, rapid in larvae. In adults (1) vigorous pulsation alternated with slow, wide dilations. Lasted few to 25 min. | Endoparasitic. Continued feeding when dissected from root suggests less dependence on external stimuli cf. ectoparasitic forms. ?No feeding by adult ♂♂ (1). No ♂♂ (2). In hosts: large syncytia, small swellings, stunting. | (2) Mankau and Linford (1960); (1) Doncaster *et al.* (1968) |
| *Meloidogyne* sp. (syn. *H. marioni*) (1); *M. javanica* (2), second-stage larvae. | (1) Cortex, epidermis growing roots *L. sativa*, *Portulaca grandiflora*, and others in black sand in observation boxes, thick root sections in water. (2) Gautheret observation chambers. | Exploration as *Heterodera* larvae. Penetration by stylet thrusts (3.6/sec); some penetrations abortive. Preference for cell elongation and root hair regions. Entry interrupted by feeds. Intra- and in- | In (2), granular secretions in subventral ducts until endoparasitic, then fluid. ?Initiated syncytia. Reservoir by each duct's outlet. Injection inferred in (1) by 15–30 sec quiescence | Rapid pump pulsation (3–4/sec) lasted 10–40 sec. | Ectoparasitic then endoparasitic. Cf. adults, injection and ingestion briefer; stylet thrusts, pump pulsation quicker. Usually ≈ 2 min between ectoparasitic feeds. Penetration holes attractive to | (1) Linford (1937c, 1942b); Gland secretions (2) Bird (1967, 1968a,b), Bird and Saurer (1967) |

| | | | | | | |
|---|---|---|---|---|---|---|
| | | tercellular migrations by cutting with stylet. | following stylet entry, then metacorpal muscular contractions, also host response. | | other larvae. In host: tunnels; cortex, etc. swelled, root elongation stopped. Syncytia in stele. | |
| *Meloidogyne* sp. (syn. *H. marioni*) (1); *M. javanica* (2), adult females | Roots *Pisum sativum;* thick sections in water preparations or nematodes dissected from roots, in 2% dextrose, 2% peptone solution. | Stylet thrust strongly when lips in contact with firm surface (about 1/sec; variable); facile head movements direct thrusts. Trial penetrations. | Quiescence (15 min +) followed penetration, suggested injection. Viscous secretions emitted in water; in (2) secretions are mainly protein, some carbohydrate, but ?enzyme synthesis in presence of host cytoplasm. Dorsal secretions increase after molting | Vigorous pump pulsation lasted 1 hr. | Endoparasitic. Continued feeding when dissected from root. Feeding by adult ♂♂ not recorded. (2) Granular dorsal secretions ?accelerate syncytial development. Roots usually heavily galled without starch. cf. *Nacobbus.* Plants stunted. | (1) Linford (1937c, 1942b); glandular secretions (2) Bird (1967, 1968a,b), Bird and Saurer (1967) |
| **Family Hoplolaimidae** | | | | | | |
| *Rotylenchus robustus* (syn. *R. uniformis*) | Roots of *L. perenne* and other spp. in polyethylene bag agar culture. Epidermis, root hairs, then cortex fed on successively. | Exploratory stylet probes. Root tips only sites avoided. Stylet thrust at 1–2/sec when lips appressed to cell wall; penetration took up to 20 min. | Quiescence (up to 20 min) during feed ?suggested injection; also metacorpal muscular contractions ?assisted secretion movement. | With stylet in cell vacuole, pump pulsation ($\approx$ 2/sec) followed quiescence and continued, with rests, for several hours. | Ectoparasitic then partly endoparasitic. Feeding sites occupied up to 5 days. Gregarious? In host: cortical tunnels, necrosis. | Klinkenberg (1963) |
| *Nacobbus serendipiticus* | Epidermis, cortex of roots *L. esculentum* in glass-ball soil substitute in observation chambers. | Stylet probing with lips slipping over epidermis; penetration for entry by vigorous stylet thrusts cutting slit (as by *Heterodera* larvae). Intracellular migrations. | Granular secretions in all ducts. Metacorpal muscular contractions (2 min) assisted injection. | Regular pump pulsation (2–2.4/sec) for few min when feeding on epidermis; up to 50 min in cortex. Some rests. | 2nd–4th larval stages penetrated and repenetrated root cortex. Ectoparasitic (2nd, 3rd? and 4th?) and endoparasitic (larvae, adult ♀♀); ♀♀ sedentary. Older larvae sluggish; they and adult ♀♀ fed deepest, on small-celled spindle. Galls, rich in starch formed (cf. *Meloidogyne*). Other pathogens entered. | Clark (1967), Doncaster (this chapter) |

TABLE I (*Continued*)

| Nematodes | Sources of food (these observations) | Behavior leading to feeding | Pharyngeal gland secretions | Ingestion | General feeding characteristics, other data, effects on host | References |
|---|---|---|---|---|---|---|
| **Family Criconematidae** | | | | | | |
| *Hemicycliophora arenaria* | Crop spp. in several families; *L. esculentum*, deep root-cortex, endodermis, root cap to elongation region. Observed in soil; closely, in agar preparations between polyethylene sheet. | Moved to roots in < 24hr, then to ½–1 mm behind root tips. Exploratory stylet probes at epidermis. Penetration by short thrusts every 5–10 sec until stylet half in root; completion by thrusting with whole body; took 15–20 min. | Injection after penetration lasted 1–2 hr interrupted by metacorpus pulsation violently moving dorsal gland. Granular secretions filled procorpus reservoir. Subventral gland secretions ingested. | Slow, irregular pump pulsation in initial food-sampling phase, during which site might be rejected; if not, stylet sealed to epidermis by polysaccharide plug. Pump with inlet, outlet valves. Pulsation, 2–4 sec intervals, continued 2–6 days. Food cell pulsed as in *H. similis*. | Ectoparasite. Feed ended by body rotating ≈ 300° around stylet axis to break adhesion; stylet retraction took 1 min. Gregarious. In host: food cell enlarged as contents adjacent cells withdrawn. Increased meristematic division maintained food supply. Galls, many branch roots. | Van Gundy (1959), Van Gundy and Rackham (1961), McElroy and Van Gundy (1968) |
| *H. similis* | (iii) Roots *L. sativa*, *Tagetes* sp., *B. oleracea*, (ii) Crop spp. in several families, but *L. sativa* and *Tagetes* reported as nonhosts. | Penetration by stylet thrusts 3–6/sec; penetrated 6–8 cell layers (½ stylet length), sometimes vascular elements. | Secretions from all glands flowed forward during pumping (ingestion); injection inferred from pulsing of food cell (see Ingestion). | Regular pump pulsation 2/sec continued for days. (iii) saw food cell contract at each pump dilation (ingestion), reexpand at pump closure (injection). | Ectoparasite. In host: root growth stopped for days; roots curved, galled at tip. | Zuckerman (1961) (i); Khera and Zuckerman (1963) (ii); Klinkenberg (1963) (iii) |
| *Criconemoides curvatum* (1); *C. xenoplax* (2). | (1) Most regions root cortex: *Dianthus caryophyllus* in root observation boxes; excised roots on micro slides. (2) Roots *Prunus persica* in peat in petri dishes. | (1,2) probed with head; lip contact stimulated slow stylet thrusts with head swinging; part of body entered cortex. In (2), stylet inserted by slow thrusts without retractions. Up to ⅔ stylet penetrated (1,2). | In (1) injection? formed granular zone around stylet tip. | Regular pump pulsations; equatorial contraction of metacorpus at each; continued several days. | Usually ectoparasitic; or partly or completely endoparasitic. In host: cortical cavities if endoparasitic. Roots stunted. | (2) Thomas (1959); (1) Streu *et al.* (1961) |

| | | | | | | |
|---|---|---|---|---|---|---|
| *Paratylenchus curvitatus* (syn. *P. dianthus*) (1); *P. projectus* (2); *P. elachistus* (syn. *P. minutus*) (3). | (1,2) Epidermis in young, mature regions of root, root hairs *T. repens, T. pratense, Nicotiana alata* var. *grandiflora;* (3) *Ananas sativa* | (1,2) Penetration by series slow stylet thrusts, alternated with rests; lips pressed against cell wall. Penetration took 5–40 min. | Granular secretions distended dorsal metacorpal and procorpal reservoirs during penetration. Injection formed granular zone 1–1¾ hr. Latterly, spasmodic metacorpal contractions ?moved secretions. | Metacorpal contractions led into regular pump pulsation (1.7–3/sec) continued up to 1 week (1 and 2). In (3), pump pulsation ≈ 1 hr) alternated with rests and spasmodic contractions. Zone oscillated by pump pulsation. | Ectoparasitic; (1,2) occasionally endoparasitic. Feeds shorter on *N. alata*. In (1) ♂♂ have no stylet and do not feed. In (2) no ♂♂ known; preadult larvae did not feed. In (3), no functional anus. Host not damaged. | (1 and 2) Rhoades and Linford (1961); (3) Linford *et al.* (1949) |
| **Family Neotylenchidae** | | | | | | |
| *Neotylenchus linfordi* | Septate hyphae *P. terrestris* and 9 other septate and nonseptate spp. | Contact between lips and hypha stimulated stylet probing along hypha. Penetration usually near septum. | At penetration granular secretions distended dorsal duct. Fluid secretions in procorpus reservoir; later replaced by granules when intestinal pulsation began after 5–15 min. Forward flow of granules continued. Secretions ?injected throughout feed. | Ingestion associated with regular contractions and redilations of anterior intestinal chamber (6–8 puls/min); at contraction, movement of intestine closed pharyngo-intestinal valve; valve opened at dilations and food entered. Ingestion depended on hyphal streaming stopping. | Ectoparasite. Pharynx without metacorpus. Feeds lasted ½–3 hr; shorter on nonseptate fungi. Penetrated cell and 5–10 adjacent became granular during feed and in some spp. died. | Hechler (1962), Pillai (1966a,b); nomenclature, Nickle (1968) |
| *Hexatylus viviparus* | Hyphae *B. cinerea* (perforated septa). | Hypha rapidly rubbed with lips, abrupt stylet probes. Stylet thick, so penetration slow (took mins). | Dorsal gland secretions flowed forward at penetration; granular, and in procorpus reservoir, fluid. Secretions ingested with food. Reverse flow after feed. | Following penetration, hyphal contents flowed rapidly through stylet and pharynx. Pharynx pumped by rapid, narrow, irregular dilations and closures of ≈ ¼ total length. No intestinal movements (cf. *N. linfordi*). | Ectoparasite. Feeds lasted up to 5 min; few, slower pharyngeal pulsations followed. | Doncaster (this chapter) |
| Superfamily Aphelenchoidea | | | | | | |
| **Family Aphelenchidae** | | | | | | |
| *Aphelenchus avenae* | (ii) Roots, root hairs of 6 spp. higher plants (15 tested). Fungal hyphae: (iv) *Thanatephorus cucumeris;* (v) *P. terrestris;* (vi) | Attraction over distances shown (i). Lips rubbed, then pressed against hypha; stylet thrusting, 6/sec (iv). Best hosts penetrated with | No injection. No secretions visibly emitted. Anterior ¼ metacorpus (i.e. where dorsal duct joins pharynx) twitched | Pump pulsation followed penetration. Pulsation quickest on best hosts (vi). Root hairs, some fungal cells partly emptied; ingestion | Ectoparasite. Endoparasite in higher plants suggested (see Christie and Arndt, 1936; Steiner, 1936) (ii). Best fungal hosts | Townshend (1964) (i); Chin and Estey (1966) (ii); Pillai (1966a) (iii); Fisher and Evans (1967) (iv); Taylor and Pillai (1967) (v); |

TABLE I (*Continued*)

| Nematodes | Sources of food (these observations) | Behavior leading to feeding | Pharyngeal gland secretions | Ingestion | General feeding characteristics, other data, effects on host | References |
|---|---|---|---|---|---|---|
| | *Amanita rubescens*, 3 *Suillus* spp. (good hosts). *Cenococcum graniforme* (moderate), *Rhizopogon roseolus* (poor). | fewest thrusts (2–4) (vi). | during stylet thrusting (function not known) (v). | lasted up to 40 sec. Cells of good hosts emptied in <5 sec (vi). | damaged most (vi). Fungal pathogens entered attacked higher plants (ii). | Sutherland and Fortin (1968) (vi) |
| **Family Aphelenchoididae** | | | | | | |
| *Aphelenchoides blastophthorus* | Hyphae of *B. cinerea*. | Hypha rubbed with lips then penetrated by stylet thrusts (4–6/sec for 2–3 sec). | No injection. Anterior $\frac{1}{4}$ metacorpus twitched during ingestion; few vigorous pump pulsations after feed passed? secretions to intestine. | Pump pulsation followed penetration (2/sec for 5–15 sec). | Ectoparasite. Host damaged little. | Doncaster (this chapter) |
| *A. ritzemabosi* | Leaf epidermis, mesophyll *S. media* in water preparations between agar slip and cover-slip- | Lips rubbed over cell wall; penetration by thrusting stylet at 4–6/sec for 1–2 sec. | No injection. Few slow, large amplitude pump pulsations after feed passed? secretions to intestine. | Rapid, vibratory, small-amplitude pump pulsation followed penetration. Lasted 10–20 sec. | Endoparasite. Slight damage to host. | Doncaster (this chapter) |
| **Family Paraphelenchidae** | | | | | | |
| *Paraphelenchus acontioides* (1); *P. myceliophthorus* (2). | Hyphae of *P. terrestris*. | (1) penetrated by ultra-rapid stylet thrusts (15–20/sec, taking <1 sec); (2) slower. | No injection. From penetration on, pharyngo-intestinal junction twitched. Metacorpal pumping (ingestion) passed secretions into pharynx. | After pause 1 sec, pump pulsation emptied penetrated cell (4 puls/sec, taking 1–2 sec). | Ectoparasitic. In intestine food and secretions formed jellylike masses, quickly digested (cf. *A. avenae* where intestinal contents remained globular). | (1) Taylor and Pillai (1967); (2) Pillai (1966a) |
| Order Dorylamida | | | | | | |
| **Subfamily Tylencholaiminae** | | | | | | |
| *Longidorus elongatus* | Roots of *Fragaria vesca* var. *semperflorens*. | Lip contact stimulated stylet probing; penetration at apical meristem. Stylet penetrated successive layers, feeding in each. Nematodes often returned to previous feeding sites. | Few sec irregular muscular contractions in pharyngeal bulb before ingestion, also host response suggested injection. | Contractions pharyngeal bulb became very regular; continued up to 70 min, stylet in one cell. | Ectoparasite. Remained at one site up to 24 hr, but stylet penetrated different cells. Gregarious in soil. 24 hr after attack, increased cell division, root swelled proximal to feeding site. | Wyss (1969) and personal communication |

| | | | | | | |
|---|---|---|---|---|---|---|
| *Xiphinema index* (1); *X. diversicaudatum* (2). | (1) Roots of *Vitis vinifera* var. Mission; (2) Roots of *Rosa* sp. | Stylet thrust immediately lips touched epidermis (2). Preferences for root hair region, then root tip (larvae (1)). Older larvae, adult ♀♀ (1) fed only about apical meristem. (1,2) Stylet penetrated successive cell layers, feeding in each. | Irregular muscular contractions in pharyngeal bulb interrupted penetration; separation of host cells suggested injection. Contractions in 2 sec periods for 5 min (iii) (2). Dorsal gland secretions injected (1) (iv). | Rhythmic contractions in pharyngeal bulb for 2 min each cell (1). In (2), irregular contractions became regular, lasted few min then usually deeper cell penetrated. | Ectoparasitic. Gregarious. Max exsertion of stylet to anterior of stylet extensions. Only growing roots penetrated. Root tips galled after 24 hr. Outer cells separated. Poor growth. | (2) Schindler (1957) (i); (1) Raski and Radewald (1958) (ii); (1,2) Fisher and Raski (1967) (iii); (1) Roggen *et al.* (1967) (iv) |
| **Family Trichodoridae** | | | | | | |
| *Trichodorus christiei* (1); *T. allius* (2); *T. viruliferus* (3). | (1) Root epidermis, cortex, (i) of *L. esculentum*, *Secale cereale;* (ii) of *V. macrocarpon;* (iii) *Brassica* spp., etc.; (iv) Isolated root cells of *Zea mays;* (v) Root epidermis, root caps, root hairs, *Triticum aestivum*. (3) (vi) Extending tree roots *Malus sylvestris*. | (3) accumulated at region of elongation (vi); (1) preferred same. Lips appressed to cell wall; penetration by rapid, direct stylet thrusts (8–10/sec) (iv), or rasping motion (i). Stylet exserted 3–4 μ (i,ii). | Injection viscous secretions (iv). Host reaction (1) suggested injection. | (1) "Sap" of food cell ingested during rapid stylet thrusting (30 sec) (ii). Convulsive, peristaltic contractions of esophagus (i,v). Stylet thrusting slowed (1/sec) as viscous cytoplasm ingested; thrusting then increased until end of feed (ii,iv). | (1) Ectoparasitic. Partly endoparasitic in root cap (v). Each cell fed on 5–10 sec (i); 1–4½ min (ii); 2–7 min (iv); 3–5 min (root hairs); endoparasitic larvae fed 5–30 min; roots penetrated 70 μ deep (v). Cyclosis in food cell stopped during feed (iv). Stubby roots. | (1) Rohde and Jenkins (1957) (i); (1) Zuckerman (1961) (ii); (1) Khera and Zuckerman (1963) (iii); (1) Chen *et al.* (1965) (iv); (1) Russell and Perry (1966) (v); (3) Pitcher (1967) (vi); Ultrastructure: (1) Hirumi *et al.* (1968) (vii); (2) Raski *et al.* (1969) (viii) |
| *T. proximus* | Root epidermis, cortex of *Stenotaphrum secundatum*. | Fed in elongating region. Lips appressed to cell wall; penetration by rapid stylet thrusts. | No injection recorded. | Stylet thrusting at 100/min; concurrent esophageal constrictions; thrusting slowed to 15/min by end. | Ectoparasite. Firmly held to roots during feeds. Stayed at each site up to 2 hr 50 min. Old sites attractive. Much feeding stopped root growth, caused lesions, swellings, browning. | Rhoades (1965) |

Addenda to Table I. *Aphelenchoides parietinus* (Christie and Arndt, 1936), *Belonolaimus longicaudatus* (Standifer and Perry, 1960), *Bursaphelenchus fungivorus* (Franklin and Hooper, 1962), *Dolichodorus heterocephalus* (Paracer *et al.*, 1967), *Helicotylenchus nanus* (Sledge, 1959), *Heterodera glycines* (Endo, 1964), *Pratylenchus crenatus* (Klinkenberg, 1963), *Tylenchus agricola*, *T. bryophilus* (Khera and Zuckerman, 1963).

[a] Injection of secretions into host.

[b] A visible zone of gland secretions within food cell.

[c] Related nematodes with similar feeding characteristics are grouped together. The following code is used for cross reference in each entry: (1), etc., to nematode; (a), etc., to source of food; (i), etc., to author.

## ACKNOWLEDGMENTS

The author wishes to thank Dr. D. W. Lawlor for much helpful discussion on pressure relationships between plants and nematodes and Dr. J. A. Currie for checking and adding to the discussion of ingestion in Tylenchida, based on Poiseuille's law.

## REFERENCES

Anderson, R. V. (1964). *Phytopathology* **54,** 1121–1126.

Barr, G. (1931). "A Monograph of Viscometry," 318 pp. Oxford Univ. Press, London and New York.

Bennet-Clark, T. A. (1959). *In* "Plant Physiology" (F. C. Steward, ed.), Vol. 2, pp. 105–191. Academic Press, New York.

Bird, A. F. (1967). *J. Parasitol.* **53,** 768–776.

Bird, A. F. (1968a). *J. Parasitol.* **54,** 475–489.

Bird, A. F. (1968b). *J. Parasitol.* **54,** 879–890.

Bird, A. F. (1969). *J. Parasitol.* **55,** 337–345.

Bird, A. F., and Saurer, W. (1967). *J. Parasitol.* **53,** 1262–1269.

Chen, T. A., Kiernan, J., and Mai, W. F. (1965). *Phytopathology* **55,** 490–491.

Chin, D. A., and Estey, R. H. (1966). *Phytoprotection* **47,** 66–72.

Christie, J. R., and Arndt, C. H. (1936). *Phytopathology* **26,** 698–701.

Clark, S. A. (1967). *Nematologica* **13,** 91–101.

Crofton, H. D. (1966). "Nematodes," 160 pp. Hutchinson, London.

Dickinson, S. (1959). *Nematologica* **4,** 60–66.

Doncaster, C. C. (1966). *Nematologica* **12,** 417–427.

Doncaster, C. C., and Shepherd, A. M. (1967). *Nematologica* **13,** 476–478.

Doncaster, C. C.. Green, C. D., and Shepherd, A. M. (1968). "Behaviour of *Heterodera* Species through the Life Cycle." Film. Brit. Film Inst., London.

Endo, B. Y. (1964). *Phytopathology* **54,** 79–88.

Fisher, J. M., and Evans, A. A. F. (1967). *Nematologica* **13,** 425–428.

Fisher, J. M., and Raski, D. J. (1967). *Proc. Helminthol. Soc. Wash.* **34,** 68–72.

Franklin, M. T., and Hooper, D. J. (1962). *Nematologica* **8,** 136–142.

Goodey, T. (1963). "Soil and Freshwater Nematodes" (rev. by J. B. Goodey from 1951 Ed.), 2nd Ed., 544 pp. Methuen, London.

Green, P. B. (1968). *Plant Physiol.* **43,** 1169–1184.

Harris, J. E., and Crofton, H. D. (1957). *J. Exp. Biol.* **34,** 116–130.

Hechler, H. C. (1962). *Proc. Helminthol. Soc. Wash.* **29,** 19–27.

Hirumi, H., Chen, T. A., Lee, K. J., and Maramorosch, K. (1968). *J. Ultrastruct. Res.* **24,** 434–453.

Hyman, L. H. (1951). "The Invertebrates: Acanthocephala, Aschelminthes and Entoprocta. The Pseudocoelomate Bilateria," Vol. III, 572 pp. McGraw-Hill, New York.

Jones, F. G. W. (1960). *Meded. Landbouwhogesch. Opzoekingssta. Staat Gent* **25,** 1009–1024.

Khera, S., and Zuckerman, B. M. (1963). *Nematologica* **9,** 1–6.

Klingler, J. (1965). *Nematologica* **11,** 4–18.

Klinkenberg, C. H. (1963). *Nematologica* **9,** 502–506.

Krusberg, L. R. (1959). *Nematologica* **4,** 187–197.
Linford, M. B. (1937a). *Proc. Helminthol. Soc. Wash.* **4,** 41–46.
Linford, M. B. (1937b). *Proc. Helminthol. Soc. Wash.* **4,** 46–47.
Linford, M. B. (1937c). *Phytopathology* **27,** 824–835.
Linford, M. B. (1942a). *Soil Sci.* **53,** 93–103.
Linford, M. B. (1942b). *Phytopathology* **32,** 580–589.
Linford, M. B., Oliveira, J. M., and Ishii, M. (1949). *Pac. Sci.* **3,** 111–119.
McElroy, F. D., and Van Gundy, S. D. (1968). *Phytopathology* **58,** 1558–1565.
Mankau, R., and Linford, M. B. (1960). *Ill. Agr. Exp. Sta. Bull.* **667,** 50 pp.
Miller, E. C. (1938). "Plant Physiology with Reference to the Green Plant," 2nd Ed., 1201 pp. McGraw-Hill, New York.
Myers, R. F. (1967). *Nematologica* **13,** 323.
Nickle, W. R. (1968). *Proc. Helminthol. Soc. Wash.* **35,** 154–160.
Paracer, S. M., Waseem, M., and Zuckerman, B. M. (1967). *Nematologica* **13,** 517–524.
Peacock, F. C. (1959). *Nematologica* **4,** 43–55.
Pillai, J. K. (1966a). Ph.D. Thesis, Univ. of Illinois, Urbana, Illinois.
Pillai, J. K. (1966b). *Diss. Abstr.* **27,** 668.
Pitcher, R. S. (1967). *Nematologica* **13,** 547–557.
Raski, D. J., and Radewald, J. D. (1958). *Plant Dis. Rep.* **42,** 941–943.
Raski, D. J., Jones, N. O., and Roggen, D. R. (1969). *Proc. Helminthol. Soc. Wash.* **36,** 106–118.
Rhoades, H. L. (1965). *Plant Dis. Rep.* **49,** 259–262.
Rhoades, H. L., and Linford, M. B. (1961). *Proc. Helminthol. Soc. Wash.* **28,** 185–190.
Roggen, D. R., Raski, D. J., and Jones, N. O. (1967). *Nematologica* **13,** 1–16.
Rohde, R. A., and Jenkins, W. R. (1957). *Phytopathology* **47,** 295–298.
Russell, C. C., and Perry, V. G. (1966). *Phytopathology* **56,** 357–358.
Schindler, A. F. (1957). *Nematologica* **2,** 25–31.
Seinhorst, J. W. (1961). *Annu. Rev. Microbiol.* **15,** 177–196.
Sledge, E. B. (1959). *Nematologica* **4,** 356.
Standifer, M. S., and Perry, V. G. (1960). *Phytopathology* **50,** 152–156.
Steiner, G. (1937). *Proc. Helminthol. Soc. Wash.* **4,** 33–38.
Streu, H. T., Jenkins, W. R., and Hutchinson, M. T. (1961). *N. J. Agr. Exp. Sta. Bull.* **800,** 32 pp.
Sutherland, J. R. (1967). *Nematologica* **13,** 191–196.
Sutherland, J. R., and Adams, R. E. (1964). *Nematologica* **10,** 637–643.
Sutherland, J. R., and Fortin, J. A. (1968). *Phytopathology* **58,** 519–523.
Taylor, D. P., and Pillai, J. K. (1967). *Proc. Helminthol. Soc. Wash.* **34,** 51–54.
Thomas, H. A. (1959). *Proc. Helminthol. Soc. Wash.* **26,** 55–59.
Townshend, J. L. (1964). *Can. J. Microbiol.* **10,** 727–737.
Van Gundy, S. D. (1959). *Proc. Helminthol. Soc. Wash.* **26,** 67–72.
Van Gundy, S. D., and Rackham, R. L. (1961). *Phytopathology* **51,** 393–397.
Wallace, H. R. (1963). "The Biology of Plant Parasitic Nematodes," 280 pp. Arnold, London.
Wright, K. A. (1965). *Can. J. Zool.* **43,** 689–700.
Wyss, U. (1969). Dissertation, Fakultaet der Technischen Universitaet, Hannover.
Yuen, P.-H. (1968). *Nematologica* **14,** 385–394.
Zuckerman, B. M. (1960). *Nematologica* **5,** 253–254.
Zuckerman, B. M. (1961). *Nematologica* **6,** 135–143.

# CHAPTER 20

# Gnotobiology

B. M. Zuckerman

*Laboratory of Experimental Biology, University of Massachusetts, East Wareham, Massachusetts*

I. Introduction . . . . . 159
Terminology . . . . . 160
II. Methodology . . . . . 161
A. Production of Axenic Nematodes . . . . . 161
B. Culturing Sterile Nematodes . . . . . 164
C. Culturing Sterile Plants . . . . . 169
III. Applications . . . . . 171
A. Studying Plant Responses to Parasitism . . . . . 171
B. Studying the Nematode . . . . . 177
IV. Conclusions . . . . . 180
References . . . . . 180

## I. INTRODUCTION

Gnotobiology, the study of a single species in the absence of other demonstrable species or in the presence of only known species, has recently received great impetus. This growth of interest is partly explained by the increasing complexity of life science research which demands the utmost control of biological variables. The space program provides but one example of contemporary research fields that have fostered the rapid growth of gnotobiotic technology.

The pure culture concepts which today receive ever-widening application to higher plants and animals involve the same principles of sterile culture that evolved during the early years of microbiology. The mycologist and bacteriologist, however, rapidly surmounted obstacles to maintaining and propagating pure cultures, whereas biologists concerned with higher animals found the problem far more complex. It is somewhat surprising that the plant nematologist, working as he was with a relatively simple organism, lagged in developing a gnotobiotic technology despite significant early progress in this direction. The techniques of Byars (1914) for the preparation and short-term maintenance of pure root-knot nematode cultures provided an excellent base for the development of phytonematode gnotobiology. Unfortunately, except for certain isolated applications, the pure culture concept lay dormant for the next 40 years. The elaboration by Mountain (1955) of a method for the continuous culture of plant parasitic nematodes under sterile conditions initiated the growth of modern phytonematode gnotobiology.

## Terminology

Gnotobiologists vigorously advocate the use of special terms to describe certain conditions and features associated with germfree studies. Dougherty (1959, 1960) emphasizes, as does Silverman (1965), that the use of these terms is justified simply because other more commonly used terms are ambiguous. To illustrate, consider the term *pure culture;* a nematologist may regard a culture of a myceliophagous nematode raised on a fungus, and in company with sundry associated microorganisms, as "pure" as long as only one species of nematode is present, but a bacteriologist, on the other hand, would not consider this culture "pure." Adherence to more precise gnotobiotic terminology resolves this difficulty. Dougherty (1959) proposed terms to eliminate nomenclatural ambiguity in regard to different types of cultural media (Table I).

TABLE I
TERMINOLOGY FOR CULTIVATION OF ORGANISMS, MOSTLY UNDER KNOWN CONDITIONS

| Term | Number of species associated with the primary species under culture | Terminology for media used in culturing |
|---|---|---|
| Gnotobiotic | Known species only, or none | Oligidic—pertaining to a medium containing crude materials of unknown chemical composition |
| Axenic | No other species | |
| Synxenic | One or more known species | Meridic—composed of a holidic base with at least one unknown substance |
| Monoxenic | One known species | |
| Dixenic | Two known species | Holidic—all intended constituents other than purified inert materials have exactly known chemical structures before compounding |
| Xenic | Unknown | |

## II. METHODOLOGY

### A. Production of Axenic Nematodes

Before plant–nematode interactions can be studied under gnotobiotic conditions it is necessary that the nematodes be axenized. Since no species of Tylenchoidea has as yet been shown to bear microorganisms in their digestive tracts, axenization involves only surface sterilization. Some methods for this sterilization are given in Table II. Some chemicals such as chloramine-T (Viglierchio and Croll, 1969), although good surface sterilants, were somewhat nematicidal. Techniques that involve but a single antibiotic have not generally been successful, whereas other methods were quite efficient. It is good practice to wash nematodes with sterile tap water following exposure to any chemical sterilant.

The problem of checking nematode sterility, as well as that of soil, plants, and other components of a gnotobiotic system, has not received sufficient attention. Most investigators (i.e., Zuckerman and Brzeski, 1965a,b) have considered it sufficient to test for fungal contaminants and bacteria. Probably most culturable organisms would be detected using this procedure, but what of those which are not readily culturable? For example, Casida (1965) described a coccoid soil microorganism that occurs in numbers greater than those of all other soil microorganisms but which can be cultured only by the most tedious and specialized procedures.

Nematodes belonging to the Dorylaimida have proved exceedingly difficult to axenize. *Trichodorus christiei* has been axenized using 0.5% hibitane diacetate diffused in water agar (Chen, 1964; Zuckerman and Brzeski, 1965b) and by a combination of three antibiotics (Chen *et al.*, 1965), although repeated trials were needed to obtain a few living nematodes. Recently, Goodman and Chen (1967) described two methods whereby large numbers of sterile *T. christiei* were obtained. In the first the nematodes were allowed to migrate through cotton held in a Melpar-Tiner storage trap filled with 0.01% Aretan. The second method involved suspending the nematodes for 20 hr in 1.5% water agar containing 0.01% Aretan. This was the first report of successful large-scale axenizing of a species belonging to the Dorylaimida. Das and Raski (1968) accomplished the first successful axenization of a *Xiphinema* species by immersion of *X. index* in either 0.01% Aretan or 0.1% dihydrostreptomycin sulfate for 1 hr. Neither treatment destroyed the infectivity of viruliferous nematodes. Species of *Longidorus* have not yet been reported as axenized.

TABLE II
SOME METHODS FOR AXENIZING PLANT NEMATODES

| Procedures | References |
|---|---|
| **I. Organic Disinfectants** | |
| A. Antibiotics or combinations including antibiotics | |
| 1. Mercuric chloride (0.01%) + streptomycin sulfate (1%) in sterile water for 2 min | Dolliver *et al.* (1962) |
| 2. 8-Hydroxyquinoline sulfate (1%) for 10 min, then streptomycin sulfate (0.2%) | DuCharme and Hanks (1961) |
| 3. Transfer through 5–6 baths of malachite green (0.002%) + streptomycin sulfate (0.1%) | Krusberg (1961) |
| 4. Sterile root-knot larvae obtained by incubating tomato roots in dihydrostreptomycin sulfate (0.1%), passed through Baermann funnel, washed 72 hr in ethoxy ethyl mercury chloride (0.0004%) + dioctyl sodium sulfosuccinate (0.01%) | Lownsbery and Lownsbery (1956) |
| 5. One to 3 days in water with actidione + aureomycin + penicillin + streptomycin (10 mg/liter each). Washed for 24 hr in methyl cellulose (4%) + ethoxy ethyl mercury chloride (0.002 to 0.005%), then 3 washes sterile water | Tiner (1960) |
| 6. *Trichodorus christiei* treated with streptomycin sulfate (0.1%) + aureomycin (0.003%) + mycostatin (35 units/ml) | Chen *et al.* (1965) |
| 7. Aureomycin + actidione + penicillin + streptomycin + neomycin + candicidin + mycostatin (10 mg/ml each) | Tiner (1960) |
| 8. Aerated for several days in novobiocin (0.1%) + streptomycin sulfate (0.1%) + ethoxy ethyl mercury chloride (0.0004%) | Townshend (1963a,b) |
| 9. *Xiphinema index* treated with dihydrostreptomycin sulfate (0.1%) or methoxy ethyl mercury chloride (Aretan) (0.013%) for 1 hr | Das and Raski (1968) |
| Different concentrations and exposure times of the above-listed antibiotics were used in other investigations | Barker (1963), Chen *et al.* (1961), Miller (1963), Mountain (1955), and Mountain and Patrick (1959) |
| B. Other organic disinfectants or combinations including organics | |
| 1. Ethoxy ethyl mercury chloride (0.000375%) for 72 hr | Crosse and Pitcher (1952a,b) |
| 2. 8-Hydroxyquinoline sulfate (0.1%) for 72 hr 8-hydroxyquinoline potassium sulfate (0.5%) for 90 hr | Crosse and Pitcher (1952a) |
| 3. Nematodes moved through peat moss soaked with malachite green (0.1%) | Hastings and Bosher (1938) |

TABLE II (*Continued*)

| Procedures | References |
|---|---|
| 4. Root-knot nematode egg masses in cetavlon (0.1%), then sterile water rinse, then hibitane diacetate (0.5%) for 15 min | Peacock (1959) |
| 5. *Meloidogyne javanica* egg masses in hydrogen peroxide (3%) for 20 min, then sterile water wash, then cetavlon (0.0%) for 5 min, then sterile water wash, then hibitane diacetate (0.5%) for 5 min, then sterile water wash. For larvae—methiolate (0.1%) for 15 min, then tetrocycline (50 mg/liter) for 15 min | Bird (1962) |
| 6. *Trichodorus christiei* suspended for 20 hr in agar (1.5%) containing Aretan (0.01%), retrieved from agar surface in distilled water | Goodman and Chen (1967) |
| 7. *Meloidogyne incognita* egg masses soaked in benzalkonium chloride (0.012%) for 3 min | Huang (1966) |
| 8. Clor-o-fen (0.25%) (an environmental disinfectant containing potassium coconut soap, isopropyl alcohol, and 4- and 6-chloro-2-phenylphenol) for 45 min or 0.5% for 20 min | Boswell (1963) |
| 9. 1000 ppm Hyamine 2389 solution (Rohm & Haas, methyldodecylbenzyltrimethyl ammonium chloride and trimethylammonium chloride) for 10 min | Hirumi *et al.* (1967) |
| Different concentrations and exposure times of the above-listed organic sterilants were used in other investigations | DuCharme (1959), Khera and Zuckerman (1962), and Zuckerman and Brzeski (1965b) |
| II. Inorganic Disinfectants | |
| 1. Egg masses of *Heterodera radicicola* in mercuric chloride (0.1%) for 20 min or hydrogen peroxide (3%) for 30 min | Byars (1914) |
| 2. Root-knot nematode eggs in Purex (10%) (a preparation with NaOCl (5.25%)) for 4 min, then sterile water wash | Loewenberg *et al.* (1960) |
| 3. Aqueous Mercurochrome (0.2 mg/ml) for 1 hr, then sterile water wash | Tiner (1961a,b) |
| Different concentrations and exposure times of the above-listed inorganic sterilants were used in other investigations | Christie and Crossman (1936), Feder (1958), Feder and Feldmesser (1957), Fenwick (1956), Myuge (1963), and Tyler (1933) |
| III. Combination Partially Employing Physical Means | |
| 1. Equal volumes of water agar (at 30°C) and disinfecting solution added directly to nematodes in sterile petri dish. Allowed to stand for 24 hr, then nematodes aseptically removed | Chen (1964) |

Axenization of bacteriophagous nematodes has not been extensively attempted by phytonematologists; however, controlled experimentation is required to substantiate or disprove the current belief held by many East European investigators that these nematodes play an important role in plant disease. Bacteriophagous nematodes are relatively difficult to axenize. For example, hibitane diacetate (0.25–0.5%), 20 vol hydrogen peroxide, malachite green (20 ppm), and streptomycin (1000 ppm) were each toxic to the eggs of a *Cephalobus* species isolated from tomato roots. Axenization was finally accomplished by treating the eggs with 1% Mercurochrome for 10 min (A. Wilski, unpublished data).

Some bacteriophagous nematodes were axenized by antibiotic solutions. Cryan (1963) reported that a combination of penicillin, streptomycin, and nystatin axenized *Rhabditis pellio, Caenorhabditis briggsae,* and *Panagrellus redivivus*. Germfree larvae of *P. redivivus* were also obtained by placing gravid females in 1% aqueous Mercurochrome for 5–7 min then transferring them to a drop of 1% Mercurochrome and rupturing them with a needle (Cryan *et al.,* 1963; as modified by Deubert and Zuckerman, 1967). Active larvae were immediately transferred to a second drop of 1% Mercurochrome for about 1 min and then collected in sterile water.

## B. Culturing Sterile Nematodes

The axenic culture of plant nematodes has proved difficult. Plant nematodes have been grown aseptically in monoxenic culture with plant or fungus tissues representing one organism of the two part system and the nematode the other. Since nematodes can be readily separated from plant tissue by one of several methods (e.g., Tiner, 1961b), monoxenic culture has proved effective for obtaining large numbers of sterile nematodes.

Metcalf (1903) was apparently the first to axenize a nematode, when by washing eggs in sterile water and then isolating uncontaminated eggs from sterile agar he obtained pure cultures of *Rhabditis brevispina,* a nematode associated with plant decay. Byars (1914), the first to culture a plant nematode monoxenically, maintained a root-knot nematode (*Meloidogyne* sp.) on cowpea and tomato seedlings in test tubes under sterile conditions for more than one month. Several others including Tyler (1933), Tanaka (1962), and Polychronopoulos and Lownsbery (1968) cultivated nematodes on seedlings within enclosed containers. However this technique often proved unsatisfactory for long-term maintenance of nematode cultures since the plant container was soon outgrown.

The culture of *Pratylenchus minyus* on sterile, excised corn roots represented the next significant advance in the propagation of germfree nematodes (Mountain, 1955). This step became possible when tissue culture techniques developed to the stage where growing, sterile plant cells and tissues could be maintained *in vitro* for many months (White, 1963). Tiner (1960, 1961a,b) established standards for efficient maintenance of nematode cultures and developed a collection trap for axenic nematodes Monoxenic culture of plant nematodes on excised roots or other plant tissue segments is summarized in Table III.

In 1958, Sayre reported that larvae of *Meloidogyne incognita* penetrated and developed within excised roots or tissue cultures of potato and tobacco grown on White's agar supplemented with auxin and kinetin. The addition of plant growth substances to a medium caused many plant tissues to produce a mass of undifferentiated parenchymatous cells termed *callus*. Soon thereafter Krusberg (1960) successfully cultured several plant parasites on alfalfa callus tissues. Using Krusberg's techniques, large quantities of germfree phytonematodes could now be accumulated for host–parasite and physiological studies. In addition, a high degree of control could be exerted in the study of nematode development.

Subsequent investigations proved that callus culture was an efficient means for maintenance and propagation of many species of plant nematodes. Studies have been made of the effects of temperature (Dolliver *et al.*, 1962; Prasad and Webster, 1967; Lownsbery *et al.*, 1967), the addition to the medium of plant growth substances (Krusberg and Blickenstaff, 1964; Webster and Lowe, 1966), and the effect of different nutrient media or varying the constitutents in the medium (Schroeder, 1963; Schroeder and Jenkins, 1963; Dolliver *et al.*, 1962) on nematode growth in callus culture. One of the significant findings of these studies was that callus derived from some plants (such as clover, Webster and Lowe, 1966) supported buildup of a large nematode population, whereas the same nematode did not reproduce on seedlings of these same plants; or if they did reproduce, population levels were not as high as those on callus (Krusberg and Blickenstaff, 1964). The observation by Schroeder and Jenkins (1963) that *Pratylenchus penetrans* usually multiplied much faster on callus than on excised roots of 11 plant species supported these conclusions. The plant growth substances are required to stimulate callus formation and also increased nematode reproduction although the effect was thought to be an indirect one. In contrast, Krusberg and Blickenstaff (1964) reported that maximum reproduction of *Ditylenchus dipsaci* occurred when kinetin was added to the medium, even though

TABLE III
NEMATODES CULTURED MONOXENICALLY ON EXCISED ROOTS OR OTHER PLANT TISSUE SEGMENTS

| Nematode | Plant[a] | | | | | | | | | | |
|---|---|---|---|---|---|---|---|---|---|---|---|
| | Alfalfa | Beet | Carrot | Clover | Cotton | Corn | Cucumber | Okra | Potato | Tobacco | Tomato |
| *Aphelenchoides asterocaudatus* | | | | 1 | | 1 | | | | | |
| *A. bicaudatus* | 2 | | | | | | | | | | |
| *A. limberi* | | | | 1 | | | | | | | |
| *A. subtenuis* | 1 | | | 1 | 1 | 1 | | | | | |
| *Aphelenchus avenae* | | | | | 1, 2 | 1 | | | | | |
| *Ditylenchus destructor* | | | 3 | 3 | | | | | 3 | 3 | |
| *Heterodera rostochiensis* | | | | | | | | | | | 4 |
| *H. schachtii* | | 5–8 | | | | | | | | | |
| *Meloidogyne incognita* | | | 9, 10 | | | | 11, 12 | | | 13 | 13–15 |
| *Nacobbus serendipiticus* | | | | | | | | | | | 16 |
| *Pratylenchus brachyurus* | | | 17 | | | 18 | | | | | |
| *P. minyus* | | | | 19, 20 | | 19, 20 | | | | 19, 20 | |
| *P. penetrans* | | | | | | 21–25 | | | | | |
| *Radopholus similis* | | | 17 | | | | | 26 | | | |

[a] The numbers refer to the following literature references: 1, Sudakova *et al.* (1965); 2, Sudakova *et al.* (1963); 3, Darling *et al.* (1957); 4, Widdowson *et al.* (1958); 5, Johnson and Viglierchio (1969a); 6, Johnson and Viglierchio (1969b); 7, Moriarty (1964); 8, Moriarty (1965); 9, Sandstedt and Schuster (1963); 10, Sandstedt and Schuster (1965); 11, McClure and Viglierchio (1966a); 12, McClure and Viglierchio (1966b); 13, Sayre (1958); 14, Dropkin (1966b); 15, Peacock (1959); 16, Prasad and Webster (1967); 17, O'Bannon and Taylor (1968); 18, Boswell (1963); 19, Mountain (1954); 20, Mountain (1955); 21, Schroeder (1963); 22, Schroeder and Jenkins (1963); 23, Tiner (1960); 24, Tiner (1961a); 25, Tiner (1961b); 26, Feder (1958).

both alfalfa seedlings and callus tissues were injured by this growth substance. They concluded that nematode reproduction was not directly correlated with callus growth.

Alfalfa has proved a versatile plant for callus culture, although other species have provided suitable substrates for the propagation of large numbers of certain plant nematodes. The maintenance of nematodes on callus is now standard procedure in many laboratories. As investigators find new plant callus–nematode combinations, the number of species cultured in this way increases (Table IV). However, as yet no species of *Belonolaimus, Criconemoides, Helicotylenchus, Hemicycliophora, Paratylenchus, Longidorus, Trichodorus,* or *Xiphinema,* as well as others, have been cultured monoxenically. Nematodes that feed on vascular tissue such as species of *Meloidogyne* and *Heterodera* do not reproduce well on callus tissue.

Nutritional and environmental factors induce mutation in free-living nematodes, and one would expect that plant nematodes serially cultured in such a highly artificial environment as represented by tissue culture to undergo morphological and physiological change. For example, Kisiel *et al.* (1969b) described a morphological variant of *Caenorhabditis briggsae* that occurred when the nematode was cultured axenically in a medium containing a bacterial growth factor. Anderson (1968) found that *Acrobeloides* sp. maintained on a mixed bacterial population developed significant morphological differences from cultures of the same nematode grown in soil. Actually, plant nematodes that were propagated on callus for many generations have been examined by several investigators, but in no case has mutation been reported. Högger (1969) detected no differences in pathogenicity between *Pratylenchus penetrans* from alfalfa callus and greenhouse cultures or nematodes collected directly from the field. Faulkner and Darling (1961) also found no differences in pathogenicity of *Ditylenchus destructor* maintained in continuous laboratory culture on callus for 4 years, nor was there evidence of variation in morphology or host preference during this period. Other investigators, however, reported that various factors affected both behavior and development of plant nematodes grown in tissue culture. In one study, the sex ratio of *Ditylenchus dipsaci* changed when the nematode was grown on different host callus tissues (Viglierchio and Croll, 1968), while in another study the rate of development of *Meloidogyne incognita* increased or decreased when the concentration of certain medium constituents were varied (McClure and Viglierchio, 1966a). Indeed, it would be surprising if, as in nature, dramatic permanent changes in structural and functional makeup did not evolve in tissue culture-reared nematodes.

Myers (1967a,b) was the first to report axenic culture of a stylet-

TABLE IV
NEMATODES CULTURED MONOXENICALLY WITH PLANT CALLUS TISSUES

| | Plant[a] | | | | | | | | | | | | | |
|---|---|---|---|---|---|---|---|---|---|---|---|---|---|---|
| | Alfalfa | Apple | Carrot | Clover | Corn | Cucumber | Marigold | Oat | Onion | Periwinkle | Potato | Rose | Tobacco | Tomato |
| *Aphelenchoides ritzemabosi* | 1–3 | 4 | 5 | 4 | | | 5 | 6 | | 5 | 4 | 4 | 5 | |
| *A. sacchari* | 25 | | | | | | | | | | | | | |
| *Aphelenchus avenae* | 7 | | 8 | | | | | | | 8 | | | 9, 8 | 8 |
| *Ditylenchus dipsaci* | 1, 2, 10, 11 | | | 10, 11 | | | | | 11 | | | | | |
| *D. destructor* | | | 12 | 12 | | | | | 12 | | 12 | | 12 | |
| *Dolichodorus heterocephalus* | | | | | 13 | | | | | | | | | |
| *Hoplolaimus coronatus* | 2 | | | | | | | | | | | | | |
| *Meloidogyne hapla* | | | | | | 14 | | | | | | | | 14 |
| *M. incognita acrita* | | | | | | 14 | | | | | 15 | | 15 | 14 |

| | Alfalfa | Cabbage | Clover | Cotton | Cucumber | Grape-fruit | Lettuce | Okra | Pea | Pepper | Snap-bean | Soy-bean | Sugar maple |
|---|---|---|---|---|---|---|---|---|---|---|---|---|---|
| *Pratylenchus brachyurus* | 16 | | 16 | 16 | | | | | | | | | |
| *P. penetrans* | 2, 17, 18 | 18 | | | 18 | | 18 | | 18 | 18 | 18 | 18 | |
| *P. vulnus* | 19 | | | | | | | | | | | | |
| *P. zeae* | 2, 17 | | | | | | | | | | | | |
| *Radopholus similis* | 20 | | | | | 20, 21 | | 20–22 | | | | | |
| *Tylenchorhynchus capitatus* | 2 | | | | | | | | | | | | |
| *T. claytoni* | 23 | | | | | | | | | | | | |
| *Tylenchus agricola* | 23 | | | | | | | | | | | | |
| *T. hexalineatus* | | | | | | | | | | | | | 24 |

[a] The numbers refer to the following literature references: 1, Krusberg (1960); 2, Krusberg (1961); 3, Webster (1967); 4, Webster and Lowe (1966); 5, Dolliver *et al.* (1962); 6, Webster (1966); 7, Sudakova and Chernyak (1967); 8, Barker and Darling (1965); 9, Barker (1963); 10, Bingefors and Eriksson (1963); 11, Viglierchio and Croll (1968); 12, Faulkner and Darling (1961); 13, Paracer and Zuckerman (1967); 14, Miller (1963); 15, Sayre (1958); 16, Chernyak (1968); 17, Krusberg and Blickenstaff (1964); 18, Schroeder and Jenkins (1963); 19, Lownsbery *et al.* (1967); 20, Myers *et al.* (1965); 21, Feder *et al.* (1962); 22, Feder (1958); 23, Khera and Zuckerman (1962); 24, Savage and Fisher (1966); 25, Myers (1967b).

bearing nematode *Aphelenchoides sacchari*, which reproduced rapidly in an oligidic medium containing horse liver extract and dextrose. The nematodes were serially subcultured four times during the following 8 months without apparent reduction in reproductive rate.

Perhaps stylet-bearing nematodes require an ingredient such as the heat-labile, proteinaceous growth factor from liver essential for reproduction of *Caenorhabditis briggsae* (Dougherty, 1951). The growth factor has been characterized as a globulin (Sayre *et al.*, 1961) that can be manipulated experimentally by certain treatments such as freezing (Hansen *et al.*, 1961) and the addition of Ficoll, a sucrose polymer (Buecher *et al.*, 1966), to increase its biological potency. Since a meridic medium has been developed for a few free-living nematodes, a similar achievement for plant parasitic nematodes is a logical goal following axenic culture in an oligidic medium.

## C. Culturing Sterile Plants

The methods for sterile culture of higher plants may be conveniently divided into three categories:

1. Small enclosed container—The entire plant is enclosed within a small container.
2. Sterile root apparatus—The roots are maintained under sterile conditions while the shoots are allowed to grow freely.
3. Flexible film isolator—Many plants are enclosed within a plastic film isolator of several cubic foot capacity.

The small enclosed container was first used by Byars (1914) when he placed nematodes in contact with a seedling grown in a tube or plate containing agar. This method has been favored by many contemporary investigators for it is simple and the equipment inexpensive. In addition, the transparent nature of all components permits direct observation of feeding, penetration, and, often, symptom development. The primary disadvantage is that the experiment generally can be of only relatively short duration since no allowance is made for continued nutrition of the plant and the seedling often outgrows the container. A further objection is that the experimental conditions are quite alien to those which occur in soil.

A variation of the system involves substitution of the agar by other substrates such as peat moss (Hastings and Bosher, 1938), sand (Townshend, 1963a), or a mixture of loam, sand, and activated charcoal (Pitcher and Crosse, 1958). While these substrates more closely approximate natural conditions, the advantage of direct observation is lost. An enclosed system wherein seedlings were grown aseptically in soil held in tubes was devised by DuCharme and Hanks (1961). A disadvantage of this device was that for purposes of watering and aeration the tubes were attached in series; therefore, if one tube became contaminated, the others soon were too. An improved apparatus was devised by Kable *et al.* (1966) for studying soil moisture relations under monoxenic conditions.

The first reported use of a sterile root apparatus (den Ouden, 1960) involved growing plants with sterile roots in thin agar layers held in polyethylene bags and injecting sterile nematodes at the proper time. An *in vitro* method devised by Feldmesser (1967) for evaluating nematicides could also be easily adapted for studying of plant–nematode interactions. The design of the apparatus is interesting and uncomplicated: The roots are implanted in agar while the seedling stem passes through a cotton-plugged hole cut in the top of a plastic petri dish. Thus the upper portions of the plant are exposed to the air, while the roots are held under sterile conditions. Plant physiologists were among the first to experiment with germ-free root culture. A typical, early, relatively simple apparatus developed by Blanchard and Diller (1950) featured a root chamber containing a liquid nutrient medium. Since few plant nematodes swim, a liquid substrate would be unsatisfactory for many phytonematological studies. In 1962, Stotsky *et al.* described an extremely complex apparatus for growing plants with sterile roots held in a quartz substrate while permitting the aseptic collection of carbon dioxide and root exudates. The advantages of the apparatus were that plants could be sustained for over 3 months within the unit, and materials within the root chamber (soil, roots, exudates, and nematodes) could be sampled at selected intervals during the experiment. This apparatus was slightly modified by Zuckerman and Brzeski (1965b) for studying of plant–nematode interactions and eventually simplified to perform the same functions except for the collection of gases and root exudates.

Biologists requiring larger facilities for sterile culture have increasingly turned to plastic film isolators of the type developed by Trexler and Reynolds (1957). Fujiwara *et al.* (1967) describe the essential features of an isolator system for the germfree culture of higher plants as follows: (1) a vinyl isolator which serves as a culture chamber; (2) a sterile lock for entry into the chamber; (3) a ventilation system which supplies germfree air; and (4) a light source and temperature control. To this should be added provision for the sterile watering of plants. The chances of contamination should be minimized by housing the entire system within a "clean" room, for if the chamber becomes contaminated the entire experiment is lost. One advantage of plastic film isolators is that many plants can be included in one test, thereby facilitating the performance of statistically designed experiments. In addition inoculation, watering, sampling, and other manipulations are more easily performed within plastic film isolators than with closed systems.

## III. APPLICATIONS

### A. Studying Plant Responses to Parasitism

The concept that gnotobiology forms an essential ingredient of any critical evaluation of pathogenesis resulting from plant–nematode interactions has been widely accepted in recent years. This acceptance evolved from a debate concerning the relation of Koch's postulates to plant nematology, a debate which brought forth strong opinions both in favor and against the feasibility of applying the postulates in this discipline (Mountain, 1960). It was natural then that the first and most prevalent objective in gnotobiotic experiments was the demonstration of pathogenesis associated with nematode parasitism. Earlier, only the physical symptoms resulting from parasitism were described; however, the past few years have witnessed highly sophisticated investigations dealing with the physiological and biochemical responses of plant tissues.

#### 1. Evaluating Physical Responses

Gnotobiotic techniques were first applied in the demonstration of pathogenesis by phytonematodes by Byars (1914), in experiments wherein *Heterodera radicicola* (probably a *Meloidogyne* sp.) was shown to induce root galls on tomato. Later gall formation under germfree conditions was demonstrated for several species of *Meloidogyne,* specifically, *M. hapla* on tomato (Dropkin and Boone, 1966; Zuckerman and Brzeski, 1965b) and bean (Zuckerman and Brzeski, 1965b), *M. incognita* on tomato (Feldmesser, 1967; Peacock, 1959; Mayol and Bergeson, 1969) and ginger Huang, 1966), *M. incognita acrita* on tomato (Pi, 1966), *M. javanica* on tomato (Bird, 1962; Dropkin and Boone, 1966) and clover (Bird, 1962), and *M. arenaria* on tomato (Dropkin and Boone, 1966). In addition, Loewenberg *et al.* (1960) reported that galling followed surface feeding without body penetration by *M. incognita.* Gnotobiotic systems have also been used to evaluate host resistance to different species of root-knot nematode (Dropkin and Boone, 1966; Dropkin and Webb, 1967; Dropkin *et al.,* 1967).

Several nematodes, other than species of *Meloidogyne,* have proved

TABLE V

SYMPTOMS CAUSED BY NEMATODES OTHER THAN *Meloidogyne* SPECIES ON WHOLE PLANTS OR PLANT TISSUES UNDER MONOXENIC CONDITIONS

| Nematode | Host | Galls | Limited hypertrophy or hyperplasia | Discolored lesions or tissue browning | Tissues cavities or necrosis | Stunting or weight reduction | Nuclear enlargement | Miscellaneous | Reference |
|---|---|---|---|---|---|---|---|---|---|
| *Aphelenchoides ritzemabosi* | Alfalfa | | | | X | X | X | | Krusberg (1961) |
| | Strawberry | | | | | | | Alaminate leaves, thick recurving petioles | Pitcher and Crosse (1958) |
| *Criconemoides curvatum* | Peach | | | X | X | | | | Hung and Jenkins (1969) |
| *Ditylenchus dipsaci* | Alfalfa | X | | X | X | | X | Same as natural symptoms | Krusberg (1960) |
| | Onion | X | | | | X | | Typical bloat symptoms | Krusberg (1960, 1961); Sayre and Mountain (1962) |
| *Dolichodorus heterocephalus* | Tomato | X | | X | | | X | Root elongation ceases | Paracer *et al.* (1967) |
| *Helicotylenchus multicinctus* | Banana | | | X | X | | X | | Blake (1966) |
| *Hemicycliophora arenaria* | Tomato | X | | | | | | Nuclei reduced and distorted; nucleoli enlarged; root hairs elongated | McElroy and Van Gundy (1968) |
| *Heterodera schachtii* | Sugar beet | | X | | X | | | Giant cells, reduced seedling emergence | Moriarty (1964), Polychronopoulus and Lownsbery (1968) |

| | | | | | | | | | |
|---|---|---|---|---|---|---|---|---|---|
| *Nacobbus serendipiticus* | Tomato | X | | | | | | Excised roots studied | Prasad and Webster (1967) |
| *Pratylenchus brachyurus* | Citrus | | | X | | | | | Feldmesser (1967) |
| *P. minyus* | Corn | | | X | X | | | | Mountain (1954) |
| | Tobacco | | | X | X | | | | Mountain (1954) |
| *P. penetrans* | Alfalfa | | | X | X | | | | Castillo and Rohde (1965) |
| | Apple | | | X | X | | | | Pitcher *et al.* (1960) |
| | Carrot | | | X | X | | | | Rohde (1963) |
| | Celery | | X | X | X | X | | | Townshend (1963a) |
| | Clover | X | | X | X | X | | Number of root hairs reduced | Chen *et al.* (1961); Kilpatrick *et al.* (1963a) |
| | Lettuce | X | | X | X | X | | | Kilpatrick *et al.* (1963a,b) |
| | Peach | | | X | | | | | Mountain and Patrick (1959) |
| | Rye grass | | | | | X | | | Troll and Rohde (1966) |
| | Soybean | | | X | | X | | | Taylor *et al.* (1968) |
| | Strawberry | | X | X | X | | | | Townshend (1963b) |
| | Tomato | | | X | X | | | | Pi (1966) |
| *P. thornei* | Wheat | | | X | X | | | | Baxter and Blake (1968) |
| *Radopholus similis* | Banana | | X | X | X | | X | | Blake (1966) |
| | Citrus | X | X | X | X | X | X | Susceptible nonresistant varieties show enlarged nuclei | DuCharme (1957, 1959); Feldmesser (1967); O'Bannon *et al.* (1967) |
| *Tylenchorhynchus claytoni* | Corn | | | | | | X | | Deubert *et al.* (1967) |
| | Fescue | | | | | X | | | Troll and Rohde (1966) |
| *Tylenchus agricola* | Corn | | | | | | X | | Deubert *et al.* (1967) |

capable of inducing by themselves detectable physical changes in whole plants or plant tissue sections. The symptoms associated with parasitism of these nematodes under monoxenic conditions have been separated into six general categories in Table V. The extensive study of a single species such as *Pratylenchus penetrans* permits some interesting comparisons. A comparison of results of several investigations involving *P. penetrans* reveals either a large difference in host susceptibility to this species or the occurrence of biological races. In fact, each of these premises has support. In greenhouse pot tests, Olthof (1968) was able to distinguish two races of *P. penetrans* by their reproductive potential and pathogenic behavior on tobacco and celery. Severe disturbances of the growth processes have been noted, as evidenced by galling on lettuce and clover, while on most hosts reaction is restricted to tissue browning and necrosis (Table V). In addition, Troll and Rohde (1966) reported that not even the usual lesions or necrosis were produced on rye grass or bluegrass, though a reduction of root weight of rye grass was noted. Mountain (1961) made a similar observation, noting that certain tobacco varieties supported large *P. minyus* populations without apparent necrosis, whereas susceptible varieties were severely damaged by comparable populations. From these observations and those of *P. penetrans* on peach, Mountain (1961) concluded that the damage associated with nematode invasion seemed to involve, at least in part, a host–response phenomenon. On the basis of these reports one may speculate that breeding plants for resistance (or to tolerance) to *P. penetrans* may prove particularly rewarding.

There have been only a few gnotobiotic studies of the interactions of bacteria or fungi and nematodes in plant disease (Table VI). This is understandable, for the complexity of the experimental design is greatly increased with the introduction of each new variable. An early study by Hastings and Bosher (1938) evaluated the role of *Pratylenchus penetrans* and *Cylindrocarpon radicicola*, both in pure and mixed cultures, on reduction of root growth of several crops. Prior to these experiments the nematodes were axenized and the fungus grown in pure culture. However, Hastings and Bosher did not clearly state how the interactions of the two organisms with the plants were evaluated. If this was accomplished in open pot experiments, then obviously gnotobiotic conditions were not maintained. The first report clearly describing a gnotobiotic study was that of Pitcher and Crosse (1958), in which was examined the relationships of *Aphelenchoides ritzemabosi* and the bacterium *Corynebacterium fasciens* to a disease of the strawberry. The results indicated that two similar but distinct diseases are found in strawberries; one caused solely by the nematode and the other predominantly a bacterial disease modified by the nematode.

TABLE VI
INVESTIGATIONS OF THE INTERACTIONS OF NEMATODES AND OTHER ORGANISMS IN DISEASE CAUSATION

| Nematode | Other organism(s) | Host | Reference |
|---|---|---|---|
| *Radopholus similis* | *Fusarium oxysporum* | *Musa ornata* | Blake (1966) |
| *Pratylenchus penetrans* | *F. oxysporum* *Trichoderma viride* | *Medicago sativa* | Edmunds and Mai (1966) |
| *Tylenchorhynchus claytoni* | *Fusarium roseum* | *Zea mays* | Kisiel *et al.* (1969a) |
| *Tylenchus agricola* | *Pythium ultimum* | | |
| *Aphelenchus avenae* | *Suillus granulatus* | *Pinus resinosa* | Sutherland and Fortin (1968) |
| *Heterodera rostochiensis* | Gray sterile fungus | *Lycoperiscon esculentum* | James (1968) |
| *Aphelenchoides ritzemabosi* | *Corynebacterium fasciens* | *Fragaria vesca* | Pitcher and Crosse (1958) |

In a review of studies of complex interactions utilizing other than a gnotobiotic approach, Powell (1963) wrote that nematodes have been shown to enhance the severity and development of several plant diseases but that nematodes also have been reported as having a depressing effect on the incidence of other plant diseases. The results of gnotobiotic experiments involving plant–nematode–fungus interactions have shown the same variability. For example, while in one study more *Pratylenchus penetrans* invaded fungus-infested roots than uninfested roots (Edmunds and Mai, 1966), James (1968) found that root penetration by another endoparasite, *Heterodera rostochiensis,* decreased in the presence of a gray, sterile fungus. James also observed that *H. rostochiensis* did not increase root susceptibility to fungus invasion. Conversely, Blake (1966) reported that *Radopholus similis* increased the invasive potential of *Fusarium oxysporum.* Kisiel *et al.* (1969a), in examining the relationships of two ectoparasitic nematodes and two fungi to root rot of corn, concluded that the fungi were the primary pathogens (Table VI). In this study increased fungus invasion of the root occurred in one nematode–fungus combination but not in three other combinations.

Blake (1968) theorized that the pathogenic capability of a nematode depends on both host sensitivity and the number of nematodes attacking the host, Thus, while a low number of nematodes may not severely affect a plant, a large number of the same species may incite severe symptoms. Indeed, even smaller numbers of a species with high pathogenic capability will cause severe damage to the plant. When considering the sensitivity of the plant to the fungus, the invasive potential of the fungus alone and in the presence of the nematode at both high and low levels of fungus inoculum and nematode population, and then adding to these factors the interplay of abiotic influences on the growth and physiological state of all organisms involved, it is clear that the isolation of variables through a gnotobiotic approach provides the most logical path to understanding the basic nature of complex interactions.

## 2. Evaluating Physiological and Biochemical Responses

Physical and biochemical changes induced in plant tissues by nematodes have been studied under aseptic conditions by several investigators. The manner in which host tissues respond to nematode injury has been rather extensively examined for species of *Pratylenchus,* particularly *P. penetrans.* In one of the earliest investigations carried out under controlled conditions, Mountain and Patrick (1959) demonstrated that *P. penetrans* hydrolyzed the glycoside amygdalin, thereby releasing the phytotoxic compounds hydrogen cyanide and benzaldehyde. It was further shown that amygdalin, which naturally occurs in peach roots, was not present within lesions and that hydrogen cyanide was given off by lesioned tissues. These workers concluded that lesion formation and necrosis resulted, at least in part, from the hydrolysis of amygdalin by nematode enzymes. They also noted that hydrogen cyanide is normally given off by crushed tissues of plants that contain cyanogenic glycosides, this action resulting from the release of certain plant enzymes. Therefore, it was concluded that mechanical damage caused by the nematode may also lead to necrosis and browning. Later investigations showed that apple root tissues which contained high concentrations of phenolic compounds (the dermal and endodermal layers) exhibited a rapid necrotic reaction when invaded by *P. penetrans,* whereas tissues that contained low amounts of phenolics (the cortical parenchyma) were quite tolerant (Pitcher *et al.,* 1960). Roots of three grass species which lacked high phenol concentrations also failed to develop discolored lesions following penetration by *P. penetrans* (Troll and Rohde, 1966). Other studies proved that lesions caused by *P. penetrans* on carrot (Rohde, 1963), celery (Townshend, 1963a), strawberry (Townshend, 1963b) alfalfa (Castillo and Rohde, 1965), and cabbage (Acedo and Rohde, 1968) contained high phenol concentrations. One of these phenolics was chlorogenic acid (Pi, 1966), a compound that commonly accumulates in diseased tissues of many plants.

Several workers studied under aseptic conditions the biochemical changes associated with other plant parasitic nematodes. Krusberg (1960,

1961) found free tryptophan in galls caused by *Ditylenchus dipsaci* on alfalfa, but not in healthy plants. He discussed the evidence provided by the presence of the free tryptophan in relation to the mechanism of galling, and suggested that the nematode might secrete indole acetic acid (IAA) or cause the plant to accumulate IAA in the vicinity of the nematode. Krusberg (1961) also found abundant free tyrosine in tissues infested by *Aphelenchoides ritzemabosi,* whereas healthy tissues or those infested by *D. dipsaci* contained little or none. Bird (1962) examined galls induced in tomato roots by *Meloidogyne javanica* and detected a growth-promoting substance that was not present in healthy roots. Bird was unable to detect IAA or free tryptophan in galled tissue, but suggested that the growth-promoting substance he found may be closely related to IAA. He further concluded that giant cell formation appeared to depend on a continuous stimulus by the nematode. Sandstedt and Schuster (1966a,b), studying the interactions of *Meloidogyne incognita* and tobacco pith, concluded that auxins were neither freed from plant tissues nor secreted by the nematode. They suggested instead that the nematode enabled the tissues to retain and use endogenous auxins that otherwise would have been transported elsewhere. In a study of *Radopholus similis* on citrus, DuCharme (1959) observed that starch disappeared from cells contiguous to lesions.

## B. Studying the Nematode

Gnotobiotic culture has not been widely utilized in the study of plant nematode development and physiology, yet this approach has at times yielded highly significant results. An example is Byars (1914), who followed and described the life cycle of the root-knot nematode from egg to egg. In later studies of the root-knot nematode on tomato seedlings, Tyler (1933) demonstrated that this nematode reproduces parthenogenetically.

Gnotobiotic techniques were also employed at an early date in behavioral studies of phytonematodes. Again we must refer to the pioneer work of Byars (1914), who was the first to observe that root-knot nematode larvae aggregate near root tips. This was the first report of root attraction to plant nematodes. Of interest is Byar's conjecture: "Thus, it seemed very probable that the root tip possessed some chemotactic stimulus which attracted the organism." A half-century later the same basic experimental design employed by Byars was used by Chen and Rich (1963a,b) to study the attraction of *Pratylenchus penetrans* to clover roots and of *Meloidogyne incognita* to excised tomato root tips by

Peacock (1959). The attractiveness of roots to nematodes was also examined by Edmunds and Mai (1967); however, the objective was to evaluate the influence of fungus infection on the finding of roots by nematodes. The three organisms in this dixenic experiment were *Pratylenchus penetrans, Fusarium oxysporum,* and alfalfa. Schuster and Sandstedt (1962), while observing *Meloidogyne* larvae in cultures containing excised roots or intact seedlings, noted that visible trails were made on the agar by the nematodes as they moved to the roots. These investigators stated that these paths provided evidence as to whether the nematodes moved in directional response to attractants emanating from the root or merely found the root while aimlessly wandering and then aggregated there. Gnotobiotic experiments of Chang and Rohde (1969) yielded significant results pertaining to host root finding by *Pratylenchus penetrans*. In these studies it was demonstrated that dark-colored callus tissues of several hosts were repellent to this nematode, whereas young, light-colored calluses were attractive. It was also shown that chlorogenic acid, the major phenolic component of tomato roots, was attractive to *P. penetrans;* but this chemical was repellent when oxidized to a yellowish brown compound. Certainly, the inclusion of gnotobiotic methodology in nematode behavioral studies is almost prerequisite to obtaining definitive results, for in no other manner can the role of the biochemical emanations in the rhizosphere be segregated and eventually understood.

A few gnotobiotic studies have had as one of their objectives the direct observation of feeding sites and the manner of feeding. Examples of such investigations were the observation of feeding by *Trichodorus christiei* on isolated root cells of corn (Chen *et al.*, 1965), *Tylenchus emarginatus* feeding on red and black pine seedling roots (Sutherland, 1967), *Tylenchorhynchus claytoni* feeding on alfalfa roots (Krusberg, 1959), and feeding of *Aphelenchus avenae* on tomato roots (Chin and Estey, 1966).

The ease with which *Ditylenchus dipsaci* can be propagated on certain plant callus tissues has favored the use of this economically important nematode as an experimental tool. Eriksson (1965) and Bingefors (1965) proved that different races of *D. dipsaci* can interbreed and produce fertile progeny on plant callus. Viglierchio and Croll (1968) observed changes in the sex ratio of *D. dipsaci* in callus culture; that is, as calluses of onion, white clover, red clover, and alfalfa were arranged in order of decreasing host suitability, the nematode populations were simultaneously arranged in order of increasing maleness. In discussing this work, Croll (1968) described an interesting experimental design wherein the responses of nematodes to susceptible and resistant tissues were evaluated by growing two callus varieties on the same medium in a common container but separated by a dialysis tubing.

Several investigators studied the development of nematodes in a monoxenic system that included the plant host. Among these Moriarty (1965) described the development of *Heterodera schachtii* on sugar beet seedlings, Bird *et al.* (1968) made observations on the embryogenesis of *Trichodorus christiei,* and Tanaka (1962) found that 28°C was optimal for the development of *Meloidogyne incognita acrita* on tobacco seedlings. Also, DuCharme and Price (1966) compared the rate of buildup of *Radopholus similis* on citrus seedlings in gnotobiotic culture with the rate of population increase in nature. In studies of a similar nature, Prasad and Webster (1967) reported that *Nacobbus serendipiticus* developed more rapidly at 25°C than at 20° or 30°C on excised tomato roots.

McClure and Viglierchio (1966b) studied the relationship between growth and nutrition of sterile, excised cucumber roots and penetration by *Meloidogyne incognita.* Their findings, though inconclusive, indicated that up to a point the concentration of certain medium constituents (i.e., sucrose) influenced the degree of root penetration but that penetration was not correlated with root growth. These investigators also found that nutrient concentration profoundly influenced both nematode development and gall formation on excised cucumber roots (McClure and Viglierchio, 1966b). Similar experiments were performed by Johnson and Viglierchio, (1969a,b) using *Heterodera schachtii* on excised sugar beet roots; specifically, the influence of environmental and nutritional factors on root penetration, nematode development, and sex ratio was studied. In studies which compared the invasive potential of *Pratylenchus penetrans* from alfalfa callus with that of nematodes from greenhouse cultures or the field, Högger (1969) detected no differences in pathogenicity between nematodes from different sources.

Physiological and morphological changes associated with parasitism were examined by Bird (1967) under monoxenic conditions. This investigator observed that granules which accumulated in the subventral esophageal gland ducts of *Meloidogyne javanica* disappeared completely within 1–3 days of entry into clover seedlings. Concurrently there was an approximate threefold enlargement of the dorsal and subventral esophageal glands. Subsequent histochemical and microspectrophotometric studies by Bird and Saurer (1967) showed that the contents of the gland ducts change their chemical composition within 2–3 days of the nematode entering the plant, becoming in the course of this change strongly periodic acid-Schiff positive. The preceding studies are illustrative of the type of careful work which will eventually lead to a more complete understanding of the complex interactions between highly specialized plant parasites and the host.

## IV. CONCLUSIONS

Unquestionably, gnotobiotic studies form an integral part of any investigation of the nematode–plant disease complex, along with surveys, greenhouse pathogenicity trials, control tests, and other procedures. Gnotobiotic concepts have been applied to practically every facet of phytonematology. This review primarily includes germfree experiments in which one of the reactants was a higher plant. Thus investigations of nematode enzymology and other biochemical studies and those dealing with the culture and study of myceliophagous nematodes have been excluded even though gnotobiotic conditions were maintained throughout.

Other reviews containing reference to phytonematode gnotobiology are by Brzeski (1966), Dropkin (1966a), Mountain (1960, 1961, 1965), Pitcher (1957, 1963, 1965), Tiner (1966), and Zuckerman (1969).

## REFERENCES


Acedo, J. R., and Rohde, R. A. (1968). *Nematologica* **14,** 1.
Anderson, R. V. (1968). *Nematologica* **14,** 2.
Barker, K. R. (1963). *Phytopathology* **53,** 870. (Abstr.)
Barker, K. R., and Darling, H. M. (1965). *Nematologica* **11,** 162–166.
Baxter, R. I., and Blake, C. D. (1968). *Nematologica* **14,** 351–361.
Bingefors, S. (1965). *NFR-Actuellt* pp. 323–330. (In Swed.; Engl. sum.)
Bingefors, S., and Eriksson, K. B. (1963). *Lantbruks-Hoegsk. Ann.* **29,** 107–118.
Bird, A. F. (1962). *Nematologica* **8,** 1–10.
Bird, A. F. (1967). *J. Parasitol.* **53,** 768–776.
Bird, A. F., and Saurer, W. (1967). *J. Parasitol.* **53,** 1262–1269.
Bird, G. W., Goodman, R. M., and Mai, W. F. (1968). *Can. J. Zool.* **46,** 292–293.
Blake, C. D. (1966). *Nematologica* **12,** 129–137.
Blake, C. D. (1968). *Proc. N.W. Nematol. Workshop, Vancouver, B. C.* pp. 17–18.
Blanchard, F. A., and Diller, V. M. (1950). *Plant Physiol.* **25,** 767–769.
Boswell, T. E. (1963). *Phytopathology* **53,** 622. (Abstr.)
Brzeski, M. W. (1966). *Ekol. Pol., Ser. B* **12,** 67–72.
Buecher, E. J., Jr., Hansen, E., and Yarwood, E. A. (1966). *Proc. Soc. Exp. Biol. Med.* **121,** 390–393.
Byars, L. P. (1914). *Phytopathology* **4,** 323–326.
Casida, L. E., Jr. (1965). *Appl. Microbiol.* **13,** 327–334.
Castillo, J. M., and Rohde, R. A. (1965). *Phytopathology* **55,** 127–128.
Chang, L. M., and Rohde, R. A. (1969). *Phytopathology* **59,** 398. (Abstr.)

Chen, T. (1964). *Phytopathology* **54,** 127.
Chen, T., and Rich, A. E. (1963a). *Phytopathology* **53,** 348. (Abstr.)
Chen, T., and Rich, A. E. (1963b). *Plant Dis. Rep.* **47,** 504–507.
Chen, T., Kilpatrick, R. A., and Rich, A. E. (1961). *Phytopathology* **51,** 799–800.
Chen, T., Kiernan, J., and Mai, W. F. (1965). *Phytopathology* **55,** 490–491.
Chernyak, E. K. (1968). *Mater. Nauch. Konf. Obshch. Gel'mintol. Uzbekistanam Sentybar'., Tashkent* pp. 135–137.
Chin, D. A., and Estey, R. H. (1966). *Phytoprotection* **47,** 66–72.
Christie, J. R., and Crossman, L. (1936). *Proc. Helminthol. Soc. Wash.* **3,** 69–72.
Croll, N. (1968). *Proc. N.W. Nematol. Workshop, Vancouver, B.C.* pp. 19–20.
Crosse, J. E., and Pitcher, R. S. (1952a). *Ann. Appl. Biol.* **39,** 475–486.
Crosse, J. E., and Pitcher, R. S. (1952b). *Annu. Rep., East Malling Res. Sta., Kent* **40,** 138–140.
Cryan, W. S. (1963). *J. Parasitol.* **49,** 351–352.
Cryan, W. S., Hansen, E., Martin, M., Sayre, F. W., and Yarwood, E. A. (1963). *Nematologica* **9,** 313–319.
Darling, H. M., Faulkner, L. R., and Wallendal, P. (1957). *Phytopathology* **47,** 7. (Abstr.)
Das, S., and Raski, D. J. (1968). *Nematologica* **14,** 55–62.
den Ouden, H. (1960). *Nematologica* **5,** 255–259.
Deubert, K. H., and Zuckerman, B. M. (1967). *Exp. Parasitol.* **21,** 209–214.
Deubert, K. H., Norgren, R. L., Paracer, S. M., and Zuckerman, B. M. (1967). *Nematologica* **13,** 56–62.
Dolliver, J. S., Hildebrandt, A. C., and Riker, A. J. (1962). *Nematologica* **7,** 294–300.
Dougherty, E. C. (1951). *Exp. Parasitol.* **1,** 34–45.
Dougherty, E. C. (1959). *Ann. N. Y. Acad. Sci.* **77,** 27–54.
Dougherty, E. C. (1960). *In* "Nematology: Fundamentals and Recent Advances with Emphasis on Plant Parasitic and Soil Forms" (J. Sasser and W. R. Jenkins, ed.), pp. 297–318. Univ. of North Carolina Press, Chapel Hill, North Carolina.
Dropkin, V. H. (1966a). *Ann. N. Y. Acad. Sci.* **139,** 39–52.
Dropkin, V. H. (1966b). *Nematologica* **12,** 89.
Dropkin, V. H., and Boone, W. R. (1966). *Nematologica* **12,** 225–236.
Dropkin, V. H., and Webb, R. E. (1967). *Phytopathology* **57,** 584–587.
Dropkin, V. H., Davis, D. W., and Webb, R. E. (1967). *Proc. Amer. Soc. Hort. Sci.* **90,** 316–323.
DuCharme, E. P. (1957). *Proc. Fla. State Hort. Soc.* **70,** 58–60.
DuCharme, E. P. (1959). *Phytopathology* **49,** 388–395.
DuCharme, E. P., and Hanks, R. W. (1961). *Plant Dis. Rep.* **45,** 742–744.
DuCharme, E. P., and Price, W. C. (1966). *Nematologica* **12,** 113–121.
Edmunds, J. E., and Mai, W. F. (1966). *Phytopathology* **56,** 1132–1135.
Edmunds, J. E., and Mai, W. F. (1967). *Phytopathology* **57,** 468–471.
Eriksson, K. B. (1965). *Nematologica* **11,** 244–248.
Faulkner, L. R., and Darling, H. M. (1961). *Phytopathology* **51,** 778–786.
Feder, W. A. (1958). *Phytopathology* **48,** 392–393.
Feder, W. A., and Feldmesser, J. (1957). *Phytopathology* **47,** 11.
Feder, W. A., Hutchins, P. C., and Whidden, R. (1962). *Proc. Fla. State Hort. Soc.* **75,** 74–76.
Feldmesser, J. (1967). *Nematologica* **13,** 141–142.
Fenwick, D. W. (1956). *Nematologica* **1,** 331–336.
Fujiwara, A., Ohira, K., Chiba, K., and Konno, I. (1967). *In* "Advances in Germfree

Research and Gnotobiology" (M. Miyakawa and T. D. Luckey, eds.), pp. 387–391. C. R. C. Press, Cleveland, Ohio.
Goodman, R. M., and Chen, T. A. (1967). *Phytopathology* **57**, 1216–1220.
Hansen, E. L., Sayre, F. W., and Yarwood, E. A. (1960). *Experientia* **17**, 32–33.
Hastings, R. J., and Bosher, J. E. (1938). *Can. J. Res., Sect. C* **16**, 225–229.
Hirumi, H., Chen, T., and Maramorosch, K. (1967). *2nd Int. Colloq. Invertebr. Tissue Cult. Como, Italy* pp. 147–152.
Högger, C. H. (1969). *J. Nematol.* **1**, 10.
Huang, C. (1966). *Phytopathology* **56**, 755–759.
Hung, C. P., and Jenkins, W. R. (1969). *J. Nematol.* **1**, 12.
James, G. L. (1968). *Ann. Appl. Biol.* **61**, 503–510.
Johnson, R. N., and Viglierchio, D. R. (1969a). *Nematologica* **15**, 129–143.
Johnson, R. N., and Viglierchio, D. R. (1969b). *Nematologica* **15**, 144–152.
Kable, P. F., Zehr, E., and Mai, W. F. (1966). *Nematologica* **12**, 175–177.
Khera, S., and Zuckerman, B. M. (1962). *Nematologica* **8**, 272–274.
Kilpatrick, R. A., Chen, T., Rich, A. E., and Rodrigues, L. (1963a). *Phytopathology* **53**, 349.
Kilpatrick, R. A., Chen, T., Rich, A. E., and Rodrigues, L. (1963b). *Plant Dis. Rept.* **47**, 497–501.
Kisiel, M., Deubert, K., and Zuckerman, B. M. (1969a). *Phytopathology* **59**, 1387–1390.
Kisiel, M., Nelson, B., and Zuckerman, B. M. (1969b). *Nematologica* **15**, 153–154.
Krusberg, L. R. (1959). *Nematologica* **4**, 187–197.
Krusberg, L. R. (1960). *Phytopathology* **50**, 643.
Krusberg, L. R. (1961). *Nematologica* **6**, 181–200.
Krusberg, L. R., and Blickenstaff, M. L. (1964). *Nematologica* **10**, 145–150.
Loewenberg, J. R., Sullivan, T., and Schuster, M. L. (1960). *Phytopathology* **50**, 322–323.
Lownsbery, B. F., and Lownsbery, J. W. (1956). *Plant Dis. Rep.* **40**, 989–990.
Lownsbery, B. F., Huang, C. S., and Johnson, R. N. (1967). *Nematologica* **13**, 390–394.
McClure, M. A., and Viglierchio, D. R. (1966a). *Nematologica* **12**, 248–258.
McClure, M. A., and Viglierchio, D. R. (1966b). *Nematologica* **12**, 237–247.
McElroy, F. D., and Van Gundy, S. D. (1968). *Phytopathology* **58**, 1558–1565.
Mayol, P. S., and Bergeson, G. B. (1969). *J. Nematol.* **1**, 17.
Metcalf, H. (1903). *Trans. Amer. Microsc. Soc.* **24**, 89–102.
Miller, C. W. (1963). *Phytopathology* **53**, 350–351.
Moriarty, F. (1964). *Parasitology* **54**, 289–293.
Moriarty, F. (1965). *Parasitology* **55**, 719–722.
Mountain, W. B. (1954). *Can. J. Bot.* **32**, 737–759.
Mountain, W. B. (1955). *Proc. Helminthol. Soc. Wash.* **22**, 49–52.
Mountain, W. B. (1960). *In* "Nematology: Fundamentals and Recent Advances with Emphasis on Plant Parasitic and Soil Forms (J. Sasser and W. R. Jenkins, eds.), pp. 422–425. Univ. of North Carolina Press, Chapel Hill, North Carolina.
Mountain, W. B. (1961). *Recent Advan. Bot.* Sect. 5, pp. 414–417.
Mountain, W. B. (1965). *In* "Ecology of Soilborne Plant Pathogens" (K. E. Baker and W. C. Snyder, eds.), pp. 285–301. Univ. of California Press, Berkeley, California.
Mountain, W. B., and Patrick, Z. A. (1959). *Can. J. Bot.* **37**, 459–470.

Myuge, S. G. (1963). "Methods for Study of Phytonematodes," pp. 126–127. *Izdat. Akad. Nauk SSSR, Moscow* (Engl. transl. from Russ.)
Myers, R. F. (1967a). *Nematologica* **13,** 323.
Myers, R. F. (1967b). *Proc. Helminthol. Soc. Wash.* **34,** 251–255.
Myers, R. F., Feder, W. A., and Hutchins, P. C. (1965). *Proc. Helminthol. Soc. Wash.* **32,** 94–95.
O'Bannon, J. H., and Taylor, A. L. (1968). *Phytopathology* **58,** 385.
O'Bannon, J. H., Myers, R. F., and Feder, W. A. (1967). *Nematologica* **13,** 147–148.
Olthof, T. H. A. (1968). *Nematologica* **14,** 482–488.
Paracer, S. M., and Zuckerman, B. M. (1967). *Nematologica* **13,** 478.
Paracer, S. M., Waseem, M., and Zuckerman, B. M. (1967). *Nematologica* **13,** 517–524.
Peacock, F. C. (1959). *Nematologica* **4,** 43–55.
Pi, C.-L. (1966). M.C. Thesis, 52 pp. Univ. of Massachusetts, Amherst, Massachusetts.
Pitcher, R. S. (1957). *Nematologica* **2** Suppl., 413–423.
Pitcher, R. S. (1963). *Phytopathology* **53,** 35–39.
Pitcher, R. S. (1965). *Helminthol. Abstr.* **34,** 1–17.
Pitcher, R. S., and Crosse, J. E. (1958). *Nematologica* **3,** 244–256.
Pitcher, R. S., Patrick, Z. A., and Mountain, W. B. (1960). *Nematologica* **5,** 309–314.
Polychronopoulos, A. G., and Lownsbery, B. F. (1968). *Nematologica* **14,** 526–534.
Powell, N. T. (1963). *Phytopathology* **53,** 28–35.
Prasad, S. K., and Webster, J. M. (1967). *Nematologica* **13,** 85–90.
Rohde, R. A. (1963). *Phytopathology* **53,** 886–887.
Sandstedt, R., and Schuster, M. L. (1963). *Phytopathology* **53,** 1309–1312.
Sandstedt, R., and Schuster, M. L. (1965). *Phytopathology* **55,** 393–395.
Sandstedt, R., and Schuster, M. L. (1966a). *Physiol. Plant.* **19,** 960–967.
Sandstedt, R., and Schuster, M. L. (1966b). *Physiol. Plant.* **19,** 99–104.
Savage, H. E., and Fisher, K. D. (1966). *Phytopathology* **56,** 898. (Abstr.)
Sayre, F. W., Hansen, E., Starr, T. J., and Yarwood, E. A. (1961). *Nature (London)* **190,** 1116–1117.
Sayre, R. M. (1958). *Diss. Abstr.* **19,** 1185.
Sayre, R. M., and Mountain, W. B. (1962). *Phytopathology* **52,** 510–516.
Schroeder, P. H. (1963). *Phytopathology* **53,** 888–889.
Schroeder, P. H., and Jenkins, W. R. (1963). *Nematologica* **9,** 327–331.
Schuster, M. L., and Sandstedt, R. M. (1962). *Nematologica* **7,** 8.
Silverman, P. H. (1965). *In* "Advances in Parasitology" (B. Dawes, ed.), Vol. 3, pp. 159–222. Academic Press, New York.
Stotzky, G., Culbreth, W., and Mish, L. B. (1962). *Plant Physiol.* **37,** 332–341.
Sudakova, I. M., and Chernyak, E. K. (1967). *Mater. Nautchnov Konf. Vses. Obshch. Gel'mintol., Moscow, 1966* pp. 315–318.
Sudakova, I. M., Chernyak, E. K., and Pimekhov, A. F. (1963). *Mater. Nauch.-Proizvod. Konf. Gel'mintol., Samarkand-Tailyak* pp. 113–115.
Sudakova, I. M., Petrovskaya, E. S., and Chernyak, E. K. (1965). *Tr. Gel'mintol. Lab. Akad. Nauk SSSR* **16,** 131–136.
Sutherland, J. R. (1967). *Nematologica* **13,** 191–196.
Sutherland, J. R., and Fortin, J. A. (1968). *Phytopathology* **58,** 519–523.
Tanaka, I. (1962). *Bull. Kagoshima Tobacco Exp. Sta.* **10,** 42 pp. (In Jap.)
Taylor, P. L., Ferris, J. M., and Ferris, V. R. (1968). *Nematologica* **14,** 17. (Abstr.)
Tiner, J. D. (1960). *Exp. Parasitol.* **9,** 121–126.
Tiner, J. D. (1961a). *J. Parasitol.* **47,** Suppl., 25.

Tiner, J. D. (1961b). *Exp. Parasitol.* **11,** 231–240.
Tiner, J. D. (1966). *Ann. N. Y. Acad. Sci.* **139,** 111–123.
Townshend, J. L. (1963a). *Can. J. Plant Sci.* **43,** 70–74.
Townshend, J. L. (1963b). *Can. J. Plant Sci.* **43,** 75–78.
Trexler, P. C., and Reynolds, L. I. (1957). *Appl. Microbiol.* **5,** 406–412.
Troll, J., and Rohde, R. A. (1966). *Phytopathology* **56,** 995–998.
Tyler, J. (1933). *Hilgardia* **7,** 373–388.
Viglierchio, D. R., and Croll, N. A. (1968). *Science* **161,** 271–272.
Viglierchio, D. R., and Croll, N. A. (1969). *J. Nematol.* **1,** 35–39.
Webster, J. M. (1966). *Nature (London)* **212,** 1472.
Webster, J. M. (1967). *Nematologica* **13,** 256–262.
Webster, J. M., and Lowe, D. (1966). *Parasitology* **56,** 313–322.
White, P. R. (1963). "The Cultivation of Animal and Plant Cells," 2nd Ed. Ronald Press, New York.
Widdowson, E., Doncaster, C. C., and Fenwick, D. W. (1958). *Nematologica* **3,** 308–314.
Zuckerman, B. M. (1969). *Helminthol. Abstr.* **38,** 1–15.
Zuckerman, B. M., and Brzeski, M. W. (1965a). *Nematologica* **11,** 46. (Abstr.)
Zuckerman, B. M., and Brzeski, M. W. (1965b). *Nematologica* **11,** 453–466.

## CHAPTER 21

# Nematodes as Vectors of Plant Viruses

C. E. TAYLOR

*Scottish Horticultural Research Institute, Invergowrie, Dundee, Scotland*

I. Introduction . . . . . . . . . . . . . . . . 185
II. The Vectors . . . . . . . . . . . . . . . . 187
A. *Xiphinema* and *Longidorus* . . . . . . . . . . 187
B. *Trichodorus* . . . . . . . . . . . . . . . 189
C. Distribution in the Soil . . . . . . . . . . . 191
III. The Viruses . . . . . . . . . . . . . . . . 192
A. NEPO Viruses . . . . . . . . . . . . . . 193
B. NETU Viruses . . . . . . . . . . . . . . 195
IV. Relationships between Viruses and Vector Nematodes . . . 196
A. Transmission . . . . . . . . . . . . . . . 196
B. Acquisition and Inoculation . . . . . . . . . . 197
C. Persistence . . . . . . . . . . . . . . . 198
D. Specificity of Transmission . . . . . . . . . . 199
E. Mechanism of Transmission . . . . . . . . . . 200
V. Ecology and Control . . . . . . . . . . . . . 204
References . . . . . . . . . . . . . . . . 207

## I. INTRODUCTION

Although nematodes had long been suspected as possible vectors of plant viruses (Percival, 1912; Johnson, 1945; Allen, 1948; Christie, 1951; McKinney *et al.*, 1957; Schindler, 1958), the first experimental evidence of such an association was not forthcoming until 1958 when Hewitt, Raski, and Goheen showed that *Xiphinema index* transmitted fanleaf virus of grapevine (Hewitt *et al.*, 1958). Discoveries of nematode vectors of other soil-borne viruses quickly followed in Europe and the United States, and at present 18 nematode species are known to transmit 11 serologically distinct viruses (Tables I and II). Only species of the

TABLE I
NEMATODE VECTORS OF NEPO VIRUSES

| Nematode | Virus | Reference |
|---|---|---|
| *Xiphinema americanum* | Tobacco ring spot—Arkansas isolate | Fulton (1962) |
| | Tomato ring spot—peach yellow bud isolate | Breece and Hart (1959) |
| | Tomato ring spot—grape yellow vein isolate | Teliz *et al.* (1966) |
| *Xiphinema coxi* | Arabis mosaic—type strain and rhubarb mosaic isolate | Fritzsche (1964) |
| | Cherry leaf roll—type strain | Fritzsche and Kegler (1964) |
| *Xiphinema diversicaudatum* | Arabis mosaic—type strain | Jha and Posnette (1959), Harrison and Cadman (1959) |
| | Brome grass mosaic-type strain | Schmidt *et al.* (1963) |
| | Carnation ring spot—German isolate | Fritzsche and Schmelzer (1967) |
| | Cherry leaf roll—type strain | Fritzsche (1964) |
| | Strawberry latent ring spot-type strain | Lister (1964) |
| *Xiphinema index* | Arabis mosaic—grapevine fanleaf strain[a] | Hewitt *et al.* (1958) |
| *Longidorus attenuatus* | Tomato black ring—English strain | Harrison *et al.* (1961) |
| *Longidorus elongatus* | Raspberry ring spot—type strain | Taylor (1962) |
| | Tomato black ring—Scottish strain | Harrison *et al.* (1961) |
| *Longidorus macrosoma* | Raspberry ring spot—English strain | Harrison (1962) |

[a] Recently *X. italiae* also has been shown to be a vector of grape fanleaf [Cohn, E. Tanne, E., and Nitzany, F. E. (1970) *Phytopathology* **60**, 181–182].

dorylaimoid genera *Xiphinema, Longidorus,* and *Trichodorus* have thus far been implicated as vectors. These genera are widely distributed throughout the world, and where host ranges have been investigated the vector species seem to be polyphagous, although they may multiply at varying rates on different hosts.

Nematode-transmitted viruses can be divided into two groups on the basis of particle shape. Cadman (1963) named those with polyhedral particles NEPO viruses, i.e., *ne*matode transmitted viruses with *po*ly-

TABLE II
NETU VIRUSES TRANSMITTED BY *Trichodorus*

| Virus | Isolate | Vector | Reference |
|---|---|---|---|
| Pea early browning | Dutch | *T. pachydermus* | van Hoof (1962) |
| | | *T. teres* | van Hoof (1962) |
| | English | *T. anemones* | Harrison (1967b) |
| | | *T. pachydermus* | Gibbs and Harrison (1964b) |
| | | *T. viruliferus* | Gibbs and Harrison (1964b) |
| Tobacco rattle | Dutch | *T. cylindricus* | van Hoof (1968) |
| | | *T. nanus* | van Hoof (1968) |
| | | *T. pachydermus* | Sol and Seinhorst (1961) |
| | | *T. teres* | van Hoof (1968) |
| | | *T. viruliferus* | van Hoof (1968) |
| | Dutch (Gladiolus notch-leaf) | *T. similis* | Cremer and Kooistra (1964) |
| | English | *T. pachydermus* | Gibbs and Harrison (1964b) |
| | | *T. primitivus* | Harrison (1961) |
| | German | *T. primitivus* | Sänger (1961) |
| | Scottish | *T. primitivus* | Mowat and Taylor (1962) |
| | United States (California) | *T. allius* | Ayala and Allen (1966) |
| | | *T. christiei* | Ayala and Allen (1966) |
| | | *T. porosus* | Ayala and Allen (1966) |
| | United States (Oregon) | *T. allius* | Jensen and Allen (1964) |
| | United States (Wisconsin) | *T. christiei* | Walkinshaw *et al.* (1961) |

hedral particles. Viruses in this group are frequently referred to as *ring spot viruses* (Harrison, 1960) and are transmitted by *Xiphinema* and *Longidorus* species. The second group includes the viruses with rod-shaped or tubular particles for which Harrison (1964a) derived the name NETU, i.e., *ne*matode transmitted viruses with *tu*bular particles. The two viruses in this group, tobacco rattle virus and pea early browning virus, are transmitted by *Trichodorus* species.

## II. THE VECTORS

### A. *Xiphinema* and *Longidorus*

Species of the genera *Xiphinema* and *Longidorus* are morphologically similar and are commonly placed together in the subfamily Longidorinae (Goodey, 1963). Species of both genera are relatively large nematodes, from about 3 to 8 mm long in the adult stage, and are characterized by the presence of a long, hollow odontostyle with which they penetrate plant roots and feed on the cell contents. Feeding is usually at the root tips and to a lesser extent along the sides of the roots, and almost invariably this causes distortion or galling (Fisher and Raski, 1967; Thomas, 1969a, 1970).

### 1. Pathogenicity

In field soils in the United States, *X. americanum* infestations have been associated with necrosis and destruction of feeder roots of laurel oak (*Quercus laurifolia*) (Christie, 1952), necrosis and swelling of roots and general symptoms of decline in maple (*Acer saccharum*) (Di Sanzo and Rohde, 1969), stunting in nursery stocks of ornamental spruce (*Picea pungens* and *P. glauca densata*) (Griffin and Epstein, 1964), black root rot of strawberry (*Fragaria* × *ananassa*) (Perry, 1958), unthriftiness of red clover (*Lolium multiflorum*) (Norton, 1967), severe decline of periwinkle (*Vinca minor*) (Epstein and Barker, 1966), and general weakening and premature decline of various tree species used in shelterbelts in South Dakota (Malek, 1968). In eastern United States, *X. diversicaudatum* was associated with severe galling on the roots of rose (*Rosa* sp.) fig (*Ficus carica*), peanut (*Arachis hypogaea*), and strawberry (Schindler, 1957; Schindler and Braun, 1957). In Great Britain this species caused galling to a greater or lesser extent on a wide variety of woody and herbaceous crop plants and weeds (Harrison and Winslow, 1961; Taylor *et al.*, 1966; Taylor and Thomas, 1968; Thomas, 1970).

Feeding by *L. elongatus* appears to cause less severe galling than that caused by *Xiphinema* (Thomas, 1969a, 1970), but in sandy soils the growth of strawberry may be affected by moderate infestations (Sharma, 1965; Seinhorst, 1966; Graham and Flegg, 1968). In Oregon (United States), *L. elongatus* caused severe necrosis or "die back" of the roots of peppermint (*Mentha piperita*) (Jensen, 1961). In Germany, Sturhan (1963) reported root galling of onions (*Allium cepa*) caused by *L. macrosoma.* In Israel, Cohn and Krikun (1966) reported similar damage to onions by *L. vineacola,* which is not known to be a virus vector.

### 2. Host Range and Life Cycle

Evidence from observations associating nematodes with root damage in the field (see Section A, 1 above) from distribution surveys (Weischer, 1966; Moore, 1967) and from laboratory studies (Lownsbery, 1964; Flores and Chapman, 1968; Thomas, 1969a) indicates that *X. americanum* and *X. diversicaudatum* have a wide host range among woody and herbaceous plants. *Xiphinema index* appears, however, to multiply successfully only on a restricted range of woody plants that includes grapevine (*Vitis vinifera*) and other *Vitis* species, fig and rose (Raski and Lider, 1959; Radewald, 1962), and to a lesser extent on *Parthenocissus* sp., peach (*Prunus persica*), Upland cotton (*Gossypium hirsutum*), *Pistacia* sp., and *Ampelopsis* sp. (Weiner and Raski, 1966; Radewald, 1962; Radewald and Raski, 1962b). In California, *X. index* reared in the laboratory on grapevine and at an optimum temperature of 30°C, completed its development from oviposition to the adult stage in 22–27 days (Radewald and Raski, 1962a); although in Israel, Cohn and Mordechai (1969) found that the life cycle extended to 7–9 months at 20°–23°C. On the other hand, *X. americanum* and *X. diversicaudatum* appear to develop more slowly, and in the field probably a year is required to complete each generation under temperate conditions (Norton, 1963; Griffin and Darling, 1964; Taylor and Thomas, 1968; Thomas, 1969b). In culture studies of *X. diversicaudatum* and *X. vuittenezi*, Flegg (1968b) found that several months may elapse between molts and that development from egg to adult took at least 2 years, with adults possibly surviving for as long as 5 years.

Large populations of *L. elongatus* have been found in association with crops of strawberry or ryegrass (*Lolium perenne*) in eastern Scotland (Taylor, 1967) and with peppermint in Oregon (United States) (Horner and Jensen, 1954). Laboratory studies have shown that *L. elongatus* is able to multiply on a large number of different host plants; but with the exception of strawberry, grasses, and clovers, herbaceous weeds such as *Stellaria media, Mentha arvensis, Poa annua,* and *Tussilago farfara* are better hosts than crop plants (Taylor, 1967; Thomas, 1969a; Yassin, 1969). In the laboratory, *L. elongatus* develops from the egg to the adult stage in 16–23 weeks at 20°C (Yassin, 1969), but in the field there is probably only one generation per year with a two- to fourfold increase of population on favored hosts (Murant and Taylor, 1965; Taylor and Murant, 1968; Thomas, 1969b). The other *Longidorus* vector species are both widely distributed in Europe (Harrison *et al.*, 1961; Debrot, 1964; Whitehead, 1965; Dalmasso and Caubel, 1966; Sturhan, 1966; Weischer, 1966; Moore, 1967), but little is known about their ecology. *Longidorus attenuatus* multiplies well on ryegrass and clover (Green, 1967; Cooke and Hull, 1967), but *L. macrosoma* probably prefers rosaceous woody plants.

### B. *Trichodorus*

*Trichodorus* species are relatively short, thick nematodes with the odontostyle in the form of a nonaxial dorsal tooth, which does not have the functional lumen seen in the odontostyle of *Xiphinema* or *Longidorus* (Hirumi *et al.*, 1968; Raski *et al.*, 1969). Feeding is probably by repeated, direct thrusts of the odontostyle into the epidermal cells of the root (Rohde and Jenkins, 1957a; Zuckerman, 1961, 1962; Chen and Mai, 1965) causing swelling of the root tips or "stubby root."

### 1. Pathogenicity

Christie and Perry (1951) were the first to recognize parasitism in *Trichodorus* when they found *T. christiei* injuring the roots of beet (*Beta vulgaris*), celery (*Apium dulce*), and sweet corn (*Zea mays*) in Florida (United States). Other crops were later found to be injured by this polyphagous species: corn (*Zea mays*), cabbage (*Brassica oleracea*), cauliflower (*B. oleracea*), fig, various grasses, onion, peach, tomato (*Lycopersicon esculentum*), and blueberry (*Vaccinium corymbosum*) (Rohde and Jenkins, 1957a; Christie, 1959; Zuckerman, 1961; Hoff and Mai, 1962; Alhassan and Hollis, 1966). In Britain, *Trichodorus* species are associated with a condition of sugar beet known as Docking disorder, which is characterized by patches of stunted plants in the crop in the early stages of growth and is common on light, sandy soils (Whitehead, 1965; Dunning and Cooke, 1967). Stunting of sugar beet grown in new polder soil in the Netherlands was associated with *T. teres* (= *flevensis*) (Kuiper and Loof, 1961). By direct observations in an underground laboratory, Pitcher and Flegg (1965) and Pitcher (1967) found that *T. viruliferus*, massing at the root tips of apple, caused necrosis, halted growth, and caused some swelling of the tips.

### 2. Host Range and Life Cycle

Many of the *Trichodorus* species that are known as vectors of tobacco rattle or pea early browning viruses have only recently been described; consequently, little is known about their biology. From their widespread distribution and frequency of occurrence it would seem that, like *T. christiei* and *T. pachydermus* which have been more fully investigated, they have a wide host range among herbaceous crops and weeds (Section B, 1 above). Apart from *T. christiei*, little is known about the life cycle of *Trichodorus* species except that some appear capable of multiplying rapidly, e.g., *T. allius* on sweet pea (Ayala and Allen, 1968). Rohde and Jenkins (1957a,b) found that *T. christiei* completed its life cycle in 21–22 days at 22°C and in 16–17 days at 30°C with a tenfold or more increase in population in 60 days. Bird and Mai (1967) found that orchard grass (*Dactylis glomerata*), tomato, and red clover (*Trifolium pratense*) are good hosts for *T. christiei*. They recorded the greatest increase in populations on pruned red clover at 25°C.

### C. Distribution in the Soil

#### 1. Vertical Distribution

Because the nematode vectors are ectoparasites feeding mainly at the root tips, it is reasonable to suppose that their distribution in depth will correspond with that of suitable roots. Most of the preferred hosts of *L. elongatus* are surface rooting; consequently, the bulk of the populations are found in the top 20 cm of the soil (Taylor, 1967). *Xiphinema diversicaudatum*, on the other hand, feeds on plants ranging from surface rooting herbaceous crops and weeds to deeper rooting woody hosts such as raspberry, and populations may be distributed to a depth of 60 cm or more (Harrison and Winslow, 1961; Taylor and Thomas, 1968). On grapevines growing in the stony, shallow soils of the Côte d'Or, France, the largest populations of *X. index* were recorded between 30–60 cm depth, but in the more fertile and deeper soils of the California Napa Valley populations were distributed to a depth of 3.6 m (Raski *et al.*, 1965). In Israel, Cohn (1969) investigated the vertical distribution of 7 species of *Xiphinema* and 2 species of *Longidorus* and found maximum numbers of all of the species in the 0–30 cm zone, with a decrease in numbers with increasing depth. In southeastern England, Flegg (1968a) found that the numbers of *X. diversicaudatum* and *X. vuittenezi* decreased with depth, but *L. macrosoma* increased with depth up to 70 cm and *L. profundorum* up to 75 cm and sometimes beyond. The depth distribution of each of the species remained more or less constant throughout the year.

Both the horizontal and vertical distribution of *Trichodorus* species in the soil appears to be more variable than that of *Xiphinema* and *Longidorus,* because many of them are capable of multiplying rapidly if suitable host plants are available, and this may give rise to relatively rapid fluctuations in their numbers. In the Netherlands, *T. teres* was found at a depth of 30–50 cm, but in the summer and autumn most were in the upper layers of the soil (Taconis and Kuiper, 1963). At one site, *T. primitivus* and *T. pachydermus* were found mainly at a depth of 20–40 cm (Taconis and Kuiper, 1963), but elsewhere *T. pachydermus* was distributed to 50 cm depth and tobacco rattle virus was detectable in soil collected from between 80 and 100 cm depth (Sol, 1963). Pitcher's (1967) observations on *T. viruliferus* feeding on apple roots indicate that *Trichodorus* species feeding on woody hosts may be very localized in their distribution. He found that *T. viruliferus* massed around the elongating zone of extending roots, the numbers increasing from few to a hundred or more in as little as 4 days; the mass of nematodes followed the growing root, remaining 1–3 mm behind the tip.

### 2. In Relation to Soil Type

The geographical distribution of many vector species appears to be related to, and limited by, soil type. Pore space and moisture are probably the most important factors of the soil that determine its suitability as a habitat for ectoparasitic nematodes. Clay soils contain more than 20% clay (particles of less than 0.002 mm) and have a texture which tends to prevent free movement of air and water. Loams contain 8–15% clay and are well aerated and easily drained, but with sufficient clay to withstand moderate periods of drought. The pore space in loam soils, as a percentage of the total volume, is about 35 compared with about 50 in clay soils. Sandy soils contain only about 5% clay and are therefore composed of relatively large particles which make them open in texture and free draining, but they are subject to drying out rapidly during periods of drought. Intermediate types between these three broad categories are usually classified as clay loam or heavy loam and sandy loam.

*Xiphinema diversicaudatum* is usually associated with clay soils or, more particularly, with soils that maintain a high moisture level (Fritzsche, 1966). In Israel, Cohn (1969) found that *X. mediterraneum* is the predominant *Xiphinema* species in poorly aerated clay soils and was almost absent in open textured soils where *X. diversicaudatum* is relatively more abundant. *Longidorus elongatus* is found in sandy to medium loams, *L. macrosoma* in clays and heavy loams, and *L. attenuatus* in sandy soils (Harrison, 1964a).

*Trichodorus* species are usually found in open textured soils. In the Netherlands, *T. teres* is found in marine, sandy soils whereas *T. pachydermus* is generally widespread, but usually in sandy soils rather than in those with much clay. In Scotland, *T. primitivus* occasionally occurs in till soils (derived from boulder clay) but is more common in soils derived from fluvioglacial and river terrace sands or from raised beach deposits, whereas *T. pachydermus* occurs only in very sandy soils (Cooper and Harrison, 1969).

## III. THE VIRUSES

Nematode-transmitted viruses infect a wide range of herbaceous and woody hosts and are commonly seed- and pollen-borne. They are readily transmitted by mechanical inoculation to herbaceous test plants, a facility which has been instrumental in the rapid advancement of our knowledge about their physical and biological properties. This, in turn, has done much to clarify the relationships between viruses and to rationalize the naming of them.

### A. NEPO Viruses

These are transmitted by species of *Xiphinema* and *Longidorus* and include arabis mosaic, strawberry latent ring spot, cherry leaf roll, tobacco ring spot, tomato ring spot, raspberry ring spot, and tomato black ring viruses (Table I). Their chemicophysical properties are similar, and all possess polyhedral particles about 27–30 nm diam, but they share no antigenic determinants and hence are readily identifiable by serological methods.

Arabis mosaic virus was first isolated from an *Arabis* plant growing in a glasshouse at Cambridge, England (Smith and Markham, 1944), and remained as a curiosity until viruses associated with raspberry yellow dwarf and strawberry yellow crinkle diseases were identified with it (Cadman, 1960) and were shown to be transmitted by *X. diversicaudatum* (Jha and Posnette, 1959; Harrison and Cadman, 1959). The type strain of arabis mosaic has so far been found only in Europe, but a distantly related strain or serotype probably occurs wherever grapevines are cultivated. In the United States the two principal diseases caused by this strain of arabis mosaic virus are called *grapevine fanleaf* and *grapevine yellow mosaic;* in Europe, the same diseases are named *arricciamento, roncet, urticado, courtnoué* or *Reisigkrankeit,* and *panachure* or *clorose infecciosa,* respectively. *Xiphinema index* was shown to be the vector in the United States (Hewitt *et al.,* 1958); subsequently, it has been found to be abundant in Europe in association with infected grapevines (Cadman *et al.,* 1960; Martelli and Raski, 1963; Vuittenez and Legin, 1964; Lamberti and Raski, 1964; Amici, 1965, 1967a,b).

Tobacco ring spot and tomato ring spot viruses are both transmitted by *X. americanum* (Breece and Hart, 1959; Fulton, 1962). Both viruses infect a wide range of crop and weed hosts, but despite the cosmopolitan distribution of the vector no outbreaks of the virus have been recorded outside the American continent. This apparent anomaly may be explained by the recent recognition of *X. americanum sensu latu* as a species complex (Tarjan, 1968, 1969; Lima, 1968) in which seven distinct species have been delineated (Lima, 1968). Most of the records of *X. americanum* in Europe should now be referred to one of the new species, *X. mediterraneum,* which commonly occurs in the south of France (Dalmasso and Caubel, 1966), in Israel (Cohn, 1969), and in central and southern Italy (Martelli and Lamberti, 1967). It has also been recorded from southern England.

*Xiphinema diversicaudatum* has been implicated as the vector of four other viruses in addition to arabis mosaic virus. Strawberry latent ring spot virus infects raspberry, cherry, black currant, and several weed hosts as well as strawberry (Lister, 1964; Taylor and Thomas, 1968). It frequently occurs with arabis mosaic. Cherry leaf roll virus (Cropley, 1961) occurs in cherry trees in England and Germany and has both *X. diversicaudatum* and *X. coxi* as vectors (Fritzsche and Kegler, 1964). Brome grass mosaic has been transmitted by *X. diversicaudatum* (Fritzsche, 1964) and carnation ring spot virus, which appears to be prevalent among carnation cultivars (Hollings and Stone, 1965), by *X. diversicaudatum,* and by *L. macrosoma* (Fritzsche and Schmelzer, 1967).

In Scotland, raspberry ring spot virus causes the economically important disease "raspberry leaf curl," but in the Netherlands an antigenically similar strain of the virus is associated with spoonleaf disease of red currant (*Ribes sanguineum*). Each is transmitted by *L. elongatus* (Taylor, 1962; van der Meer, 1965). Strains of raspberry ring spot virus isolated from blackberry (*Rubus fruticosus*) and raspberry in England, and serologically distinguishable from the Scottish strain, are transmitted by *L. macrosoma* (Harrison, 1964b). As with raspberry ring spot virus, there are several strains of tomato black ring virus that have different *Longidorus* species as vectors. The Scottish strain which causes beet ring spot disease is transmitted by *L. elongatus,* whereas the strain causing celery yellow vein and lettuce ring spot in England and potato bouquet and potato pseudo-aucuba in Germany is transmitted by *L. attenuatus* (Harrison *et al.,* 1961).

Diseases caused by NEPO viruses typically show the "shock" and "recovery" phases that are characteristic of tobacco ring spot disease in tobacco, i.e., the first parts of the plants to become systemically infected show severe symptoms but later growth, although infected, appears healthy. Grapevines newly infected with grapevine fanleaf virus show ring spot patterns on their leaves (Vuittenez, 1961), but leaves produced later show only the typical fanleaf symptoms (Hewitt, 1950). Chlorotic rings precede the development of rasp leaf symptoms in sweet cherry (Blumer, 1954, 1955); Lister (1960) found that strawberry infected with raspberry ring spot virus developed ring spot symptoms each year on the new foliage, whereas leaves produced later in the summer were normal in appearance.

Transmission of virus through the seeds of infected plants is characteristic for the NEPO group of viruses, but the relevance of this to the ecology of the individual viruses varies. Tomato black ring and raspberry ring spot viruses transmitted by *L. elongatus* are commonly present in seeds of various weed species in soil from outbreak areas, but this is rarely so with other NEPO viruses transmitted by *Xiphinema* spp. (Lister and Murant, 1967). Infected weed seeds appear to form an important reservoir of infection for raspberry ring spot and tomato black ring viruses and serve as a means whereby these viruses perennate in soils (Murant and Taylor, 1965; Murant and Lister, 1967; Taylor and Murant, 1968). These viruses do not persist for more than about 8 weeks in their vector, *L. elongatus*, whereas the viruses transmitted by *Xiphinema* (e.g., arabis mosaic, strawberry latent ring spot, grapevine fanleaf viruses) are rarely found in weed seeds from infested soils but persist in their vectors for 8 months or more (Taylor, 1968a).

### B. NETU Viruses

There are only two viruses in this group: tobacco rattle virus and pea early browning virus. Tobacco rattle virus causes a disease in tobacco which was first described and named *mauche* by Behrens (1899), but it is nowadays better known as the cause of economically important diseases in bulbous crops in the Netherlands and one cause of internal necrosis or spraing disease of potato tubers in Britain, continental Europe, and the United States. The virus infects a large number of crop and weed hosts in Europe and in North and South America. Preparations of the virus contain tubular particles of two main lengths, about 75 and 185 nm, but only the long particles are infective (Harrison and Nixon, 1959; Harrison and Woods, 1966). Although the shorter particles on their own are not infective they nevertheless appear to be essential for the production of stable virus (Frost *et al.*, 1967; Lister, 1966).

Pea early browning virus occurs naturally in pea and alfalfa (*Medicago sativa*) crops in the Netherlands and in Britain, but has not been found in North America (Bos and van der Want, 1962; Gibbs and Harrison, 1964b). This virus also has straight tubular particles of two main lengths, about 210 and 105 nm, and like tobacco rattle virus it does not readily infect plants systemically. In 1961, three *Trichodorus* species were shown to be vectors of tobacco rattle virus: *T. pachydermus* in the Netherlands (Sol and Seinhorst, 1961), *T. primitivus* in England (Harrison, 1961), and *T. christiei* in the United States (Walkinshaw *et al.*, 1961). Since then, many more *Trichodorus* species have been added to the list of vectors of both tobacco rattle and pea early browning viruses in Europe (Table II), and in the United States *T. allius* and *T. porosus* have been implicated as vectors of the Californian isolate of tobacco rattle virus (Jensen and Allen, 1964; Ayala and Allen, 1966, 1968).

## IV. RELATIONSHIPS BETWEEN VIRUSES AND VECTOR NEMATODES

In assessing the value of the experimental evidence that has accumulated on virus vector relationships since the initial discovery of *X. index* as a virus vector (Hewitt *et al.*, 1958), it must be recognized that the techniques and conditions of experimentation have varied considerably; consequently, the data are not strictly comparable. The ability of a nematode species to transmit virus may be affected by the technique of extraction from the soil (Teliz, 1967), by the potting mixture used in test pots (Ayala and Allen, 1968), by the size of the test pots (Fritzsche, 1967), or by the age and kind of "bait plants" which are exposed to the nematodes (Cadman and Harrison, 1960). Furthermore, the tendency has been to follow the experimental approach and terminology used in work on insect-borne viruses which may be misleading, restrictive, and possibly largely irrelevant to nematode studies.

### A. Transmission

The adult and juvenile stages of most nematode vector species transmit virus, and probably with equal efficiency. Grapevine fanleaf is transmitted by all stages of *X. index* (Raski and Hewitt, 1960) and tomato ring spot by all stages of *X. americanum* (Teliz *et al.*, 1966). Arabis mosaic and strawberry latent ring spot viruses are also transmitted by adult and juvenile stages of *X. diversicaudatum* (Jha and Posnette, 1961; Harrison, 1967a), but males appear to transmit strawberry latent ring spot less frequently than females (Harrison, 1967a; Taylor and Thomas, 1968). Harrison *et al.* (1961) reported that they obtained transmissions of tomato black ring virus only with juvenile stages of *L. elongatus*, but later experiments (Yassin, 1968; Taylor and Murant, 1969) showed that both adults and juveniles can transmit the virus. Taylor (1962) transmitted raspberry ring spot virus by female and juvenile *L. elongatus*. Juvenile and adult stages of *T. pachydermus* (Sol, 1963; Gibbs and Harrison, 1964a,b), *T. teres* (van Hoof, 1964a), and *T. allius* (Ayala and Allen, 1968) can transmit tobacco rattle virus; Ayala and Allen (1968) obtained transmissions with single specimens of all stages of *T. allius*, except the first-stage juvenile, which was not tested. Tobacco ring spot virus has been transmitted by single *X. americanum* (Teliz, 1967) and strawberry latent ring spot and arabis mosaic viruses by single *X. diversicaudatum* (Lister, 1964; Harrison, 1967a).

## B. Acquisition and Inoculation

The minimum periods variously recorded for nematodes to acquire viruses from plants and then to transmit them to other plants refer to access periods and not to feeding periods. Teliz *et al.* (1966), in their experiments with *X. americanum* and tomato ring spot virus, emphasize this and propose the terms *acquisition access threshold* to express the minimum time required for a nematode to have access to a virus source in order to become viruliferous and *inoculation access threshold* for the minimum time that a viruliferous nematode must have access to a healthy plant for transmission to occur. Neither definition implies that the nematode feeds continuously during this time.

In some early experiments with *X. diversicaudatum* and arabis mosaic virus (Jha and Posnette, 1961) and *X. index* and grapevine fanleaf virus (Hewitt and Raski, 1962), transmissions were obtained after an acquisition access period of 24 hr, the minimum time tried. More recently, acquisition access thresholds of an hour have been established for *X. americanum* and tomato ring spot virus (Teliz *et al.*, 1966) and for *T. allius* and tobacco rattle virus (Ayala and Allen, 1968) and 15 min for *X. index* and grapevine fanleaf virus (Das and Raski, 1968). With several vector–virus combinations, it has been demonstrated that an increase in the acquisition access period results in a proportionate increase in the frequency of transmission obtained; for example, *X. americanum* and tomato ring spot virus (Teliz *et al.*, 1966), *X. americanum* and tobacco ring spot virus (McGuire, 1964), *X. index* and grapevine fanleaf virus (Das and Raski, 1968), and *T. allius* and tobacco rattle virus (Ayala and Allen, 1968). These results may, however, reflect the efficiency of the nematodes to find the roots of the bait plant rather than their efficiency as vectors per se.

Inoculation access threshold periods of less than 24 hr have been established for several nematode vectors, and as with acquisition of virus, frequency of transmission increases with time. Using tobacco ring spot virus-infected field soil containing large numbers of *X. americanum,* Bergeson *et al.* (1964) obtained no transmissions at 4 or 8 hr, but 4 out of 5 transmissions after 24 hr and 5 out of 5 at 72 hr. Teliz *et al.* (1966) obtained few transmissions of tomato ring spot virus by *X. americanum* after 1 hr inoculation access but 100% transmission after 4 days. Similarly, the frequency of transmission of tobacco rattle virus by *T. allius* increased as the inoculation access time was extended from 1 hr up to 48 hr. In addition to experimental variations such as the numbers of nematodes used per test, the physiological state and kind of infector and bait plants used, temperature, soil type, and other environmental factors, these results are affected by the intermittent feeding behavior of some, if not all, species. This is apparent in the experiments of van Hoof (1964b) in which single *T. pachydermus,* carrying tobacco rattle virus, were serially transferred each day to fresh *Nicotiana rustica* bait seedlings, but transmissions with each individual occurred intermittently up to 5 days. Harrison (1967a) also found that transmissions were rarely in serial succession when he transferred individual *X. diversicaudatum* carrying strawberry latent ring spot virus to fresh cucumber bait seedlings at 2-day intervals.

### C. Persistence

Nematode vectors kept in fallow soil usually retain infective virus for periods which much exceed the *in vitro* longevity of the virus in plant sap (Taylor, 1968a). In *Chenopodium* sap at 18°–20°C grapevine fanleaf virus remains infective for 10–28 days (Cadman *et al.,* 1960) but the virus can persist in its vector, *X. index,* for up to 8 months (Taylor and Raski, 1964). *Xiphinema americanum* was shown to transmit tobacco ring spot virus after a starvation period of 49 weeks (Bergeson and Athow, 1963) and *X. diversicaudatum* to transmit strawberry latent ring spot and arabis mosaic viruses after 84 and 112 days, respectively (Taylor and Thomas, 1968). When virus-carrying *Xiphinema* are allowed to feed on virus-immune hosts, persistence of virus in the vectors may also be for relatively long periods: *Xiphinema index* retained grapevine fanleaf virus for 12 weeks when allowed to feed on virus-immune fig (Das and Raski, 1968); and *X. diversicaudatum* allowed to feed on Malling Jewel raspberry, which is immune to arabis mosaic virus, apparently retained the virus for 8 months (Harrison and Winslow, 1961). *Longidorus* vectors may not retain infectivity for as long as *Xiphinema.* Raspberry ring spot and tomato black ring viruses persist in *L. elongatus* for a maximum period of about 8 weeks (Lister and Murant, 1967; Taylor, 1968a) compared with *in vitro* longevities of 15 to 21 and 50 days, respectively (Cadman, 1963). Debrot (1964) found that *L. macrosoma* retained the English strain of raspberry ring spot virus for 34 days but did not experiment further. Sol (1963) found that the Dutch strain of tobacco rattle virus was retained by *T. pachydermus* for at least 36 days, and van Hoof (1964b) reported transmission of the virus with *T. pachydermus* in soil stored for 10 months at temperatures fluctuating between 3° and 20°C. *Trichodorus allius* remained infective with the Californian strain of tobacco rattle virus for 20 weeks when maintained without a host, but for 27 weeks when feeding on virus-immune sweet pea (Ayala and Allen, 1968).

Despite the long retention of viruses in their nematode vectors, the few experimental data available indicate that nematode-transmitted viruses do not persist through the egg or through the molt. Taylor and Raski (1964) support this hypothesis with *X. index* and grapevine fanleaf virus and Harrison and Winslow (1961) with *X. diversicaudatum* and arabis mosaic virus, but both series of experiments were done with too few nematodes to exclude the possibility that infectivity is occasionally retained, particularly through the molt. Das and Raski (1968) maintained a population of *X. index* on virus-immune fig for 12 weeks, which is normally sufficient time for all juveniles to reach the adult stage, and still obtained transmission of grapevine fanleaf virus with preadult stages, suggesting that the virus was retained through the molt or possibly even through the egg stage. Retention of virus by the vector is also discussed later with reference to the mechanism of transmission (Section IV, E).

### D. Specificity of Transmission

Although unrelated NEPO viruses may share a common nematode vector species, experimental evidence suggests that strains of a single virus that differ serologically require different nematode species of the same genus for their transmission (Harrison, 1964b). Thus, whereas *X. diversicaudatum* transmits both arabis mosaic and strawberry latent ring spot viruses, it does not transmit the grapevine fanleaf strain of arabis mosaic, which is transmitted by *X. index* (Dias and Harrison, 1963). Scottish strains of raspberry ring spot and tomato black ring viruses are transmitted by *L. elongatus*, but the English strains, which share only about half of their antigenic groups with the Scottish strains, have *L. macrosoma* and *L. attenuatus*, respectively, as vectors (Table I). On the other hand, strains of virus showing only slight serological differences but distinguishable in host range do not necessarily have different vectors. *Xiphinema americanum* can transmit both the type strain and the serologically distinguishable grapevine yellow vein strain of tomato ring spot virus (Teliz *et al.*, 1966). Sauer (1966) transmitted two serologically distinguishable strains of tobacco ring spot virus with a single specimen of *X. americanum* and found no evidence of protection of one strain against the other in the nematode. Murant *et al.* (1968) found a variant of raspberry ring spot virus infecting Lloyd George raspberry, which is immune to the type Scottish strain, and showed that it was readily transmitted by *L. elongatus*. Later, in laboratory experiments, Taylor and Murant (1969) found that *L. elongatus* transmitted English isolates of raspberry ring spot virus almost as efficiently as the Scottish strain, but it transmitted English and German strains of tomato black ring virus only occasionally. Since *L. elongatus* is widespread throughout Britain, it would appear from the results of these experiments that the natural distribution of the English and Scottish strains of raspberry ring spot virus is not dependent on vector specificity of transmission per se but may lie in some combination of ecological factors (Murant *et al.*, 1968). Different ecological requirements for tomato and tobacco ring spot viruses may also explain their geographical separation in the United States where the common vector, *X. americanum*, is widely distributed.

American and European cultures of tobacco rattle virus are serologically distinguishable (Cadman, 1963; Oswald and Bowman, 1958), and they are transmitted by different *Trichodorus* species. However, many isolates of tobacco rattle virus are each transmitted by more than one vector species, and conversely, a single *Trichodorus* species can transmit more than one isolate (Table II; van Hoof *et al.*, 1966; Ayala and Allen, 1968). Nevertheless, in comparing transmission of tobacco rattle virus by nine *Trichodorus* species from the Netherlands, van Hoof (1968) concluded that at least one, *T. pachydermus*, shows specificity in transmission by becoming "viruliferous only when a virus isolate is available that 'suits' it." Isolates of pea early browning are serologically distinguishable and are transmitted by several species of *Trichodorus* (Table II) but not enough is known to establish any conclusions about specific relationships.

### E. Mechanism of Transmission

Some viruses may be detected in their vectors by cutting infective nematodes in a drop of water and using this suspension to inoculate suitable assay plants. Tobacco rattle virus was detected in this way in *Trichodorus* sp. (Sänger *et al.*, 1962), *T. pachydermus* (Sol, cited in Raski and Hewitt, 1963), and *T. allius* (Ayala and Allen, 1968). Grapevine fanleaf virus was similarly detected in *X. index* (Raski and Hewitt, 1963; Das and Raski, 1968), but Taylor and Thomas (1968) failed to detect arabis mosaic or strawberry latent ring spot viruses in *X. diversicaudatum*. Raspberry ring spot and tomato black ring viruses were readily detected in *L. elongatus* (Taylor, 1964; Taylor and Murant, 1969), but Debrot (1964) claimed little success with this technique with *L. macrosoma* and the English strain of raspberry ring spot virus. The virus detected in this way is probably that in the intestine, and because of the "one-way" action of the esophago-intestinal valve it is unlikely to play any part in the transmission process. Moreover, detection of virus in a nematode is not necessarily indicative of any specific relationship between virus and vector since viruses have been detected in nematodes that do not transmit them, viz., tobacco rattle virus in *X. diversicaudatum*, *X. coxi*, and *Pratylenchus penetrans* (van Hoof, 1967) and arabis mosaic and strawberry latent ring spot viruses in *L. elongatus* (Taylor, 1968b; Taylor and Robertson, 1969a).

Apparent differences in the retention of virus by *X. index* and *L. elongatus* led Taylor and Raski (1964) to suggest that there may be at

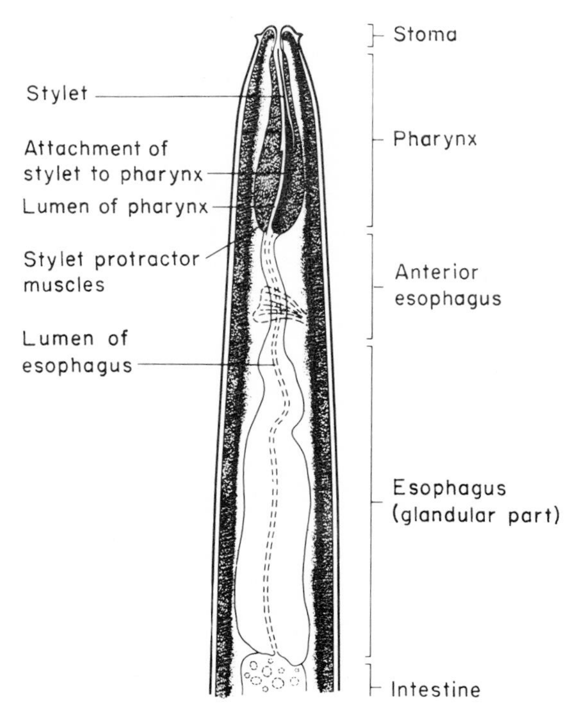


FIG. 1. Diagram of the esophageal region of *Trichodorus pachydermus* (× 400). (to use in conjunction with electronmicrographs of tobacco rattle virus in *T. pachydermus*). [Taken from *The Journal of General Virology* (1970).]

least two different mechanisms of transmission among the NEPO viruses, a close biological association existing between virus and vector in *Xiphinema* species, whereas in *Longidorus* the virus is mechanically retained. Taylor and Robertson (1969a) showed by means of electron microscopy of thin sections that the inner surface of the guiding sheath of *L. elongatus* is probably the main site of accumulation and retention of raspberry ring spot and tomato black ring viruses. They suggested that particles of these viruses are selectively and specifically adsorbed onto the cuticular guiding sheath from the sap flow, whereas other viruses such as arabis mosaic and strawberry latent ring spot, which may be ingested but are not normally transmitted by *L. elongatus,* pass with the

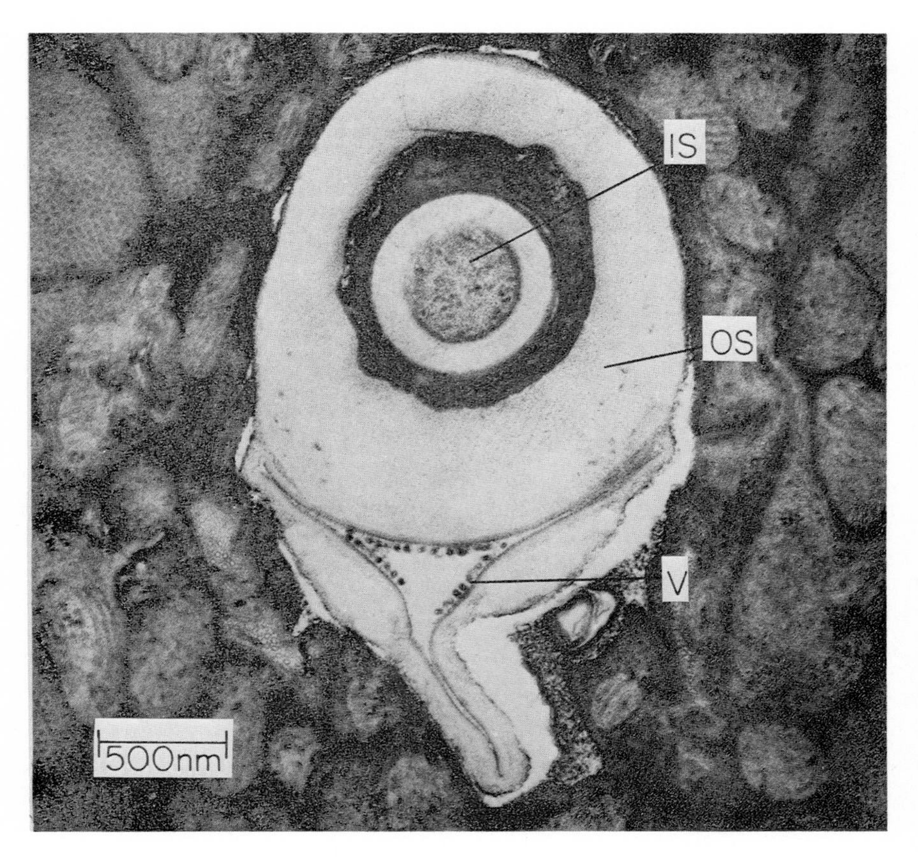


FIG. 2. Cross section of the pharynx of *Trichodorus pachydermus* in the region where the inner spear (IS) is enclosed within the outer spear (OS). Cross-sectioned particles of tobacco rattle virus (V) are present in close association with the cuticle lining the pharyngeal lumen.

sap flow to the intestine. Particles of tobacco rattle virus have also been found in close association with the cuticle lining the lumina of the pharynx and the esophagus of *T. pachydermus* (Taylor and Robertson, 1969b, 1970a), again suggesting a mechanical but specific mechanism of retention of the virus (see Figs. 1–4).

Comparing viruliferous with virus-free *X. index*, Roggen (1966) observed morphological and physiological differences in the hypodermal chords; but in what way, if any, these differences are associated with virus infection is unclear. Betto and Raski (1966) injected suspensions of grapevine fanleaf virus into *X. index*, but those nematodes that sur-

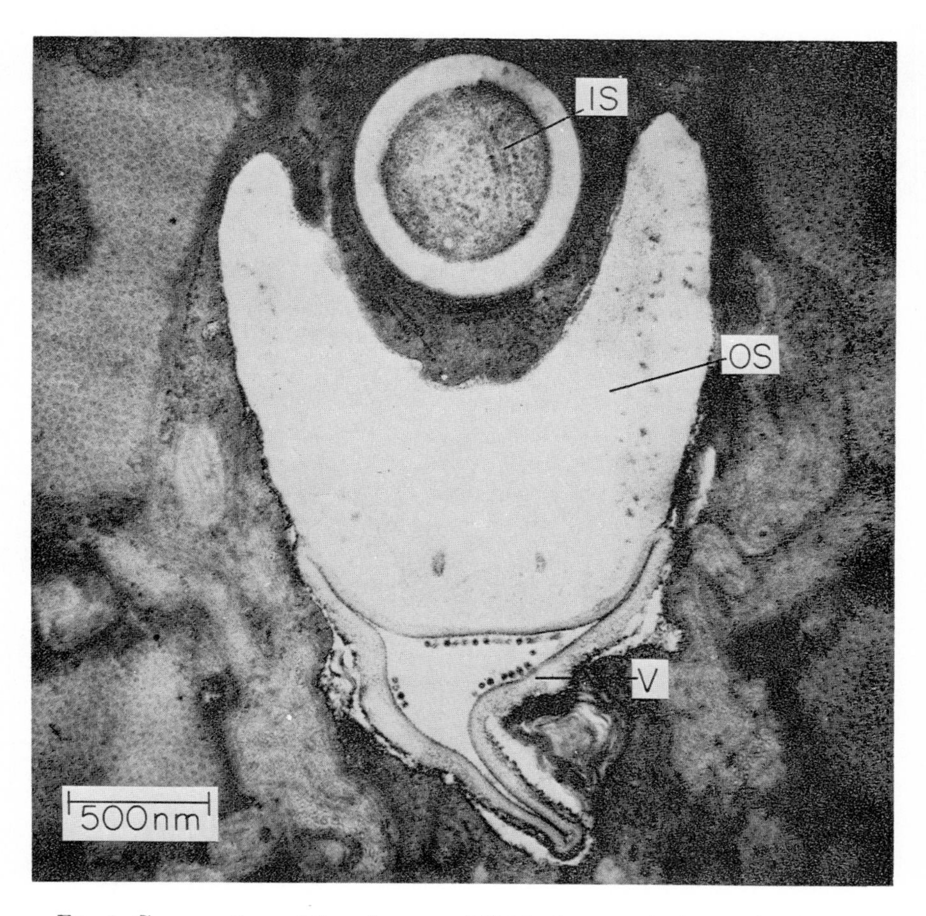


FIG. 3. Cross section of the pharynx of *Trichodorus pachydermus* in the region of attachment of the outer spear (OS) to the pharyngeal wall. Cross-sectioned particles of tobacco rattle virus (V) are shown in close association with the cuticle lining the pharyngeal lumen. IS, inner spear.

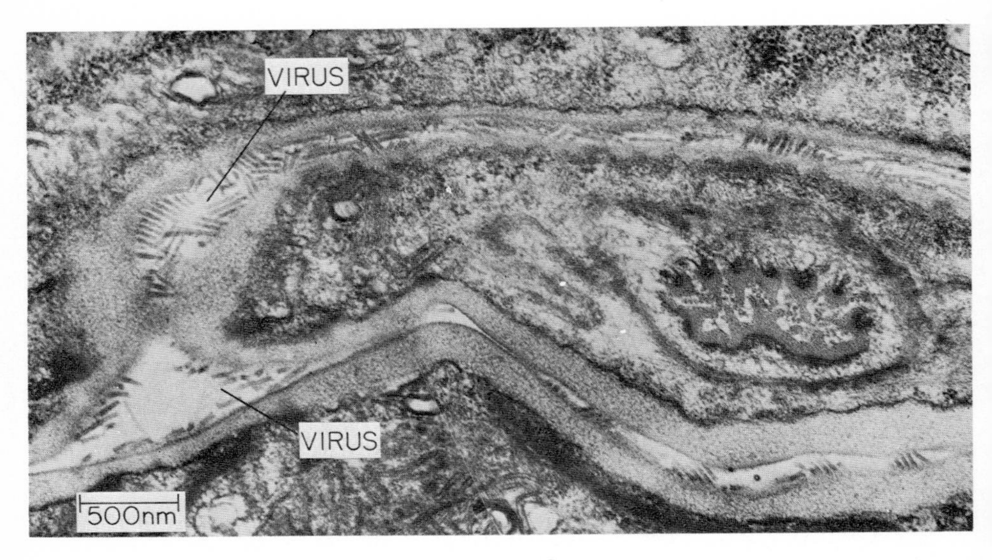


FIG. 4. Longitudinal section through the triradiate lumen of the esophagus of *Trichodorus pachydermus* showing the presence of particles of tobacco rattle virus.

vived the operation did not transmit the virus, nor was there any observed effect on their feeding or multiplication. Earlier work on electron microscopy of *Xiphinema* species failed to provide any clues to the location of virus retained for transmission (Wright, 1965; Roggen *et al.*, 1967; López-Abella *et al.*, 1967). Recently, Taylor and Robertson (1970b) have observed virus particles apparently specifically adsorbed onto the cuticle lining the stylet extension and the esophagus in both *X. index* and *X. diversicaudatum*, and they suggested that transmissible virus is retained at these loci and not within the tissues.

## V. ECOLOGY AND CONTROL

The distribution and use of infected planting material is probably the most important way by which nematode-transmitted viruses are disseminated. Grapevine fanleaf virus has almost certainly been distributed in this way from Europe to new areas such as North America and Australia. In California the virus is widespread in grapevine plantings where *Vitis rupestris* St. George rootstocks were used in earlier years, indicating that the nematode and virus were transported to these fields as a contaminant of planting material (Raski and Hewitt, 1963). Ornamental bulbs are an effective way of disseminating tobacco rattle virus and probably also other nematode-transmitted viruses that appear to commonly infect commercial *Narcissus* stocks (Brunt, 1966).

Weed seeds may provide an effective means of disseminating viruses and enable them to become established in areas where vectors are present. Whether or not this happens in nature is a matter of conjecture because specific evidence such as finding virus-infected seeds or seedlings in soils without the vectors, both near to and remote from infected sites, is difficult to obtain and no detailed investigations have been made (Murant and Lister, 1967). However, there is little doubt that infected weed seeds are the most important, if not the only, overwintering sources of raspberry ring spot and tomato black ring viruses in a disease outbreak (Murant and Taylor, 1965; Murant and Lister, 1967; Taylor and Murant, 1968).

Infected weed seeds appear to play no comparable role in the survival and dissemination of *Xiphinema*-transmitted viruses. Nevertheless, in common with all the NEPO viruses, soils remain infested for long periods. During limited periods without infector plants the viruses may persist in the long-lived adults, but over long periods reservoirs of virus may be maintained in root pieces of plants remaining after a crop has been removed. Pieces of grapevine roots not only act as a reservoir of fanleaf infection but also help to maintain populations of *X. index* in the absence of suitable host plants (Raski *et al.*, 1965).

Apart from measures taken to prevent virus and vectors reaching new sites, by using only virus-free planting material, for example, control measures are mainly concerned with the eradication of virus from land already infected.

Most viruses and vectors have such a wide host range that crop rotation is largely precluded as a means of control. Nematode populations decrease during periods of fallow but most vector species can survive for months, if not years, without feeding. Moreover, reservoirs of infection remain in virus-infected weed seeds and root pieces. With *L. elongatus,* the use of fallow or rotations of nonhost crops is difficult to implement as a means of control because of the importance of weeds as hosts for the nematode and the viruses it transmits (Thomas, 1969a). In Scotland, weed control resulted in some reduction of infection in a strawberry crop grown on raspberry ring spot virus-infected soil but compared unfavorably with the control obtainable by using nematicides (Taylor and Murant, 1968). Weed control may, however, be more effective in preventing the spread of virus infection when the crop itself is a poor host for the vector; the raspberry crop and *L. elongatus* is an example (Taylor *et al.,* 1965; Taylor, 1967). Not only does *L. elongatus* fail to reproduce on raspberry, but also when crop residues of canes and shoots are plowed into the soil they apparently release a nematicidal chemical (Taylor and Murant, 1966). The eradication of root pieces remaining after the removal of a crop is dependent on their natural decay and takes time; 5 years was advocated as a minimum break between grapevine crops to ensure a complete removal of sources of infection (Hewitt *et al.*, 1962; Raski *et al.*, 1965).

With long-term crops such as grapevines, stone fruits, and raspberry, the planting of disease-escaping or virus-immune cultivars is an attractive solution to the problem of nematode-transmitted viruses. However, although a crop may be immune from infection it may remain as a suitable host for the vector and allow it to multiply, thus increasing the chances of infection in subsequent susceptible crops once a virus source has been reintroduced. Moreover, most vector species cause some direct damage to crops by their feeding, and their increase is therefore to be discouraged.

Nematicides have so far proved the most satisfactory way of dealing with nematode vectors. Although their application will rarely result in the complete elimination of a nematode population in the soil, particularly where the population is distributed to some depth, for most virus-vector situations such treatment can be long lasting because of the low rate of multiplication of the nematodes. Good control of *X. index* has been obtained with D–D (dichloropropane–dichloropropene), carbon bisulfide, and methyl bromide (Vuittenez, 1961; Boubals, 1968). In California, Raski and Schmitt (1964) obtained significant increases in yields of Tokay grapevines by applications of DCBP (dichlorobromopropane) metered into irrigation water, and the beneficial effect was apparent for up to 3 years. Dichloropropane–dichloropropene proved effective against *X. diversicaudatum* and *L. elongatus* infesting strawberry crops in Britain (Harrison *et al.*, 1963; Murant and Taylor, 1965; Taylor and Murant, 1968) and against *L. attenuatus* and *Trichodorus* sp. causing Docking disorder of sugar beet in eastern England (Whitehead and Tite, 1967). Quintozene, which is normally used as a fungicide for the control of *Rhizoctonia,* also proved effective against *L. elongatus* and prevented the infection of strawberry crops with raspberry ring spot virus (Taylor and Murant, 1968).

The problem of nematode-transmitted viruses can, of course, be circumvented by not planting susceptible crops on land containing virus-carrying nematodes. The presence of the vector can be detected by means of soil sampling, and the extent of virus infection can be assayed by bait testing. If the use of virus-infested land cannot be avoided, then the results of such tests, in conjunction with knowledge of the ecology of the vector and its association with the viruses it transmits, can help to decide whether chemical treatment of the soil or other control measures need be undertaken.

## REFERENCES


Alhassan, S. A., and Hollis, J. P. (1966). *Phytopathology* **56**, 573–574.
Allen, M. W. (1948). *Phytopathology* **38**, 612–627.
Amici, A. (1965). *Riv. Patol. Veg., Ser. IV* **1**, 109–128.
Amici, A. (1967a). *Riv. Patol. Veg., Ser. IV* **3**, 85–88.
Amici, A. (1967b). *Riv. Patol. Veg., Ser. IV* **3**, 99–104.
Ayala, A., and Allen, M. W. (1966). *Nematologica* **12**, 87. (Abstr.)
Ayala, A., and Allen, M. W. (1968). *J. Agr. Univ. P. R.* **52**, 101–125.
Behrens, J. (1899). *Landwirt. Vers.-Sta.* **52**, 442–447.
Bergeson, G. B., and Athow, K. L. (1963). *Phytopathology* **53**, 871. (Abstr.).
Bergeson, G. B., Athow, K. L., Laviolette, F. A., and Thomasine, Sister M. (1964). *Phytopathology* **54**, 723–728.
Betto, E., and Raski, D. J. (1966). *Nematologica* **12**, 453–461.
Bird, G. W., and Mai, W. F. (1967). *Phytopathology* **57**, 1368–1371.
Blumer, S. (1954). *Schweiz. Z. Obst- Weinbau* **63**, 516–519, 525–529.
Blumer, S. (1955). *Schweiz. Z. Obst- Weinbau* **64**, 2–11.
Bos, L., and van der Want, J. P. H. (1962). *Tijdschr. Plantenziekten* **68**, 368–390.
Boubals, D. (1968). *C. R. 8th Symp. Int. Nematol., Antibes, 1965* pp. 112–113.
Breece, J. R., and Hart, W. H. (1959). *Plant Dis. Rep.* **43**, 989–990.
Brunt, A. A. (1966). *Plant. Pathol.* **15**, 157–160.
Cadman, C. H. (1960). *Virology* **11**, 653–664.
Cadman, C. H. (1963). *Annu. Rev. Phytopathol.* **1**, 143–172.
Cadman, C. H., and Harrison, B. D. (1960). *Virology* **10**, 1–20.
Cadman, C. H., Dias, H. F., and Harrison, B. D. (1960). *Nature (London)* **187**, 577–579.
Chen, T. A., and Mai, W. F. (1965). *Phytopathology* **55**, 128. (Abstr.)
Christie, J. R. (1951). *Proc. Fla. State Hort. Soc.* **64**, 120–122.
Christie, J. R. (1952). *Soil Sci. Soc. Fla., Proc.* **12**, 30–39.
Christie, J. R. (1959). "Plant Nematodes. Their Bionomics and Control," 256 pp. Agr. Exp. Sta., Univ. of Florida, Gainesville, Florida.
Christie, J. R., and Perry, V. G. (1951). *Science* **113**, 491–493.
Cohn, E. (1969). *Nematologica* **15**, 179–192.
Cohn, E., and Krikun, J. (1966). *Plant. Dis. Rep.* **50**, 711–712.
Cohn, E., and Mordechai, M. (1969). *Nematologica* **15**, 295–302.
Cooke, D. A., and Hull, R. (1967). *Rep. Rothamsted Exp. Sta.* 1966, pp. 283–284.
Cooper, J. I., and Harrison, B. D. (1969). *Rep. Scot. Hort. Res. Inst.* **15**, 48–49.
Cremer, M. C., and Kooistra, G. (1964). *Nematologica* **10**, 69–70. (Abstr.)
Cropley, R. (1961). *Ann. Appl. Biol.* **49**, 524–529.
Dalmasso, A., and Caubel, G. (1966). *C. R. Acad. Agr. Fr.* **52**, 440–445.
Das, S., and Raski, D. J. (1968). *Nematologica* **14**, 55–62.
Debrot, E. A. (1964). *Ann. Appl. Biol.* **54**, 183–191.
Dias, H. F., and Harrison, B. D. (1963). *Ann. Appl. Biol.* **51**, 97–105.

Di Sanzo, C. P., and Rohde, R. A. (1969). *Phytopathology* **59,** 279–284.
Dunning, R. A., and Cooke, D. A. (1967). *Brit. Sugar Beet Rev.* **36,** 23–29.
Epstein, A. H., and Barker, K. R. (1966). *Plant Dis. Rep.* **50,** 420–422.
Fisher, J. M., and Raski, D. J. (1967). *Proc. Helminthol. Soc. Wash.* **34,** 68–72.
Flegg, J. J. M. (1968a). *Nematologica* **14,** 189–196.
Flegg, J. J. M. (1968b). *Nematologica* **14,** 197–210.
Flores, H., and Chapman, R. A. (1968). *Phytopathology* **58,** 814–817.
Fritzsche, R. (1964). *Wiss. Z. Univ. Rostock, Math.-Naturwiss. Reihe* **13,** 343–347.
Fritzsche, R. (1966). *Nachrichtenbl. Deut. Pflanzenschutzdienst (Berlin),* **20,** 8–11.
Fritzsche, R. (1967). *Biol. Zentralbl.* **86,** 753–759.
Fritzsche, R., and Kegler, H. (1964). *Naturwissenschaften* **51,** 299.
Fritzsche, R., and Schmelzer, K. (1967). *Naturwissenschaften* **54,** 498–499.
Frost, R. R., Harrison, B. D., and Woods, R. D. (1967). *J. Gen. Virol.* **1,** 57–70.
Fulton, J. P. (1962). *Phytopathology* **52,** 375.
Gibbs, A. J., and Harrison, B. D. (1964a). *Plant Pathol.* **13,** 144–150.
Gibbs, A. J., and Harrison, B. D. (1964b). *Ann. Appl. Biol.* **54,** 1–11.
Goodey, T. (1963). "Soil and Freshwater Nematodes" (rev. by J. B. Goodey from 1951 Ed.), 2nd Ed., 544 pp. Methuen, London.
Graham, C. W., and Flegg, J. J. M. (1968). *Plant Pathol.* **17,** 191.
Green, C. D. (1967). *Rep. Rothamsted Exp. Sta.,* 1966, p. 148.
Griffin, G. D., and Darling, H. M. (1964). *Nematologica* **10,** 471–479.
Griffin, G. D., and Epstein, A. H. (1964). *Phytopathology* **54,** 177–180.
Harrison, B. D. (1960). *Advan. Virus Res.* **7,** 131–161.
Harrison, B. D. (1961). *Rep. Rothamsted Exp. Sta.* 1960, p. 118.
Harrison, B. D. (1962). *Rep. Rothamsted Exp. Sta.* 1961, p. 105.
Harrison, B. D. (1964a). *In* "Plant Virology" (M. K. Corbett and H. D. Sisler, eds.), pp. 118–147. Univ. of Florida Press, Gainesville, Florida.
Harrison, B. D. (1964b). *Virology* **22,** 544–550.
Harrison, B. D. (1967a). *Ann. Appl. Biol.* **60,** 405–409.
Harrison, B. D. (1967b). *Rep. Rothamsted Exp. Sta.* 1966, p. 115.
Harrison, B. D., and Cadman, C. H. (1959). *Nature (London)* **184,** 1624–1626.
Harrison, B. D., and Nixon, H. L. (1959). *J. Gen. Microbiol.* **21,** 569–581.
Harrison, B. D., and Winslow, R. D. (1961). *Ann. Appl. Biol.* **49,** 621–633.
Harrison, B. D., and Woods, R. D. (1966). *Virology* **28,** 610–620.
Harrison, B. D., Mowat, W. P., and Taylor, C. E. (1961). *Virology* **14,** 480–485.
Harrison, B. D., Peachey, J. E., and Winslow, R. D. (1963). *Ann. Appl. Biol.* **52,** 243–255.
Hewitt, W. B. (1950). *Calif. Dep. Agr., Bull.* **39,** 62–63.
Hewitt, W. B., and Raski, D. J. (1962). *In* "Biological Transmission of Disease Agents" (K. Maramorosch, ed.), pp. 63–72. Academic Press, New York.
Hewitt, W. B., Raski, D. J., and Goheen, A. C. (1958). *Phytopathology* **48,** 586–595.
Hewitt, W. B., Goheen, A. C., Raski, D. J., and Gooding, G .W., Jr. (1962). *Vitis* **3,** 57–83.
Hirumi, H., Chen, T. A., Lee, K. J., and Maramorosch, K. (1968). *J. Ultrastruct. Res.* **24,** 434–453.
Hoff, J. K., and Mai, W. F. (1962). *Plant Dis. Rep.* **46,** 24–25.
Hollings, M., and Stone, O. M. (1965). *Ann. Appl. Biol.* **56,** 73–86.
Horner, C. E., and Jensen, H. J. (1954). *Plant Dis. Rep.* **38,** 39–41.
Jensen, H. J. (1961). *Bull. Agr. Exp. Sta. Oregon State Univ., Corvallis* **579,** 34 pp.

Jensen, H. J., and Allen, Jr., T. C. (1964). *Plant Dis. Rep.* **48,** 333–334.
Jha, A., and Posnette, A. F. (1959). *Nature (London)* **184,** 962–963.
Jha, A., and Posnette, A. F. (1961). *Virology* **13,** 119–123.
Johnson, F. (1945). *Ohio J. Sci.* **45,** 125–128.
Kuiper, K., and Loof, P. A. A. (1961). *Versl. Meded. Plantenziektenk. Dienst Wageningen* **136,** 193–200.
Lamberti, F., and Raski, D. J. (1964). *Phytopathol. Mediter.* **3,** 41–43.
Lima, M. B. (1968). *C. R. 8th Symp. Int. Nematol., Antibes, 1965* p. 30.
Lister, R. M. (1960). *Agriculture (London)* **67,** 25–29.
Lister, R. M. (1964). *Ann. Appl. Biol.* **54,** 167–176.
Lister, R. M. (1966). *Rep. Scot. Hort. Res. Inst.* **12,** 45–46.
Lister, R. M., and Murant, A. F. (1967). *Ann. Appl. Biol.* **59,** 49–62.
López-Abella, D., Jiménez-Millán, F., and Garcia-Hidalgo, F. (1967). *Nematologica* **13,** 283–286.
Lownsbery, B. F. (1964). *Plant Dis. Rep.* **48,** 218–221.
McGuire, J. M. (1964). *Phytopathology* **54,** 799–801.
McKinney, H. H., Paden, W. R., and Koehler, B. (1957). *Plant Dis. Rep.* **41,** 256–266.
Malek, R. B. (1968). *Plant Dis. Rep.* **52,** 795–798.
Martelli, G. P., and Lamberti, F. (1967). *Phytopathol. Mediter.* **6,** 65–85.
Martelli, G. P., and Raski, D. J. (1963). *Informatore Fitopathol.* **13,** 416–420.
Moore, J. F. (1967). *Sci. Proc. Roy. Dublin Soc.* **2,** 81–86.
Mowat, W. P., and Taylor, C. E. (1962). *Rep. Scot. Hort. Res. Inst.* **9,** 67–68.
Murant, A. F., and Lister, R. M. (1967). *Ann. Appl. Biol.* **59,** 63–76.
Murant, A. F., and Taylor, C. E. (1965). *Ann. Appl. Biol.* **55,** 227–237.
Murant, A. F., Taylor, C. E., and Chambers, J. (1968). *Ann. Appl. Biol.* **61,** 175-186.
Norton, D. C. (1963). *Phytopathology* **53,** 66–68.
Norton, D. C. (1967). *Phytopathology* **57,** 1390–1391.
Oswald, J. W., and Bowman, T. (1958). *Phytopathology* **48,** 396. (Abstr.)
Percival, J. (1912). *J. Southeast. Agr. Coll., Wye, Engl.* **1,** 5–9.
Perry, V. G. (1958). *Phytopathology* **48,** 420–423.
Pitcher, R. S. (1967). *Nematologica* **13,** 547–557.
Pitcher, R. S., and Flegg, J. J. M. (1965). *Nature (London)* **207,** 317.
Radewald, J. D. (1962). Ph.D. Thesis, Univ. of California, Davis, California.
Radewald, J. D., and Raski, D. J. (1962a). *Phytopathology* **52,** 748. (Abstr.)
Radewald, J. D., and Raski, D. J. (1962b). *Phytopathology* **52,** 748–749 (Abstr.)
Raski, D. J., and Hewitt, W. B. (1960). *Nematologica* **5,** 166–170.
Raski, D. J., and Hewitt, W. B. (1963). *Phytopathology* **53,** 39–47.
Raski, D. J., and Lider, L. (1959). *Calif. Agr.* **13**(g), 13–15.
Raski, D. J., and Schmitt, R. V. (1964). *Amer. J. Enol. Viticult.* **15,** 199–203.
Raski, D. J., Hewitt, W. B., Goheen, A. C., Taylor, C. E., and Taylor, R. H. (1965). *Nematologica* **11,** 349–352.
Raski, D. J., Jones, N. O., and Roggen, D. R. (1969). *Proc. Helminthol. Soc. Wash.* **36,** 106–118.
Roggen, D. R. (1966). *Nematologica* **12,** 287–296.
Roggen, D. R., Raski, D. J., and Jones, N. O. (1967). *Nematologica* **13,** 1–16.
Rohde, R. A., and Jenkins, W. R. (1957a). *Phytopathology* **47,** 295–298.
Rohde, R. A., and Jenkins, W. R. (1957b). *Phytopathology* **47,** 29. (Abstr.)
Sänger, H. L. (1961). *Proc. 4th Conf. Potato Virus Dis., Braunschweig, Germany, 1960* pp. 22–28.

Sänger, H. L., Allen, M. W., and Gold, A. H. (1962). *Phytopathology* **52,** 750. (Abstr.)
Sauer, N. I. (1966). *Phytopathology* **56,** 862–863.
Schindler, A. F. (1957). *Nematologica* **2,** 25–31.
Schindler, A. F. (1958). *Plant Dis. Rep.* **42,** 1348–1350.
Schindler, A. F., and Braun, A. J. (1957). *Nematologica* **2,** 91–93.
Schmidt, H. B., Fritzsche, R., and Lehmann, W. (1963). *Naturwissenschaften* **50,** 386.
Seinhorst, J. W. (1966). *Nematologica* **12,** 275–279.
Sharma, R. D. (1965). *Meded. Landbouwhogesch. Opzoekingssta. Staat Gent* **30,** 1437–1443.
Smith, K. M., and Markham, R. (1944). *Phytopathology* **34,** 324–329.
Sol, H. H. (1963). *Neth. J. Plant Pathol.* **69,** 208–214.
Sol, H. H., and Seinhorst, J. W. (1961). *Tijdschr. Plantenziekten* **67,** 307–309.
Sturhan, D. (1963). *Z. Angew. Zool.* **50,** 129–193.
Sturhan, D. (1966). *Gesunde Pfl.* **18,** 93–96.
Taconis, P. J., and Kuiper, K. (1963). *Versl. Meded. Plantenziektenk. Dienst Wageningen* **141,** 177–178.
Tarjan, A. C. (1968). *C. R. 8th Symp. Int. Nematol., Antibes, 1965* pp. 11–12.
Tarjan, A. C. (1969). *Nematologica* **15,** 241–252.
Taylor, C. E. (1962). *Virology* **17,** 493–494.
Taylor, C. E. (1964). *Rep. Scot. Hort. Res. Inst.* **11,** 65.
Taylor, C. E. (1967). *Ann. Appl. Biol.* **59,** 275–281.
Taylor, C. E. (1968a). *C. R. 8th Symp. Int. Nematol., Antibes 1965* pp. 109–110.
Taylor, C. E. (1968b). *Rep. Scot. Hort. Res. Inst.* **14,** 66.
Taylor, C. E., and Murant, A. F. (1966). *Nematologica* **12,** 488–494.
Taylor, C. E., and Murant, A. F. (1968). *Plant Pathol.* **17,** 171–178.
Taylor, C. E., and Murant, A. F. (1969). *Ann. Appl. Biol.* **64,** 43–48.
Taylor, C. E., and Raski, D. J. (1964). *Nematologica* **10,** 489–495.
Taylor, C. E., and Robertson, W. M. (1969a). *Ann. Appl. Biol.* **64,** 233–237.
Taylor, C. E., and Robertson, W. M. (1969b). *Rep. Scot. Hort. Res. Inst.* **15,** 58.
Taylor, C. E., and Robertson, W. M. (1970a). *J. Gen. Virol.* **6,** 179–182.
Taylor, C. E., and Robertson, W. M. (1970b). *Rep. Scot. Hort. Res. Inst.* **16,** 63–64.
Taylor, C. E., and Thomas, P. R. (1968). *Ann. Appl. Biol.* **62,** 147–157.
Taylor, C. E., Chambers, J., and Pattullo, W. I. (1965). *Hort. Res.* **5,** 19–24.
Taylor, C. E., Thomas, P. R., and Converse, R. H. (1966). *Plant Pathol.* **15,** 170–174.
Teliz, D. (1967). *Nematologica* **13,** 177–185.
Teliz, D., Grogan, R. G., and Lownsbery, B. F. (1966). *Phytopathology* **56,** 658–663.
Thomas, P. R. (1969a). *Plant Pathol.* **18,** 23–28.
Thomas, P. R. (1969b). *Nematologica* **15,** 582–590.
Thomas, P. R. (1970). *Ann. Appl. Biol.* **65,** 169–178.
van der Meer, F. A. (1965). *Neth. J. Plant Pathol.* **71,** 33–46.
van Hoof, H. A. (1962). *Tijdschr. Plantenziekten* **68,** 391–396.
van Hoof, H. A. (1964a). *Neth. J. Plant Pathol.* **70,** 187.
van Hoof, H. A. (1964b). *Nematologica* **10,** 141–144.
van Hoof, H. A. (1967). *Neth. J. Plant Pathol.* **73,** 193–194.
van Hoof, H. A. (1968). *Nematologica* **14,** 20–24.

van Hoof, H. A., Maat, D. Z., and Seinhorst, J. W. (1966). *Neth. J. Plant Pathol.* **72**, 253–258.

Vuittenez, A. (1961). "Les Nématodes," pp. 55–78. C.N.R.A., Versailles.

Vuittenez, A., and Legin, R. (1964). *Etud. Vir. Appl.* **5**, 59–62.

Walkinshaw, C. H., Griffin, G. D., and Larson, R. H. (1961). *Phytopathology* **51**, 806–808.

Weiner, A., and Raski, D. J. (1966). *Plant Dis. Rep.* **50**, 27–28.

Weischer, B. (1966). *Mitt. Biol. Bundesanst. Land-Forstwirt. Berlin-Dahlem* **118**, 100–106.

Whitehead, A. G. (1965). *Brit. Sugar Beet Rev.* **34**, 77–78, 83–84, 92.

Whitehead, A. G., and Tite, D. J. (1967). *Plant Pathol.* **16**, 107–109.

Wright, K. A. (1965). *Can. J. Zool.* **43**, 689–700.

Yassin, A, M. (1968). *Nematologica* **14**, 419–428.

Yassin, A. M. (1969). *Nematologica* **15**, 169–178.

Zuckerman, B. M. (1961). *Nematologica* **6**, 135–143.

Zuckerman, B. M. (1962). *Phytopathology* **52**, 1017–1019.

*Biochemistry and Physiology*

CHAPTER 22

# Chemical Composition of Nematodes*

L. R. Krusberg

*Department of Botany, University of Maryland, College Park, Maryland*

I. Introduction . . . . . . . . . . . . . . . 213
II. Inorganic Substances . . . . . . . . . . . . 214
III. Carbohydrates . . . . . . . . . . . . . . 215
IV. Amino Acids and Proteins . . . . . . . . . . . 216
V. Lipids . . . . . . . . . . . . . . . 222
A. Fatty Acids . . . . . . . . . . . . . 223
B. Sterols . . . . . . . . . . . . . . 227
VI. Plant Growth Regulators . . . . . . . . . . . 231
VII. Other . . . . . . . . . . . . . . . 232
VIII. Conclusions . . . . . . . . . . . . . . 233
References . . . . . . . . . . . . . . 233

## I. INTRODUCTION

The chemical composition of phytoparasitic nematodes has received little attention in general; hence, information is limited. The composition of animal parasitic nematodes is far better known, but the information was recently reviewed in the excellent book by von Brand (1966) and thus will not be included here with a few exceptions. This chapter will deal primarily with the composition of plant parasitic and free-living nematodes, excluding enzymes which are covered elsewhere in this volume.

* Scientific Article No. A1550, Contribution No. 4257 of the Maryland Agricultural Experiment Station.

## II. INORGANIC SUBSTANCES

Relatively little is known about the inorganic constituents of phytoparasitic nematodes.

In a study of the composition of egg shells of the potato cyst nematode, *Heterodera rostochiensis*, Clarke *et al.* (1967) found that the shells contained 3% ash. Elements detected in the ash by spectrographic analysis included Ca, Al, Cu, Mg, Si, Fe, and Mn although the relative quantities of each were not determined. Using similar methods in a later study, Clarke (1968) found that the cyst wall contained 23 elements and about 5% ash.

Ellenby (1968) investigated the water content of nematodes by measuring microscopically the displacement of interference fringes in individual nematode specimens. The free-living nematode, *Panagrellus redivivus*, and the marine nematode, *Enoplus brevis*, each contained about 80% water. The rate of water uptake by *H. rostochiensis* larvae just freed from cysts was measured in the same manner. Larvae increased from 19 to 50% water content during the first 5 min. The rate of water uptake then slowed down so that after 30 min the water content was 60%, and after 4 hr it was about 75%.

Ammonia was detected in homogenates of the fungus-feeding nematode, *Ditylenchus triformis*, by Myers and Krusberg (1965) accounting for about 0.04% of the dry weight. The total nitrogen content of this nematode was found to average 5.3% of the dry weight (range: 4.1–6.7%). Ammonia accounted for 27–44% of the total nitrogen discharged by freshly harvested *D. triformis*. Since nematodes discharged more ammonia into solution in 24 hr than they contained, it was concluded that they must eliminate ammonia from the body as it is formed. The total nitrogen content of dried egg shells of *H. rostochiensis* was found to be 9.5% (Clarke *et al.*, 1967).

Osmoregulation in the nematodes *Panagrellus redivivus* and *Aphelenchus avenae* was investigated by Myers (1966). Incubation in Fenwick's salt solution which contains NaCl, KCl, $CaCl_2$, and $MgCl_2$ resulted in an increase in every ion measured in the nematodes as compared to those incubated in distilled water. In nematodes of both species Na levels seemed to fluctuate most readily and consistently. In salt solution the Na content of nematodes was 3–3.5 times as high as in nematodes incubated in distilled water. Other elements that occurred in high levels in nematodes incubated in salt solution were K in *A. avenae* and K, Ca, P, Mg, and Cl in *P. redivivus*. In general, it was discovered that as the concentration of a salt in the incubation solution varied, the concentration of elements of the salt in *P. redivivus* varied directly but not always proportionately. At higher salt concentrations in the medium, proportionately less salt entered the nematode. When compared to specimens of *P. redivivus* incubated in Fenwick's salt solution, nematodes incubated for 25 hr in distilled water or in NaCl solution at 6 atm gained water whereas those incubated in NaCl or glucose solutions at 14 atm lost water. Nematodes incubated in glucose solution at 6 atm gained water between 7 and 15 hr and then lost water to the original level by 25 hr. Thus it appears that both the water and ion content of nematodes is somewhat dependent upon the concentration of solutes in the medium. This nematode has at least a limited capacity to regulate these substances in the body, and the nematode body expands or contracts during changes in salt content.

## III. CARBOHYDRATES

Data concerning carbohydrate composition of nematodes are particularly lacking.

Dried eggshells of *H. rostochiensis* contained 6% carbohydrate based on estimation of total reducing and nonreducing sugars in acid hydrolysates (Clarke *et al.*, 1967). Individual sugars were not identified, and uronic acids were not detected. Eggshells contained about 9% chitin, whereas none was detected in the cuticle of larvae or in walls of cysts. Chitin was identified both by color tests of van Wisselingh and by presence of glucosamine in hydrolysates. In another study, Clarke (1968) found that the cyst wall contained 0.5% carbohydrate. A small amount of glucosamine was detected in cyst wall hydrolysate, but again no chitin was detected in the cyst walls. Only a trace of uronic acid was detected.

Investigating changes associated with parasitism in the root-knot nematode, *Meloidogyne javanica,* Bird and Saurer (1967) noted that ducts of subventral glands of second-stage larvae became strongly periodic acid-Schiff positive 2–3 days after they entered plant roots, indicating carbohydrate. Because this material was Alcian blue negative it was suggested to be mucopolysaccharide. These same histochemical tests failed to detect carbohydrate in glands of preparasitic larvae, suggesting to the investigators that the subventral glands may have more than one function depending upon the stage of parasitism. Stylet exudate from living females contained carbohydrate (Bird, 1968). Cells in female nematodes which produce the gelatinous matrix stained positive for acid-mucopolysaccharide, and the matrix also contained carbohydrate (Bird and Rogers, 1965b).

Glucose and fructose were detected in extracts of homogenized *Ditylenchus dipsaci* and *Aphelenchoides ritzemabosi* by paper chromatography, but the quantities of each were not measured (Krusberg, 1961). An unidentified fungus-feeding *Aphelenchoides* sp. was found to contain 2–9% carbohydrate of which 85% was glycogen (R. Petriello, personal communication). Trehalose and mannose were also identified from this nematode.

## IV. AMINO ACIDS AND PROTEINS

The adult female cuticle of *M. hapla* and *M. javanica* consisted of a thin, tanned lipoprotein outer layer covering a thicker, less-resistant homogeneous protein layer split by two bands of darker tissues (Bird, 1958). The egg sac (gelatinous matrix) consisted of a tanned glycoprotein of uniform composition. Proteins of both the egg sac and female cuticle resembled secreted collagens. Pepsin partially digested both egg sac and cuticle, whereas trypsin hydrolyzed only the cuticle. Acid hydrolysates of egg sacs of both *Meloidogyne* spp. contained the same 14 amino acids as seen in Table I. In a later study of the ultrastructure and formation of the cuticle in *M. javanica,* Bird and Rogers (1965a) detected both ribonucleic acid and protein in the hypodermis histochemically and found additional evidence that the cuticle originated from the hypodermis. In females of this nematode both the cells producing the gelatinous matrix and the matrix itself stained positive for protein in histochemical tests (Bird and Rogers, 1965b).

Eggshells of *H. rostochiensis* were 59% protein based on nitrogen content according to Clarke *et al.* (1967) and accounted for virtually all the amino acids present in the shells. Following acid hydrolysis, 18 amino acids were identified from eggshells with proline most abundant (36% of the total) and aspartic acid, glycine, and serine also present in large amounts (Table I). The small amounts of aromatic and sulfur-containing amino acids and the presence of hydroxyproline suggested that the eggshell protein was collagen. Eggshell collagen was less organized in structure than is usual for collagen since x-ray diffraction studies failed to reveal any crystalline structure. On the other hand, collagen of *Ascaris* cuticle is crystalline by x-ray diffraction. The authors suggest that collagen is a structural material in nematodes. Somewhat similar results were obtained by Clarke (1968) in a study of the cyst wall. The wall was 70% protein based on nitrogen content, and hydrolysates of cyst walls contained 19 amino acids (Table I). Proline, glycine, and alanine were the predominant amino acids found. The eggshell and cyst wall differed in amino acid content.

Krusberg (1961) investigated qualitatively the free amino acids and the amino acids in acid hydrolysates following extraction of free amino acids of *Ditylenchus dipsaci* and *Aphelenchoides ritzemabosi* (Table I). Differences occurred in the 15 free amino acids among the two nematodes, but hydrolysates contained the same 17 amino acids. The 13 protein amino acids found in an unidentified fungus-feeding *Aphelenchoides* sp. by M. Balasubramanian (personal communication) are also presented in Table I. Thus nematodes contain most or all of the more common amino acids. Relative quantities of amino acids among nematode species might reveal relationships which could be utilized in biochemical studies and taxonomy.

Total amino acids of *Ditylenchus triformis* accounted for 13–18% of the dry weight of nematodes according to Myers and Krusberg (1965). In the same study surface sterilized nematodes of *Ditylenchus myceliophagus, D. triformis, D. dipsaci, Pratylenchus penetrans,* and *Meloidogyne incognita* larvae also discharged from 8–23 amino acids into 1% glucose solution in which they were held for 24 hr at 30°C (Table I). When live *D. triformis* or *Meloidogyne* larvae were incubated in solution containing Na acetate-2-$^{14}$C, several amino acids both within and discharged into the solution by nematodes became radioactive; more amino acids were labeled when nematodes were incubated in solution containing labeled acetate than labeled glucose. These results demonstrated that stylet-bearing nematodes take up substances from ambient solution and metabolize them in the absence of living plant tissue substrate.

Label from Na acetate-2-$^{14}$C was incorporated into 7 of 15 amino acids identified from *Caenorhabditis briggsae* in experiments of Nicholas *et al.* (1960). Nematodes were incubated for 14 days at 20°C in medium to which the $^{14}$C-acetate was added. The 15 amino acids identified are presented in Table I.

Utilizing microweighing and extraction methods and histochemical staining, Van Gundy *et al.* (1967) determined that freshly hatched larvae of *Meloidogyne javanica* and *Tylenchulus semipenetrans* contained around 24% protein which they found was probably metabolized partially to sustain the nematodes during periods of starvation. Similar results were obtained by Wilson (1965) during storage of infective larvae of the animal parasite, *Nippostrongylus brasiliensis.*

Changes in the composition of free amino acids in cysts of *Heterodera*

TABLE
AMINO ACID

| Amino acid | *M. hapla* and *M. javanica*[a] | | *H. rostochiensis* | | | *D. dipsaci*[e] | |
|---|---|---|---|---|---|---|---|
| | Egg sac | Female cuticle | Eggshell[b] | Cyst wall[c] | Maturing cysts[d] | Free | Hydrolysates |
| Alanine | + | + | 3 | 13 | + | + | + |
| Arginine | + | + | 2 | 2 | | + | + |
| Aspartic acid | + | + | 11 | 6 | + | + | + |
| Glutamic acid | + | + | 5 | 8 | + | + | + |
| Glycine | + | + | 10 | 18 | + | + | + |
| Histidine | + | + | 2 | 1 | | + | + |
| Leu/Isoleu | + | + | 3 | 3 | + | + | + |
| Lysine | + | + | 4 | 2 | | | + |
| Hydroxyproline | + | + | 5 | 5 | | | + |
| Proline | + | + | 34 | 19 | + | | + |
| Serine | + | | 8 | 4 | + | + | + |
| Cystine | | + | | | | | |
| Cysteic acid | | + | | | | + | + |
| Threonine | + | + | 2 | 2 | + | + | + |
| Tyrosine | + | + | 3 | 1 | | + | + |
| Valine | + | + | 2 | 2 | + | + | + |
| Cysteine | | | 3 | 6 | | | |
| Phenylalanine | | | 2 | 1 | | | + |
| Methionine | | | 1 | 3 | + | | + |
| γ-Aminobutyric | | | | | | + | |
| Asparagine | | | | | | | |
| Citrulline | | | | | | | |
| Glutamine | | | | | | | |
| Methionine sulfoxide | | | | | | | |
| Ornithine | | | | Trace | | | |

[a] Bird (1958), hydrolysates.
[b] Clarke *et al.* (1967), hydrolysates, percent by weight.
[c] Clarke (1968), hydrolysates, percent by weight.
[d] Smith and Ellenby (1967), free amino acids.
[e] Krusberg (1961).
[f] Myers and Krusberg (1965).
[g] Balasubramanian (personal communication, 1969), protein hydrolysates.
[h] Nicholas *et al.* (1960), protein hydrolysates.

*rostochiensis* during maturation were investigated by Smith and Ellenby (1967). The amino acids identified are listed in Table I. Aspartic acid, glutamic acid, and serine increased during maturation while proline, glycine, and valine decreased, and alanine, methionine, and isoleucine virtually disappeared (Table I). During these studies the yellow pigment of cysts was examined briefly. Although its composition was not

I
COMPOSITION OF NEMATODES

| Amino acid | *A. ritzemabosi*[e] Free | *A. ritzemabosi*[e] Hydrolysates | *D. triformis*[f] Discharged free amino acids | *D. dipsaci*[f] Discharged free amino acids | *D. myceliophagus*[f] Discharged free amino acids | *P. penetrans*[f] Discharged AA | *M. incognita*[f] Discharged AA | *Aphelenchoides* sp.[g] Discharged AA | *C. briggsae*[h] Discharged AA |
|---|---|---|---|---|---|---|---|---|---|
| Alanine | + | + | + | + | + | + | + | + | + |
| Arginine | + | + | + | + | + | | + | + | + |
| Aspartic acid | + | + | + | + | + | + | + | + | + |
| Glutamic acid | + | + | + | + | + | + | + | + | + |
| Glycine | + | + | + | + | + | + | + | + | + |
| Histidine | | + | + | | | | + | + | + |
| Leu/Isoleu | | + | + | + | + | | + | + | + |
| Lysine | + | + | + | + | + | | + | + | + |
| Hydroxyproline | | + | + | | | | | | |
| Proline | | + | + | | | | + | + | + |
| Serine | + | + | + | + | + | + | + | + | + |
| Cystine | | | | | | | | | |
| Cysteic acid | + | + | + | + | | | | | + |
| Threonine | + | + | + | + | + | + | + | + | + |
| Tyrosine | | + | + | + | + | | Trace | + | + |
| Valine | + | + | + | + | + | | Trace | | + |
| Cysteine | | | + | + | | | | | |
| Phenylalanine | | + | + | + | | | + | + | |
| Methionine | + | + | | + | + | | | | + |
| γ-Aminobutyric | + | | + | Trace | | | Trace : | | |
| Asparagine | | | + | + | + | | + | | |
| Citrulline | | | Trace | Trace | Trace | | Trace | | |
| Glutamine | | | + | + | | | + | | |
| Methionine sulfoxide | | | + | + | + | + | + | | |
| Ornithine | | | + | + | + | + | + | | |

determined, evidence was obtained that the pigment is composed of a quininoid derivative of tyrosine.

Protein was detected by histochemical and microspectrophotometric methods in the ducts of subventral glands of second-stage *Meloidogyne javanica* larvae by Bird and Saurer (1967), and in the ampulla and stylet exudations of females and dorsal esophageal glands of both larvae and females (Bird, 1968). However, after second-stage larvae penetrated the plant tissues this protein could no longer be detected. Tests for activity of six enzymes were negative. The protein in stylet exudates of females was histonelike, and Bird suggested that it may possess plant growth regulating influence which plays a role in the plant tissue changes occurring in root-knot nematode infections (Bird, 1968).

Interest has recently been increasing concerning the protein composition of nematodes. Electrophoresis, particularly using acrylamide gels, has been demonstrated to be a powerful tool for separating nematode proteins, including enzymes. At least 20 proteins were separated from crude homogenates of *Panagrellus redivivus* and 16 proteins from homogenates of *Ditylenchus triformis* by Benton and Myers (1966). Thirteen of the protein bands from preparations of *P. redivivus* and 8 bands from *D. triformis* showed esterase or acid phosphatase activities. Recently, Chow and Pasternak (1969) investigated the protein patterns of second-, third-, and fourth-stage larvae and adults of *P. silusiae* propagated in synchronous cultures. They found the same 16 bands on acrylamide gel electrophoresis of protein preparations from the nematodes of various ages, although varying intensities of certain bands in preparations from different aged nematodes indicated quantitative differences. Unique enzymic patterns on gels for the various aged nematodes suggested to them that the physiology and biochemistry of nematodes changes with development. They also studied proteins in *P. redivivus* and obtained results which disagreed with those of Benton and Myers (1966), especially concerning enzyme bands. Similarly, Dickson *et al.* (1968) separated proteins of adult female *Meloidogyne javanica, M. incognita,* and *M. hapla* and demonstrated that qualitative differences in protein composition allowed the species to be distinguished from one another. Culturing nematodes on different host plants had no influence on the protein pattern of a nematode species. Many of these proteins had enzymic activities.

Gysels (1968), using agar gel electrophoresis, investigated the protein contents of crude extracts of *Panagrellus silusiae, Aphelenchoides fragariae, Pelodera teres, Rhabditis terricola,* and *Caenorhabditis dolichura.* Of the 15 proteins found in extracts of *P. silusiae,* 2 were characterized as glycoproteins and 1 as a lipoprotein. Preparations from *A. fragariae* contained 11; *P. teres,* 10; *R. terricola,* 10; and *C. dolichura,* 8 protein components. From studies with *P. silusiae* protein preparations, it was found that freshly prepared samples or those refrigerated for 18 hr or less contained the largest number of separable proteins; longer refrigeration, freezing, or lyophilization resulted in loss of proteins and changed mobilities for certain others. Amylase and protease activities were detected in certain protein bands for 4 of the 5 nematode species studied. Each species of nematode could be distinguished by the qualitative and quantitative differences in protein patterns on agar gels. Thus it appears that in the future, studies of nematode proteins by electrophoresis and other means will lead to elucidation of differences and similarities among nematodes such as taxonomic relationships among genera, species, and populations. This will also provide an approach for investigating questions of susceptibility and resistance in plants to nematodes as well as contributing basic information on the protein and enzyme composition of nematodes.

Serological techniques have been used to determine differences and similarities among populations and species of nematodes. Webster and Hooper (1968) injected crude homogenates or extracts of nematodes into rabbits to induce formation of antibodies. Using agar gel diffusion plates they were able to separate *Heterodera rostochiensis, H. trifolii,* and *H. schachtii* into one group and *H. cruciferae, H. carotae,* and *H. goettingiana* into another group based on serological differences. *Ditylenchus dipsaci, D. myceliophagus,* and *D. destructor* were found to be serologically distinct. Only slight serological differences were detected among the narcissus, oat, red clover, tulip, and "giant" populations of *D. dipsaci* and between the Colyton and Woburn pathotypes of *H. rostochiensis.* Gibbins and Grandison (1968) also investigated serological differences among four populations of *D. dipsaci* (obtained from *Trifolium pratense, T. repens, Medicago sativa,* and *Narcissus* sp.) and *Aphelenchoides ritzemabosi* from *Lilium* sp. using antiserum to *D. dipsaci* (*T. pratense* population). Agar gel diffusion gave better results than immunoelectrophoresis. They observed serological differences among the populations of *D. dipsaci* and *A. ritzemabosi.* However, the significance of these differences was obscured by the fact that differences also occurred in *D. dipsaci* from *T. pratense* harvested during the same season but from adjacent sites in the field, making the differences detected among populations meaningless. They suggested that this variability resulted from different proportions of adults, juveniles, and eggs in the various samples of *D. dipsaci* from *T. pratense* or that the nematodes exist in the field as a mixture of genotypes. El-Sherif and Mai (1968) tested antisera of *Aphelenchus avenae, Panagrellus redivivus,* and *Diplogaster* sp. against antigens of the same nematodes using agar gel diffusion plates. Results indicated that *P. redivivus* and *Diplogaster* sp. are closely related (slight reactions of antibodies with antigens) whereas *A. avenae* is not closely related (no reactions with antigens or antibodies of the two other nematodes). Therefore a start has been made to use serology in nematode taxonomy, but how useful or reliable a tool it will be remains to be demonstrated by future investigations.

## V. LIPIDS

The total lipid content of a number of plant parasitic and free-living nematodes has now been determined. Comenga Gerpe (1955) reported that the vinegar eelworm, *Turbatrix aceti,* contained 41% lipid on a dry weight basis, whereas more recent measurements using nematodes propagated axenically indicated that the lipid content is about 24%, range 20.2–28.7% (Krusberg, unpublished). Tracey (1958) reported that *Ditylenchus dipsaci, D. destructor,* and *D. myceliophagus* each contained about 33% chloroform-soluble material (lipid) on a dry weight basis; Krusberg (1967) found that *D. dipsaci* contained 38% lipid. Krusberg (1967) also determined that *Ditylenchus triformis* contained about 23%, *Pratylenchus penetrans* about 27%, *Tylenchorhynchus claytoni* about 37%, and *Aphelenchoides ritzemabosi* about 11% total lipid. An unidentified fungus-feeding species of *Aphelenchoides* contained 24–30% lipid (R. F. Myers, personal communication). *Panagrellus redivivus* contained 24% lipid according to Sivapalan and Jenkins (1966). Van Gundy *et al.* (1967) found that second-stage larvae of *Meloidogyne javanica* contained about 30% lipid by extraction. By cutting transverse sections through the intestinal region of larvae and measuring the area of lipid globules, they estimated that 43% was lipid and in *Tylenchulus semipenetrans* 53% of this area was lipid. They found that this lipid disappeared as the nematodes were starved and concluded that lipid was a primary source of energy during periods of food stress. Similar results were recorded by Wilson (1965) for stored infective larvae of *Nippostrongylus brasiliensis.* Dried eggshells of *H. rostochiensis* contained 7% lipid (Clarke *et al.,* 1967), whereas the cyst wall contained 2% lipid (Clarke, 1968). Bird and Saurer (1967) were unable to detect any lipid in the ducts of subventral glands of *M. javanica* second-stage larvae nor in the ampulla or stylet exudation of females (Bird, 1968). However, Bird and Rogers (1965b) did detect lipid histochemically in the cells which produce the gelatinous matrix in females. With the exception of *A. ritzemabosi* those plant parasitic nematodes investigated thus far contain relatively large amounts of lipid. On the other hand, animal parasitic nematodes contain much less lipid, from 3.5 to 8.9% according to von Brand (1966).

### A. Fatty Acids

Further studies of the lipid of *Panagrellus redivivus* propagated under xenic conditions were reported by Sivapalan and Jenkins (1966). Phospholipids accounted for about 33% of the total lipid and those detected by thin-layer chromatography were phosphatidic acid, phosphatidyl ethanolamine, phosphatidyl choline, and phosphatidyl inositol. The fatty acid content of the total lipid, neutral lipid, and phospholipid was also investigated. Of the 21 fatty acids detected in the total lipid, 14 were identified as to chain length and degree of unsaturation and are presented in Tables II and III. Of the total fatty acids, 80% contained 18 or 20 carbon atoms, and 90% were unsaturated. The predominant fatty acids identified in the total lipid were 18:1 and 18:2, and these were also major components of both the phospholipids and neutral lipids. The phospholipids tended to contain a great proportion of 20-carbon fatty acids and the neutral lipids more of the 18-carbon fatty acids.

Krusberg (1967) investigated the fatty acid composition of several plant parasitic nematodes. For these studies *Ditylenchus triformis* was propagated on cultures of the fungus *Pyrenochaeta terrestris,* whereas *Ditylenchus dipsaci, Pratylenchus penetrans, Aphelenchoides ritzemabosi,* and *Tylenchorhynchus claytoni* were all propagated monxenically on tissue cultures of alfalfa. The fatty acids identified in these nematodes are listed in Tables II and III. Several fatty acids present in lesser quantities were also detected in these nematodes but were not identified and are not included here. The unsaturated fatty acids of *D. triformis* and *D. dipsaci* were also analyzed to determine the positions of the double bonds, and these data are also presented in Table III. Although all 5 nematode species contained essentially the same fatty acids, there was a distinctive relative quantitative pattern of fatty acids in each nematode. It is interesting that 18:1 was the major fatty acid fraction in every nematode examined, and that in 3 species it accounted for 19% of the dry weight of nematodes and in *D. triformis* for 7.5% of the dry weight. A further surprise was that when the double bond positions of the 18:1 in *D. triformis* and *D. dipsaci* were determined it consisted of 75 and 92% vaccenic acid ($18{:}1^{\omega 7}$),* respectively, with the rest oleic acid ($18{:}1^{\omega 9}$). The 18:1 from the fungus and alfalfa tissue was only about 3% vaccenic acid, with the rest oleic acid. Only *D. triformis* lacked $20{:}4^{\omega 3}$ and $20{:}5^{\omega 3}$, but this nematode contained traces of $22{:}1^{\omega 9}$ which was lacking in the other nematodes. In addition, the total lipid

* $18{:}1^{\omega 7}$=18 carbons: 1 unsaturation, $\omega^7$ 7 carbons from methyl end of chain to first unsaturation. Additional unsaturations separated by units of 3 carbons.

TABLE II
SATURATED FATTY ACID COMPOSITION OF NEMATODES

| Fatty acid | *Panagrellus redivivus*[a] | *Ditylenchus triformis*[b] | *Ditylenchus dipsaci*[b] | *Pratylenchus penetrans*[b] | *Aphelenchoides ritzemabosi*[b] | *Tylenchorhynchus claytoni*[b] | *Turbatrix aceti*[c] |
|---|---|---|---|---|---|---|---|
| 10:0 | | | | | | | Trace |
| 11:0 | | | | | | | Trace |
| 12:0 | 0.1 | Trace | Trace | Trace | Trace | Trace | 0.2 |
| 13:0 | | | | | | | 0.1 |
| 14:0 | 0.5 | 1.1 | 6.5 | 6.7 | 0.7 | 8.0 | 2.4 |
| 15:0 | | | | | | | 4.6 |
| 16:0 | 3.9 | 3.1 | 16.6 | 10.1 | 2.4 | 19.5 | 8.6 |
| 17:0 | | | | | | | 2.5 |
| 18:0 | 0.6 | 3.0 | 6.4 | 2.2 | 4.4 | 5.2 | 13.8 |
| 20:0 | | 1.6 | 0.3 | 2.9 | Trace | 4.5 | 0.4 |

[a] Sivapalan and Jenkins (1966), relative percent composition.
[b] Krusberg (1967), mg/g dry weight.
[c] Krusberg (unpublished), relative percent composition.

TABLE III
UNSATURATED FATTY ACID COMPOSITION OF NEMATODES

| Fatty acid | *Panagrellus redivivus*[a] | *Ditylenchus triformis*[b] | *Ditylenchus dipsaci*[b] | *Pratylenchus penetrans*[b] | *Aphelenchoides ritzemabosi*[b] | *Tylenchorhynchus claytoni*[b] | *Turbatrix aceti*[c] |
|---|---|---|---|---|---|---|---|
| 12:1 | 0.2 | | | | | | |
| 14:1 | 4.0 | | | | | | |
| 16:1 | 3.7 | 4.7, $^{\omega}7$, 38%<br>$^{\omega}5$, 62% | 13.3, $^{\omega}9$, 43%<br>$^{\omega}7$, 32%<br>$^{\omega}5$, 25% | 4.6 | 0.4 | 35.3 | 0.3, $^{\omega}7$, 90%<br>$^{\omega}5$, 10% |
| 18:1 | 16.3 | 75.4, $^{\omega}9$, 25%<br>$^{\omega}7$, 75% | 191.5, $^{\omega}9$, 8%<br>$^{\omega}7$, 92% | 189.7 | 11.5 | 188.7 | 29.6, $^{\omega}9$, 50%<br>$^{\omega}7$, 50%<br>$^{\omega}5$, Trace |
| 20:1 | | 7.4, $^{\omega}9$, 30%<br>$^{\omega}7$, 70% | 8.3, $^{\omega}9$, 29%<br>$^{\omega}7$, 71% | 15.7 | 5.5 | 6.3 | 2.6, $^{\omega}11$, 3%<br>$^{\omega}9$, 23%<br>$^{\omega}7$, 74% |
| 22:1 | | Trace, $^{\omega}9$, 100% | | | | | |
| 18:2 | 20.7 | 6.7, $^{\omega}6$, 100% | 22.6, $^{\omega}6$, 100% | 2.0 | 8.3 | 7.2 | 11.0, $^{\omega}6$, 100% |
| 20:2 | 0.6 | 0.8, $^{\omega}6$, 100% | 1.3, $^{\omega}6$, 100% | 1.5 | 0.6 | 2.8 | 0.9, $^{\omega}6$, 100% |
| 18:3 | 1.5 | Trace, $^{\omega}3$, 100% | 0.4, $^{\omega}3$, 100% | Trace | 1.4 | 1.5 | 2.1, $^{\omega}6$, 85%<br>$^{\omega}3$, 15% |
| 20:3 | 4.7 | 5.7, $^{\omega}6$, 100% | 3.7, $^{\omega}6$, 100% | 3.3 | 1.8 | 1.0 | 6.5, $^{\omega}6$, 100% |
| 20:4 | 5.2 | 7.0, $^{\omega}6$, 100% | 9.0, $^{\omega}6$, 24%<br>$^{\omega}3$, 76% | 6.5 | 1.4 | 1.8 | 6.2, $^{\omega}6$, 50%<br>$^{\omega}3$, 50% |
| 20:5 | 4.5 | | 7.7, $^{\omega}3$, 100% | 1.8 | Trace | 1.8 | 8.1, $^{\omega}3$, 100% |
| Unident | 33.4 | | | | | | |

[a] Sivapalan and Jenkins (1966), relative percent composition.
[b] Krusberg (1967), mg/g dry weight.
[c] Krusberg (unpublished), relative percent composition.

and fatty acid contents of the fungus *P. terrestris* and alfalfa tissues were determined. Based on the quantities and kinds of fatty acids found in the plant tissues it was concluded that the nematodes must be synthesizing most of their component fatty acids rather than obtaining them from host tissues. A less likely explanation is that nematode feeding stimulates synthesis of additional fatty acids in the host tissues which they then consume and incorporate into nematode lipid.

Recently, the fatty acid composition of *Turbatrix aceti* was investigated (Krusberg, unpublished). Use of a nematode propagated axenically such as this one, precludes the activity of another organism influencing the lipid composition of the medium or the nematode. Fatty acids accounted for slightly over half of the total lipid in *T. aceti* and the 27 identified are presented in Tables II and III. Saturated fatty acids made up about 34%, monounsaturates about 33%, and polyunsaturates about 33% of the total fatty acids. Several fatty acids in addition to those identified occurred in small quantities in *T. aceti*. In this nematode 18:1 was also the major fatty acid fraction and consisted of equal amounts of oleic and vaccenic acids with just a trace of $18{:}1^{\omega 5}$. Surprisingly the 18:3 fraction turned out to be primarily $\gamma$-linolenic acid rather than the more common $\alpha$-linolenic acid, the latter being the only form found in the other nematodes examined. Only 16 fatty acids were detected in the medium used to propagate *T. aceti*, and only 2 of these ($16{:}1^{\omega 7}$ and $18{:}3^{\omega 3}$) occurred in quantities sufficient to account for the amounts of these fatty acids found in the nematodes. By measuring the difference in the quantity of each fatty acid in the medium before and after nematode propagation and comparing these figures with the quantities of corresponding fatty acids found in the nematodes, it was concluded that the nematode synthesized from some to all of each of its component fatty acids with the exception of $16{:}1^{\omega 7}$ and $18{:}3^{\omega 3}$.

Rothstein and Gotz (1968), in a study of fatty acid biosynthesis in *T. aceti*, found essentially the same fatty acids in this nematode as did Krusberg (unpublished), although they made no quantitative measurements and determined double bond positions in only 3 unsaturated fatty acid fractions. Nematodes were propagated for 2–3 weeks with $^{14}C$-labeled acetate or one of several $^{14}C$-labeled fatty acids used, and then the pattern of labeling in the nematode fatty acids was determined. Label was found in both the methyl and carboxyl ends of oleic, linoleic, and $20{:}3^{\omega 6}$ acids, and Rothstein and Gotz therefore claimed that *T. aceti* was capable of *de novo* fatty acid biosynthesis. Varying amounts of label were detected in every nematode fatty acid examined regardless of the labeled substrate used, which indicated that the nematodes did take up these substrates from the medium and incorporated the label from the substrates into fatty acids.

From these studies it is seen that the nematodes tested thus far contain the same kinds of fatty acids although certain qualitative and quantitative differences occur among species. Nematodes certainly possess a rich variety of fatty acids with unsaturated ones predominating. It seems likely that nematodes can and do synthesize most if not all of their fatty acids, but it remains to be proved if any fatty acids are "essential" for nematodes. The metabolic role of these fatty acids in nematodes also remains to be determined.

## B. Sterols

Several recent studies have concerned the composition and biosynthesis of sterols in nematodes. Cole and Krusberg (1967a), in a preliminary study of *Ascaris lumbricoides*, found that males contained almost twice as much total sterol as females. Males contained 0.38% total sterols on a dry weight basis with 60% of the total sterols in ester form, whereas females contained 0.21% total sterols with 40% in ester form (Table IV). Cholesterol was the major sterol detected with lesser quantities of several plant sterols. The plant sterols were probably derived from the diet materials of the pig hosts of these nematodes.

The sterol composition of *Ditylenchus triformis*, *D. dipsaci*, and their

TABLE IV
STEROL COMPOSITION OF NEMATODES AS PERCENT OF TOTAL STEROL

| Sterol | *Ascaris lumbricoides*[a] | *Ditylenchus triformis*[b] | *Ditylenchus dipsaci*[b] | *Turbatrix aceti*[c] |
|---|---|---|---|---|
| Cholesterol | 40 | 40 | 50 | 46 |
| Lathosterol | | 60 | 50 | 10 |
| 7-Dehydrocholesterol | | | | 44 |
| Cholestanol | 20 | | | |
| Campesterol | 7 | | | |
| $\beta$-Sitosterol | 13 | | | |
| Stigmastanol | 11 | | | |
| Campestanol | 7 | | | |
| $\alpha$-Spinasterol | | | Trace | |
| $\Delta$7-Stigmastene-3$\beta$-ol | | | Trace | |

[a] Cole and Krusberg (1967a).
[b] Cole and Krusberg (1967b).
[c] Cole and Krusberg (1968).

host tissues was also examined by Cole and Krusberg (1967b). *Ditylenchus triformis* contained 0.09% total sterols on a dry weight basis of which 48% was in ester form, and *D. dipsaci* contained 0.06% sterols of which 58% was in ester form. The major sterols in both nematodes were identified as cholesterol and lathosterol (Table IV), although traces of the phytosterols $\alpha$-spinasterol and $\Delta$7-stigmastene-3$\beta$-ol were also sometimes found in *D. dipsaci.* These two phytosterols were also found in the alfalfa tissues. The major sterol of the fungus *P. terrestris,* host for *D. triformis,* was ergosterol. Traces of a sterol tentatively identified as 22-dehydroergosterol also occurred. Traces of cholesterol were found in the media used to propagate both the fungus and alfalfa tissues. Interestingly, cholesterol and lathosterol were also the two major sterols in the insect parasitic nematode DD-136 (Dutky *et al.,* 1967a). Since the host plants and the media lacked sufficient quantities of specific sterols to account for those present in the plant parasitic nematodes, it appears that these nematodes must at least be able to convert plant sterols to nematode sterols if they cannot synthesize them from smaller precursors.

Investigations were made on sterol metabolism of a nematode which was propagated axenically enabling the pathways of sterol synthesis and transformations to be followed more exactly. In preparation for such studies, Cole (1967) identified the sterols in *Turbatrix aceti* as cholesterol, 7-dehydrocholesterol, and lathosterol (tentative identification) as shown in Table IV. The nematode was propagated axenically on medium containing fresh beef liver extract, yeast extract, soy peptone, and acetic acid. The total sterol content of the nematode on a dry weight basis was only 0.02–0.03%. The culture medium contained 2.5–3.0 $\mu$g/ml of cholesterol, mainly contributed by the liver extract. This amount of cholesterol was ample to account for the amount of sterols found in the nematodes. However, by supplementing the medium with an additional 10 $\mu$g/ml of cholesterol the total sterol content of the nematodes rose to around 0.15% of the dry weight, although there was little increase in yield of nematodes (Cole and Krusberg, 1968). The structures of cholesterol, 7-dehydrocholesterol, and lathosterol are presented in Fig. 1.

Sterol metabolism in *T. aceti* was then investigated using nematodes propagated in medium containing mevalonic acid-2-$^{14}$C lactone, a sterol precursor (Cole and Krusberg, 1968). No radioactivity was detected in nematode sterols, suggesting that the nematode was incapable of *de novo* sterol biosynthesis. However, when nematodes were propagated in the presence of cholesterol-4-$^{14}$C, label was incorporated into nematode 7-dehydrocholesterol. In addition, when nematodes were propagated in the presence of $^{3}$H-$\beta$-sitosterol, tritium was incorporated into nematode cholesterol and 7-dehydrocholesterol. Propagating nematodes in medium

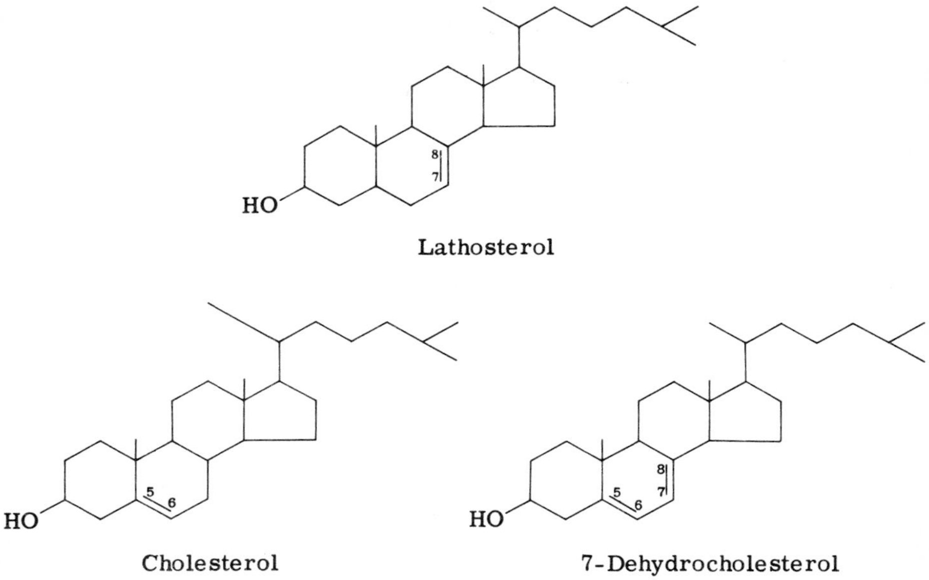

Lathosterol

Cholesterol

7-Dehydrocholesterol

FIG. 1. Structures of the three principal sterols found in free-living and plant parasitic nematodes.

supplemented with 10 $\mu$g/ml of fucosterol led to a 10-fold increase in the amount of cholesterol and lathosterol in nematodes, although total yield of nematodes was again not increased. When nematodes were propagated in the presence of Triparanol succinate, a known inhibitor of sterol biosynthesis, and fucosterol or $^{3}$H-$\beta$-sitosterol, desmosterol accumulated in the nematodes. Desmosterol did not accumulate in nematodes grown in medium supplemented with Triparanol succinate but not supplemented with plant sterols. These results suggested that the inhibitor interfered with dealkylation of plant sterols being metabolized by the nematode rather than interfering with *de novo* sterol biosynthesis in the nematode.

Somewhat similar studies were carried out by Rothstein (1968) using *Turbatrix aceti*, *Caenorhabditis briggsae*, and *Panagrellus redivivus*, all propagated axenically. None of these nematodes incorporated $^{14}$C from acetate-2-$^{14}$C or mevalonate-2-$^{14}$C into nematode cholesterol. When cholesterol-4-$^{14}$C was added to the medium in which *T. aceti* or *C. briggsae* were grown, no $^{14}CO_2$ was released, indicating that the cholesterol was not broken down. However, the author was unable to clearly demonstrate 7-dehydrocholesterol, either labeled or cold, in either nematode, although he obtained circumstantial evidence that *T. aceti* esterified cholesterol-4-$^{14}$C.

Two attempts have been made to determine whether or not nematodes require sterol. Hieb and Rothstein (1968) found that *C. briggsae* could reproduce on cells of *Escherichia coli* in phosphate buffer only when the cells were supplemented with sterols. Cholesterol, 7-dehydrocholesterol, ergosterol, $\beta$-sitosterol, and stigmasterol were used individually and in a mixture, and cholestane, with Tween 80 added to obtain solution, all supported nematode reproduction on *E. coli* cells. Squalene, vitamin A, or a mixture of vitamin E, linoleic acid, and arachidonic acid did not support nematode reproduction on cells of *E. coli*. The lower limit of sterol which allowed consistent nematode reproduction was 1.3 $\mu$g/ml of bacterial medium. However, the authors did not determine if the bacterial cells might be altering the sterols or if impurities were present in the sterols, except for cholesterol and cholestane, which could influence nematode reproduction. Cole and Dutky (1969) did similar studies with *T. aceti* and *P. redivivus* using a sterol-deficient test system consisting of endospores of *Bacillus subtilis* on peptone-glucose agar slants. The medium contained minute amounts of cholesterol and $\beta$-sitosterol, and the endospores a small amount of cholesterol. The bacteria did not alter cholesterol-4-$^{14}$C added to the medium and all was recovered after 14 days incubation. In the same test system, nematodes grew and reproduced only when cholesterol was added. *Panagrellus redivivus* seemed to require less sterol than *T. aceti* since it had to be subcultured twice in the sterol-deficient test system to demonstrate a distinct sterol requirement for reproduction. In this test system using highly purified sterols, it was demonstrated that *T. aceti* could utilize $\Delta$7-cholestenol, cholestanol, $\beta$-sitosterol, fucosterol, 24-methylene-cholesterol, 25-norcholesterol, cholest-4-ene-3-one, cholest-5-ene-3-one, campesterol, stigmasterol, desmosterol, and 7-dehydrocholesterol for growth and reproduction, but not coprostanol or coprost-7-ene-ol. The latter two compounds are stereoisomers of cholestanol and $\Delta$7-cholestenol, respectively, both of which supported nematode reproduction. Included here are plant sterols, animal sterols, and synthetic sterols. A sterol requirement was also demonstrated for the insect parasitic nematode DD-136 (Dutky *et al.*, 1967b). Thus a requirement for exogenous sterol has been demonstrated in several nematodes and may be a general nutritional requirement among nematodes. On the other hand, nematodes seem to have the capacity to convert many sterols into nematode sterols.

## VI. PLANT GROWTH REGULATORS

The similarities between deformations in plants induced by certain nematodes and deformations produced by application of particular plant growth regulators to plants have led a few investigators to examine nematodes for such growth regulators. Yu and Viglierchio (1964), using three species of root-knot nematodes, found that the nematodes contained characteristic indolic growth regulators. These regulators were identified by relative $R_f$ values on paper chromatograms of substances from nematode extracts, color reactions, fluorescence, and activity in the *Avena* first internode bioassay. Indoleacetic acid, indoleacetic acid ethyl ester, and indoleacetonitrile were detected in second-stage larvae and egg masses of *Meloidogyne hapla*. Only indoleacetic acid was detected in egg masses of *M. javanica,* and no growth regulator was detected in second-stage larvae. Only indolebutyric acid was found in egg masses of *M. incognita* and again, no growth regulator in larvae. The same growth regulators were found in tomato root galls induced by the respective nematodes. Bird (1966) found that distilled water in which *M. javanica* second-stage larvae had been incubated for 24 hr contained a dialyzable, ether-soluble substance which inhibited growth of *Avena* coleoptile segments. No auxin, kinin, or gibberellin activity was detected in such exudates or in extracts of homogenized larvae.

When Viglierchio and Yu (1968) examined extracts of cysts and second-stage larvae from cysts of *Heterodera schachtii* raised on sugar beets or brussel sprouts, and extracts of cysts of *H. trifolii* raised on white clover, no auxin activity was detected using the *Avena* first internode bioassay. Later, however, Johnson and Viglierchio (1969) found what they called indoleacetic acid in extracts of *H. schachtii* larvae.

Using extracts of *Ditylenchus triformis* and *D. dipsaci,* Cutler and Krusberg (1968) isolated an indolic plant growth promoter with the following structural formula:

$(CH_2)_n—C(=O)OCH_3$ (attached to position 3 of indole, N–H)

(I)

The *Avena* first internode bioassay was used to determine biological activity. The fungus *Pyrenochaeta terrestris,* host for *D. triformis,* contained a growth promoter with characteristics identical to indoleacetic acid, whereas alfalfa tissues, host for *D. dipsaci,* contained no detectable growth regulators. Neither *D. triformis* nor *D. dipsaci* would take up from liquid medium or esterify $\alpha$-$^{14}$C-indoleacetic acid.

While it has been demonstrated that nematodes contain plant growth regulators, the origin of these substances (whether they are taken from plant tissues or synthesized by the nematodes) and whether or not they are injected into plant tissues by nematodes and play a role in symptom production remain to be determined.

## VII. OTHER

As with other animals, the organic acid content of nematodes is low (Castillo, 1968). Total organic acids accounted for about 0.8% of the dry weight of *Ditylenchus triformis* and 1.2% of *Turbatrix aceti.* Fumaric acid made up 55% of the total organic acids in *D. triformis* and 70% in *T. aceti.* Other acids identified in both nematodes included succinic, pyruvic, citric, lactic, malic, and $\alpha$-ketoglutaric acids. Considerable variation occurred in levels of particular acids in one nematode species when compared with the other. When *T. aceti* or *D. triformis* were incubated in media containing pyruvate-3-$^{14}$C, glucose-6-$^{14}$C, or glucose-UL-$^{14}$C, label was incorporated into every organic acid.

Using histochemical techniques, polyphenols were detected in the exo- and endocuticle of the wall of cysts of *Heterodera rostochiensis* by Ellenby (1963). When the cyst wall was not methylated chemically, polyphenols were detected in the exocuticle only, but, when methylated, both layers of cuticle were observed to contain phenols. Dried eggshells of *H. rostochiensis* were aslo found to consist of 3% polyphenols (Clarke *et al.,* 1967), and the cyst wall 2% (Clarke, 1968). The phenols were identified from acid hydrolysates of the eggshells or cyst walls.

Cytochromes $a + a_3$, b, and c were detected spectroscopically in massed specimens of *Ditylenchus triformis* and also by preparing absorption spectra of sodium cholate extracts of this nematode (Krusberg, 1960).

Using microchemical tests, Myers and Krusberg (1965) found that live *D. triformis* discharged into the incubation solution 1,2-dicarboxylic acids and aldehydes in addition to a number of other compounds mentioned earlier. However, tests were negative for the following types of compounds; primary aliphatic amines, formaldehydes, formic acid, methanol, alcohols, acetic acid, sulfonic and sulfinic acids, sulfonamides, sulfones, *o*-dioxomethylene and oxomethylene compounds, and urea.

## VIII. CONCLUSIONS

Although information on the chemical composition of free-living and plant parasitic nematodes is limited and fragmentary, the findings thus far indicate that they contain a rich variety of substances. Perhaps unique biochemical and physiological processes are operating in these nematodes. Results of future investigations not only will provide us with a better understanding of nematode composition but also perhaps make it possible to better relate these primitive organisms to other organisms in the evolutionary sequence.

## REFERENCES

Benton, A. W., and Myers, R. F. (1966). *Nematologica* **12,** 495–500.
Bird, A. F. (1958). *Nematologica* **3,** 205–212.
Bird, A. F. (1966). *Nematologica* **12,** 471–482.
Bird, A. F. (1968). *J. Parasitol.* **54,** 879–890.
Bird, A. F., and Rogers, G. E. (1965a). *Nematologica* **11,** 224–230.
Bird, A. F., and Rogers, G. E. (1965b). *Nematologica* **11,** 231–238.
Bird, A. F., and Saurer, W. (1967). *J. Parasitol.* **53,** 1262–1269.
Castillo, J. M. (1968). Ph.D. Thesis, 51 pp. Univ. of Maryland, College Park, Maryland.
Chow, H. H., and Pasternak, J. (1969). *J. Exp. Zool.* **170,** 77–84.
Clarke, A. J. (1968). *Biochem. J.* **108,** 221–224.
Clarke, A. J., Cox, P. M., and Shepherd, A. M. (1967). *Biochem. J.* **104,** 1056–1060.
Cole, R. J. (1967). Ph.D. Thesis, 33 pp. Univ. of Maryland, College Park, Maryland.
Cole, R. J., and Dutky, S. R. (1969). *J. Nematol.* **1,** 72–75.
Cole, R. J., and Krusberg, L. R. (1967a). *Comp. Biochem. Physiol.* **21,** 109–114.
Cole, R. J., and Krusberg, L. R. (1967b). *Exp. Parasitol.* **21,** 232–239.
Cole, R. J., and Krusberg, L. R. (1968). *Life Sci.* **7,** 713–724.
Comenga Gerpe, M. (1955). *Rev. Espan. Fisiol.* **11,** 181–186.
Cutler, H. G., and Krusberg, L. R. (1968). *Plant Cell Physiol.* **9,** 479–497.
Dickson, D. W., Sasser, J. N., and Huisingh, D. (1968). *Nematologica* **14,** 5. (Abstr.)
Dutky, S. R., Kaplanis, J. N., Thompson, M. J., and Robbins, W. E. (1967a). *Nematologica* **13,** 139–140. (Abstr.)
Dutky, S. R., Robbins, W. E., and Thompson, J. V. (1967b). *Nematologica* **13,** 140. (Abstr.)
Ellenby, C. (1963). *Experientia* **19,** 256–257.
Ellenby, C. (1968). *Experientia* **24,** 84–85.
El-Sherif, M., and Mai, W. F. (1968). *Nematologica* **14,** 593–595.

Gibbins, L. N., and Grandison, G. S. (1968). *Nematologica* **14,** 184–188.
Gysels, H. (1968). *Nematologica* **14,** 489–496.
Hieb, W. F., and Rothstein, M. (1968). *Science* **160,** 778–780.
Johnson, R. N., and Viglierchio, D. R. (1969). *Nematologica* **15,** 159–160.
Krusberg, L. R. (1960). *Phytopathology* **50,** 9–22.
Krusberg, L. R. (1961). *Nematologica* **6,** 181–200.
Krusberg, L. R. (1967). *Comp. Biochem. Physiol.* **21,** 83–90.
Myers, R. F. (1966). *Nematologica* **12,** 579–586.
Myers, R. F., and Krusberg, L. R. (1965). *Phytopathology* **55,** 429–437.
Nicholas, W. L., Dougherty, E. C., Hansen, E. L., Hansen, O. H., and Moses, V. (1960). *J. Exp. Biol.* **37,** 435–443.
Rothstein, M. (1968). *Comp. Biochem. Physiol.* **27,** 309–317.
Rothstein, M., and Gotz, P. (1968). *Arch. Biochem. Biophys.* **126,** 131–140.
Sivapalan, P., and Jenkins, W. R. (1966). *Proc. Helminthol. Soc. Wash.* **33,** 149–157.
Smith, L., and Ellenby, C. (1967). *Nematologica* **13,** 395–405.
Tracey, M. V. (1958). *Nematologica* **3,** 179–183.
Van Gundy, S. D., Bird, A. F., and Wallace, H. R. (1967). *Phytopathology* **57,** 559–571.
Viglierchio, D. R., and Yu, P. K. (1968). *Exp. Parasitol.* **23,** 88–95.
von Brand, T. (1966). "Biochemistry of Parasites," 429 pp. Academic Press, New York.
Webster, J. M., and Hooper, D. J. (1968). *Parasitology* **58,** 879–891.
Wilson, P. A. G. (1965). *Exp. Parasitol.* **16,** 190–194.
Yu, P. K., and Viglierchio, D. R. (1964). *Exp. Parasitol.* **15,** 242–248.

# CHAPTER 23

# Respiration

R. A. Rohde

*Department of Plant Pathology, University of Massachusetts, Amherst, Massachusetts*

I. Introduction . . . . . . . . . . . . . . 235
Comparison with Other Organisms . . . . . . . . . 236
II. Factors Influencing Respiration . . . . . . . . . 237
A. Oxygen . . . . . . . . . . . . . . 237
B. Reduced Oxygen Levels . . . . . . . . . . 237
C. Carbon Dioxide . . . . . . . . . . . . 238
D. Temperature . . . . . . . . . . . . 239
E. Osmotic Pressure . . . . . . . . . . . 240
F. Metabolic Cycles . . . . . . . . . . . 241
G. Fermentation . . . . . . . . . . . . 242
H. Hexose Monophosphate Shunt . . . . . . . . 242
I. Tricarboxylic Acid Cycle . . . . . . . . . . 243
J. Terminal Oxidation . . . . . . . . . . . 243
K. Other Cycles . . . . . . . . . . . . 243
III. Methods of Measuring Respiration . . . . . . . . 244
References . . . . . . . . . . . . . . 245

## I. INTRODUCTION

The beginning student of nematology soon learns that nematodes stop moving when left too long in the Baermann funnel but that they soon start to wriggle again when placed in a dish of water. In more formal terms, we may state that soil nematodes are aerobic but well able to survive periods of partial or complete anaerobiosis. These conclusions have been confirmed by the sophisticated techniques of respirometry, but the fact remains that we know relatively little else about the respiration of nematodes. There has not been a complete study of the respiration of even a single species of plant parasitic nematode, and some fundamental questions have yet to be answered. Much more is known about animal parasitic nematodes although caution must be used in drawing conclusions from species highly adapted for habitats such as the large intestine of a pig.

The reader is referred to several recent books and reviews on respiration of nematodes, dealing mainly with animal parasites but applicable to plant parasites as well (von Brand, 1960; Fairbairn, 1960; Lee, 1965; Saz, 1969). Respiration measurements were used by a number of authors who were not measuring respiration per se but rather using oxygen uptake as an index of other life processes. Free energy for all activities of an organism is extracted by cells from chemical bonds through the process of respiration, and respiration measurements therefore provide a means of quantifying the influence of external factors on the overall well being of the organism.

No special respiratory organs are known in nematodes. Gas exchange takes place by diffusion through the body wall.

## Comparison with Other Organisms

Respiration rates ($QO_2$) of plant parasitic nematodes fall within the ranges of similar measurements for animal parasitic forms as well as other small invertebrates. Only very general conclusions are valid in regard to any given $QO_2$, since a number of external factors will induce a severalfold change. For the same reason, comparison of rates quoted by various authors for a given species is subject to considerable error.

Small nematodes respire at a faster rate than large ones following the same pattern whereby mice respire at a higher rate than elephants. Overgaard Nielsen (1949) provided data that indicate, in plant parasitic nematodes, that increase in oxygen uptake is proportional to increase in body surface rather than body weight. The data of Wallace and Greet (1964) on two species of *Tylenchorhynchus* support this contention, although all authors point out the risk of relying too heavily on such data. Volume and length are quite variable within an individual living nematode (Crofton, 1966) and, consequently, so is weight if weight is measured by one of the commonly used methods employing linear measurements times specific gravity.

Overgaard Nielsen (1949) attempted to use respiration measurement of nematodes to assess their relative importance in the soil. He showed that nematodes are a major factor in mineralization of soil nitrogen. Conservative estimates put the weight of nematodes at 15–20 gm/m$^2$ of a cultivated field. Considering their high $QO_2$, nematodes consume relatively large amounts of food and are important in the recycling of organic compounds in the soil.

## II. FACTORS INFLUENCING RESPIRATION

### A. Oxygen

Plant parasitic nematodes are exposed to high levels of oxygen during most of their life. Even the roots of rice and other marsh grasses growing in anaerobic mud are oxygen rich (Sculthorpe, 1967). All soil nematodes investigated thus far have shown an ability to utilize $O_2$ when it is available, although it must be emphasized that ability to take up $O_2$ does not in itself prove that $O_2$ is essential or is used in respiration.

The soil environment undergoes periods when $O_2$ is depleted or absent and nematodes survive these periods quite well. Water-saturated soils and rotting plant tissues are low in $O_2$, but nematodes are found in these habitats. Van Gundy *et al.* (1968) found that immediately after irrigation there was no detectable $O_2$ in the top 61 cm of soil around citrus trees, and $O_2$ did not return to normal levels in the top 15 cm for 12 hr. It was 7 days before $O_2$ returned to the 61-cm depth. Presumably, every rain brings a similar temporary $O_2$ depletion in soils.

All nematodes exist in habitats that are characterized by periodic oxygen loss, and they vary in ability to survive under these conditions (cf. Chapter 14 by Cooper and Van Gundy). *Tylenchorhynchus martini* withstands 18 days under $N_2$ (Johnston, 1957), *Dorylaimus* sp. from African swamps survive 68–86 days under anaerobic conditions (Banage, 1966). There is no question, however, that $O_2$ is necessary for growth, reproduction, and other normal activities. Flooding of fields has been used to control some species. As little as 12 hr of anaerobic conditions every 3 days significantly reduced *Hemicycliophora arenaria* populations around citrus roots (Cooper *et al.*, 1970).

### B. Reduced Oxygen Levels

Overgaard Nielsen (1949) found that the nematodes he studied respired well in 10% $O_2$ with no apparent reduction in respiration rate. He made no observations at lower $O_2$ levels, but predicted that very low levels would reduce $QO_2$. Rogers (1962) calculated that at about 2%

$O_2$, diffusion rate of $O_2$ into the body became a limiting factor in nematodes over 50 $\mu$ in diameter. With few exceptions, soil nematodes are well below this size.

The influence of low levels of $O_2$ would be in part dependent on terminal oxidation, and the little evidence available on this subject will be discussed later. If a system with a high affinity for $O_2$, such as the cytochrome system, is operative, $O_2$ uptake would continue at very low concentrations of $O_2$ (2%). Other vital processes with lower affinities for $O_2$ could, of course, raise the lower limit of $O_2$ concentration at which respiration occurs.

Baxter and Blake (1969) discussed the influence of low levels of $O_2$ on hatch, development, and migration of *Meloidogyne javanica* larvae. They reasoned that the $O_2$ concentration required for half of maximum egg hatch ($K_{50}$) was equivalent to the Michaelis constant ($K_m$) of an $O_2$-requiring enzyme. If this is the case, observed $K_{50}$ values for hatch were around 3%, which is much higher than recorded $K_m$ values for most terminal oxidases.

Since development, movement, and other activities are curtailed at low $O_2$ levels, it is reasonable to expect that demand for $O_2$ is reduced as well. It is not possible to state whether reduced respiration supplies less energy for activity or that less activity requires less energy. In natural habitats, the confusion is further compounded because reduced $O_2$ is usually associated with high $CO_2$ and increased activity of antagonistic microorganisms. Oxygen itself appears to be of primary importance in that as little as 4 hr of $N_2$ every 3 days will reduce reproduction *in vitro* of *Aphelenchus avenae* and *Caenorhabditis* sp. (Cooper *et al.*, 1970).

### C. Carbon Dioxide

Carbon dioxide produced by biological activities in soil tends to accumulate and soil nematodes are often exposed to concentrations up to several percent. Roots, as well, are high in $CO_2$, with concentrations reaching 8% in meristems of fibrous roots (Fadeel, 1963). Nematodes observed in habitats with high $CO_2$ concentrations behave normally and some are even attracted to regions of high $CO_2$ concentration (Klingler, 1959).

Concentrations of $CO_2$ up to 2% have no apparent inhibitory effect on respiration of *Aphelenchoides ritzemabosi*, *Ditylenchus dipsaci*, and *Pratylenchus penetrans* (Bhatt and Rohde, 1970). There appeared to be some correlation between habitat and tolerance to $CO_2$ in that *A.*

*ritzemabosi*, a foliar parasite, respired at a higher rate in air (0.03% $CO_2$) than in higher concentrations of $CO_2$, whereas the others, from habitats often high in $CO_2$, did not. Very high concentrations of $CO_2$ probably inhibit respiration since other activities are curtailed, but this has not been investigated.

It is doubtful that nematodes encounter $CO_2$-free environments in nature. This is a factor that must be considered, however, since many of the common methods of measuring $O_2$ uptake require atmospheres wherein $CO_2$ has been removed with alkali. The plant parasitic nematodes studied thus far consume $O_2$ at higher rates in the presence of $CO_2$ than in its absence (Rohde, 1960; Bhatt and Rohde, 1965). Why this happens is not known. The depression is not permanent and is reversed by exposing nematodes to air. Collis-George and Wallace (1968) found that $O_2$ uptake of eggs and larvae of *M. javanica* declined during storage in a Warburg respirometer. They explained this in terms of decrease in amount of available $O_2$, decreased permeability of membranes to $O_2$, or an abnormally high initial respiration rate induced by handling. The influence of removing $CO_2$ from the respirometer vessels could be added. Santmyer (1956) attributed the reduction of $O_2$ uptake by *Panagrellus redivivus* during 24 hr in the respirometer to starvation. The data of Overgaard Nielsen (1949) show an initial reduction of respiration, but the rate then becomes level.

### D. Temperature

The increase in $QO_2$ with increase in temperature of plant parasites deviates somewhat from the normal curve of Krogh, particularly at higher temperatures (Fig. 1). Maximum respiration of *D. dipsaci* occurs at 22°C. This agrees with other characteristics of this cool-climate species. Maximum respiration of *A. ritzemabosi* and *P. penetrans* occurs at 30°–35°C, indicating an adaptation to warmer environments.

Von Brand (1960) discussed the use of the Arrhenius equation to determine the temperature characteristic ($\mu$) of $O_2$ uptake. Within a given temperature range, a constant temperature characteristic indicates a single limiting reaction. When the temperature characteristic changes, it indicates a change in the limiting reaction. Von Brand, using the data of Santmyer (1956), showed a change in temperature characteristic of respiration at 34.7°C. This is the temperature at which $\mu$ for respiration of *P. penetrans* and *A. ritzemabosi* also changes, a possible indication that the same temperature-dependent reactions are involved (Bhatt and Rohde, 1970). Data given by Santmyer (1956) can be interpreted as

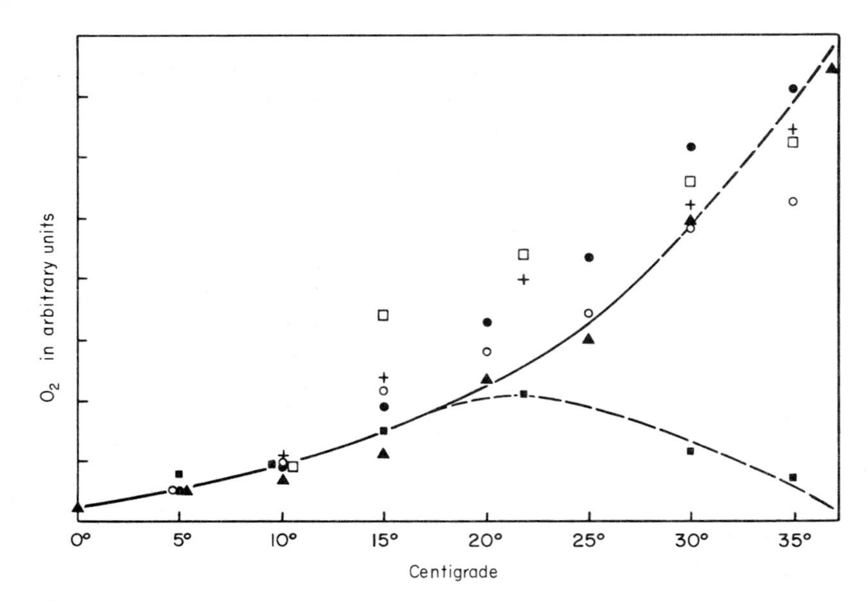


FIG. 1. Oxygen consumption of nematodes projected from Krogh's normal curve. (▲) Larval *Eustrongylides*, (○) *Panagrellus redivivus*, fresh; (●) *Panagrellus redivivus*, starved 24 hr; (+) *Pratylenchus penetrans;* (□) *Aphelenchoides ritzemabosi;* and (■) *Ditylenchus dipsaci*. (After Von Brand, 1960.) *Panagrellus* data from Santmyer (1956), *Pratylenchus, Aphelenchoides,* and *Ditylenchus* data from Bhatt and Rohde (1970).

showing a change in respiratory quotient (RQ) of *P. redivivus* at 34.7°C as well.

Current methods for measuring respiration cannot be used at temperatures below freezing and no information is available on respiration of nematodes at these temperatures. Such data would be of ecological value, particularly for species such as *Ditylenchus radicicola* that may be active during periods of freezing temperatures (Chiaravalle and Stessel, 1962).

## E. Osmotic Pressure

The influence of osmotic pressure (OP) on nematodes is discussed in detail elsewhere in this book, but brief mention should be made of the influence of OP on respiration.

The plant parasitic nematodes investigated thus far respire well in urea, mannitol, or sodium chloride solutions up to 44.8 atm OP (Wallace and Greet, 1964; Bhatt and Rohde, 1970). When compared with respira-

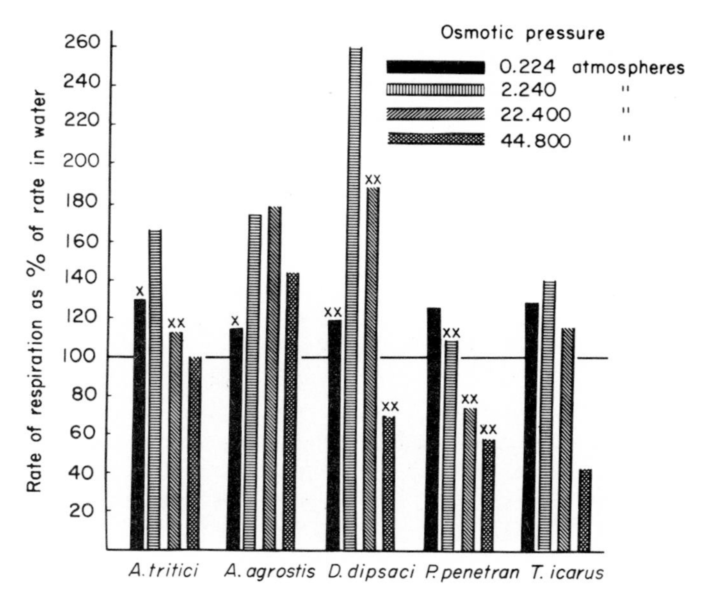


FIG. 2. Respiratory rates of plant parasitic nematodes in urea solutions. Crosses over bars indicate significant differences between adjacent bars ($x = 0.05$ level, $xx = 0.01$ level). Values for *Tylenchorhynchus icarus* calculated from Wallace and Greet (1964). (From Bhatt and Rohde, 1970.)

tion in distilled water, $QO_2$ in dilute salt solutions is usually higher, possibly because of work done by an osmoregulatory system to prevent water loss and maintain body turgor.

There is an apparent relationship between resistance to desiccation and ability to respire at high OP (Fig. 2). *Anguina tritici* and *A. agrostis* respire as well at high OP as they do in distilled water, whereas $QO_2$ values of *P. penetrans,* not resistant to desiccation, are reduced at high OP. *Ditylenchus dipsaci* and *Tylenchorhynchus icarus,* moderately resistant to desiccation, are intermediate in reaction.

## F. Metabolic Cycles

Saz (1969) thoroughly summarized the information available on the metabolism of animal parasitic nematodes. All consume $O_2$ when it is available, although some utilize glucose and survive just as well in an atmosphere of nitrogen, and some are actually poisoned by high levels of $O_2$ because they cannot break down the high amounts of $H_2O_2$ produced. In general, carbohydrates are incompletely oxidized to various acids and relatively little $CO_2$ is released, indicating limited importance of the Kreb's cycle, even for aerobic forms. Glycolysis appears to be similar to that of other organisms. Fixation of $CO_2$ into succinate is common, and there are probably many other major and minor differences between the metabolic pathways of nematodes and higher animals.

### G. Fermentation

All nematodes studied thus far have been found to anaerobically ferment hexoses to pyruvate. A typical glycolytic cycle similar to that of other animals is presumed to exist in plant parasitic nematodes, and Krusberg (1960) demonstrated several glycolytic enzymes in *Ditylenchus*. Labeled glucose is utilized by this species and converted to $CO_2$ (Castillo, 1968). Fermentation is sufficient to satisfy the limited need for energy of animal parasites in an environment containing no $O_2$ and abundant food. Survival of plant parasites in the absence of $O_2$ also points to the occurrence of active fermentation.

Many animal parasitic nematodes convert pyruvate to a variety of organic acids that are excreted or combined with $CO_2$ to form succinate. In this way, no $O_2$ debt is built up. This process has not been demonstrated in plant parasites, although some of the acids have been found. The observation that $CO_2$ stimulates respiration may also have some significance here.

### H. Hexose Monophosphate Shunt

Two of the enzymes of the hexose monophosphate (HMP) pathway, glucose-6-phosphate and 6-phosphogluconic dehydrogenases, were found in *D. dipsaci* and *D. triformis* by Krusberg (1960). More complete evidence exists for the HMP pathway in *Ascaris* (Entner and Gonzalez, 1959). Experimental observations indicating the relative importance of this pathway as opposed to glycolysis (C6/C1 ratio) have not been made.

Galls induced by *M. incognita* showed an increased utilization of HMP shunt when compared with normal tomato roots, as demonstrated by reduction of tetrazolium dyes and a C6/C1 ratio less than one (DeMott, 1965). An increase in the HMP shunt is typical of most infected plant tissue, and there is no evidence that the nematodes themselves contributed to the increase.

### I. Tricarboxylic Acid Cycle

Most of the enzymes (Krusberg, 1960) and organic acids (Castillo, 1968) of the tricarboxylic acid (TCA) cycle have been demonstrated in species of *Ditylenchus*. In addition, homogenates of *D. triformis* were able to utilize labeled glucose and pyruvate and convert them to $CO_2$, although in reduced amounts (Castillo, 1968).

Further work is necessary before it can be concluded that the TCA cycle is an important feature of respiration of plant parasitic nematodes. Available evidence points to its importance, but stoichiometric measurements with TCA intermediates, preferably on isolated mitochondria, will be needed.

Tricarboxylic acid intermediates are involved in amino acid synthesis, lipid oxidation, and other processes, and the presence of enzymes and intermediates is not proof of function in respiration. Several animal parasitic nematodes, both aerobic and anaerobic, possess all or most of the enzymes, yet the cycle is not functional or is limited (Saz, 1969).

### J. Terminal Oxidation

The same cautions expressed about the TCA cycle hold true for terminal oxidation as well. Enzymes of the cytochrome system have been demonstrated in *D. triformis* (Krusberg, 1960; Deubert and Zuckerman, 1968). Available evidence is incomplete but points to this system as being functional.

The aerobic bacterial feeder, *Caenorhabditis briggsae,* is insensitive to $1 \times 10^{-4}$ cyanate, indicating atypical terminal oxidation. Also, as previously mentioned, Baxter and Blake (1969) have indirect evidence that the $O_2$-dependent enzyme of *M. javanica* differs from typical cytochromes.

Terminal oxidation pathways other than cytochromes are common in animal parasitic forms, but these have not been investigated in plant parasites.

### K. Other Cycles

As Saz (1969) pointed out, differences between metabolic patterns of nematodes and their hosts are the basis for effective use of anthelmintics. Rothstein and Mayoh (1965, 1966) found isocitric lyase and malate synthetase in the bacterial feeders *P. redivivus, Turbatrix aceti,* and

*Rhabditis anomala*. The significance of these findings is not certain, but these enzymes are active in the glyoxylate cycle, previously known only in bacteria and higher plants. One might speculate that the pathway from isocitrate to glyoxalate to malate, providing a mechanism for the introduction of breakdown products of lipids into the TCA cycle, would be of great value to nematodes. Acetyl-CoA from β-oxidation of lipids condenses with glyoxylic acid to form malic acid. Lipids are stored in great quantity by nematodes, up to one-fourth of their body weight, and utilized during periods of starvation (cf. Krusberg, Chapter 23 and Cooper and Van Gundy, Chapter 14).

## III. METHODS OF MEASURING RESPIRATION

The major reason for our inadequate knowledge of nematode respiration has been a lack of simple and precise methods. Most plant parasitic nematodes weigh 1–2 μg, and an individual will consume perhaps $10^{-3}$ μl $O_2$/hr. Given these minute values, the investigator is faced with a choice of either using ultramicrorespirometers or producing large numbers of animals. Both approaches have limitations, but both have been productive.

A number of sensitive methods, suitable for measurement of $QO_2$ of small numbers of nematodes, are discussed in detail by Glick (1949). The theory behind most of the methods described is that uptake of $O_2$ from an enclosed space produces a decrease in either volume or pressure, and either can be measured when the other is held constant.

A simple respirometer can be made from a length of capillary tubing closed at one end and containing the test organism. The tube is sealed with a drop of alkali which absorbs $CO_2$ produced by respiration. As $O_2$ is consumed, pressure is reduced, and the alkali drop is drawn into the tube. The distance the drop moves is a measure of the amount of $O_2$ consumed. This method is reported to have a sensitivity of $5 \times 10^{-5}$ μl $O_2$/hr.

Other methods utilize different principles. As little as 1 μl of gas can be analyzed for either $O_2$ or $CO_2$ content by either chemical methods or gas–liquid chromatography. Platinum and other electrodes can be made to measure $O_2$ concentration in single drops of solutions.

Special mention should be made of the Cartesian diver respirometer because it is one of the most reliable and versatile of the ultrasensitive methods (Holter and Linderstrom-Lang, 1943; Boell, 1960). Most of the techniques of the more familiar Warburg respirometer, including indirect and Pardee's methods (Umbreit *et al.*, 1964), can be utilized with this technique which will measure changes of 0.01 μl/hr.

For the most part, instruments and equipment for these ultrasensitive methods are not commercially available, and considerable skill is required for their manufacture. Calibration and use require a great deal of time, manual dexterity, and patience. Unless an investigator plans to make extensive use of the equipment, he is apt to become discouraged.

One of the more important breakthroughs in plant nematology in the last 10 years was the publication by Krusberg (1961) of a relatively simple method of producing large numbers of aseptic plant parasitic nematodes grown on plant callus tissue. Large numbers of nematodes can be produced in culture and used in conventional respirometers such as the Warburg or Gilson (Umbreit *et al.*, 1964).

Many problems still exist, even with nematodes produced in culture. A minimum of 10,000 individuals of most species are needed for each 7 ml respirometer vessel in order to obtain good readings in a reasonable time period, and such numbers are difficult to produce on a daily basis. Relatively few nematode species have been successfully cultured and methods for criconematoid, dorylaimoid, and other nematodes are not available.

Populations obtained from culture contain all life stages and individuals in a variety of metabolic states. Synchronous cultures would be of great value. Large numbers of second-stage larvae, all in a similar physiological state, could be obtained from *Anguina* galls or by hatching *Heterodera* cysts. Hatch of *Meloidogyne* eggs can be delayed in solutions with high OP or in dry soil. Development continues, larvae develop, and when water is added, large numbers of eggs hatch (Dropkin *et al.*, 1958).

Respiration measurements made on nematodes freshly extracted from soil are subject to the criticism that associated microorganisms influence these data. If nematodes are rinsed in aerated sterile water, no significant respiration can be measured from the rinsings; and if water is the suspension medium, it is not likely that respiration of associated microbes will affect respiration measurements made on nematodes (Rohde, 1960). Tylenchid nematodes have no intestinal flora, and methods used to prepare them for axenic culture could be used to prepare them for respiration measurements (cf. Zuckerman, Chapter 20). Care must be taken, however, to determine the effect of the cleaning process itself on respiration.

## REFERENCES

Banage, W. B. (1966). *Oikos* **17,** 113–120.
Baxter, R. I., and Blake, C. D. (1969). *Ann. Appl. Biol.* **63,** 191–203.
Bhatt, B. D., and Rohde, R. A. (1965). *Phytopathology* **55,** 1283. (Abstr.)

Bhatt, B. D., and Rohde, R. A. (1970). *J. Nematol.* **2,** 277–285.
Boell, E. J. (1960). *In "Nematology:* Fundamentals and Recent Advances with Emphasis on Plant Parasitic and Soil Forms" (J. N. Sasser and W. R. Jenkins, eds.), pp. 109–121. Univ. of North Carolina Press, Chapel Hill, North Carolina.
Castillo, J. (1968). Ph.D. Thesis, 51 pp. Univ. of Maryland, College Park, Maryland.
Chiaravalle, P. D., and Stessel, G. J. (1962). *Phytopathology* **52,** 923. (Abstr.)
Collis-George, N., and Wallace, H. R. (1968). *Aust. J Biol. Sci.* **21,** 21–35.
Cooper, A. F., Jr., Van Gundy, S. D., and Stolzy, L. H. (1970). *J. Nematol.* **2,** 182–188.
Crofton, H. D. (1966). "Nematodes," 160 pp. Hutchinson Univ. Library, London.
DeMott, H. E. (1965). Ph.D. Thesis, 63 pp. Univ. of Virginia, Charlottesville, Virginia. (*Diss. Abstr.* **27,** 4257-B.)
Deubert, K. H., and Zuckerman, B. M. (1968). *Nematologica* **14,** 453–455.
Dropkin, V. H., Martin, G. C., and Johnson, R. W. (1958). *Nematologica* **3,** 115–126.
Entner, N., and Gonzalez, C. (1959). *Exp. Parasitol.* **8,** 471–479.
Fadeel, A. A. (1963). *Physiol. Plant.* **16,** 870–888.
Fairbairn, D. (1960). *In "Nematology:* Fundamentals and Recent Advances with Emphasis on Plant Parasitic and Soil Forms" (J. N. Sasser and W. R. Jenkins, eds.) pp. 267–296. Univ. of North Carolina Press, Chapel Hill, North Carolina.
Glick, D. (1949). "Techniques of Histo- and Cytochemistry," 531 pp. Wiley (Interscience), New York.
Holter, H., and Linderstrom-Lang, K. (1943). *C. R. Trav. Lab. Carlsberg, Ser. Chim.* **24,** 333–478.
Johnston, T. (1957). *Phytopathology* **47,** 525–526.
Klingler, J. (1959). *Mitt. Schweiz. Entomol. Ges.* **32,** 311–316.
Krusberg, L. R. (1960). *Phytopathology* **50,** 9–22.
Krusberg, L. R. (1961). *Nematologica* **6,** 181–200.
Lee, D. L. (1965). "The Physiology of Nematodes," 154 pp. Freeman, San Francisco, California.
Overgaard Nielsen, C. (1949). *Natura Jutlandica* **2,** 1–131.
Rogers, W. P. (1962). "The Nature of Parasitism," 287 pp. Academic Press, New York.
Rohde, R. A. (1960). *Proc. Helminthol. Soc. Wash.* **27,** 160–164.
Rothstein, M., and Mayoh, H. (1965). *Comp. Biochem. Physiol.* **16,** 361–365.
Rothstein, M., and Mayoh, H. (1966). *Comp. Biochem. Physiol.* **17,** 1181–1188.
Santmyer, P. H. (1956). *Proc. Helminthol. Soc. Wash.* **23,** 30–36.
Saz, H. J. (1969). *In* "Chemical Zoology" (M. Florkin and B. T. Scheer, eds.), Vol. 3, 329–360. Academic Press, New York.
Sculthorpe, C. D. (1967). "The Biology of Aquatic Vascular Plants," p. 161. Arnold, London.
Umbreit, W. W., Burris, R. H., and Stauffer,, J. F. (1964). "Manometric Techniques," 305 pp. Burgess, Minneapolis, Minnesota.
Van Gundy, S. D., McElroy, F. D., Cooper, A. F., and Stolzy, L. H. (1968). *Soil Sci.* **106,** 270–274.
von Brand, T. (1960). *In "Nematology:* Fundamentals and Recent Advances with Emphasis on Plant Parasitic and Soil Forms" (J. N. Sasser and W. R. Jenkins, eds.), pp. 233–266. Univ. of North Carolina Press, Chapel Hill, North Carolina.
Wallace, H. R., and Greet, D. N. (1964). *Parasitology* **54,** 129–144.

# CHAPTER 24

# Mating and Host Finding Behavior of Plant Nematodes

C. D. GREEN

*Rothamsted Experimental Station, Harpenden, Herts., England*

I. Introduction . . . . . . . . . . . . . . . 247
II. Sources of Stimulants . . . . . . . . . . . . . 248
A. Intraspecific Communication . . . . . . . . . . 248
B. Interspecific Stimuli . . . . . . . . . . . . 250
III. Dissemination of Stimuli . . . . . . . . . . . 253
A. Distances Nematodes Move to a Root . . . . . . . 253
B. Loss of Stimuli (Disappearance by Degradation and Absorption) 254
C. Dispersal of Stimuli . . . . . . . . . . . . 255
IV. Responses to Stimuli . . . . . . . . . . . . 258
A. Changes in Rate of Movement, Arousal, and Orthokinesis . . 259
B. Attraction . . . . . . . . . . . . . . . 260
C. Retention at a Source of Stimulus . . . . . . . . 262
V. Discussion . . . . . . . . . . . . . . . 263
References . . . . . . . . . . . . . . . 264

## I. INTRODUCTION

Behavior is a characteristic sequence of responses to internal or external stimuli. Each response may be exchanged, suppressed, or rearranged to form patterns appropriate to the species of nematode in a particular condition at a particular age. The natural selection of behavioral characteristics is as rigorous as that of morphological characters, and behavioral variation becomes increasingly important when morphological variation is limited. Responses that are irrelevant are usually dangerous or uneconomic in terms of energy and are suppressed. The distribution of plant nematodes in time, space, and habitat must be related to their hosts, but the consummation of each stage is that usual in animals, development of young and reproduction of adults.

Of the many organs suggested as sensory receptors in nematodes, none has been confirmed (Lee, 1965). Cilia have recently been found in some of the suggested sensory organs (Roggen *et al.*, 1966; Yuen, 1967, 1968), showing their similarity to most other animal receptors. Amphids and phasmids may be secretory or chemosensory but not at the same time. The elaborate structure of some amphids seemed better adapted to measure cuticular stress than to detect chemicals (Inglis, 1964) and phasmids often seemed to be secretory or excretory (Paramonov, 1954; Poinar, 1965). The occurrence of cholinesterases (Bird, 1966; Ramisz, 1966; Rohde, 1960) usually signifies nervous activity, but in invertebrates acetylcholine in unusual concentrations or at nonnervous sites often seemed to aid secretion, possibly by changing membrane permeability to salts (Kerkut, 1967) as in mammal erythrocytes (Lindvig *et al.*, 1951).

The following sections outline the known and likely sources of stimulants that relate the life of plant nematodes to the organisms around them. Relationships are discussed qualitatively, but this is no great disadvantage since the systems are very variable and dynamic.

## II. SOURCES OF STIMULANTS

Only chemical, galvanic, thermal, vibratory, and tactile stimuli can be received in the soil environment. The tactile sense is restricted and, except as a pressure sense, cannot operate at a distance; nor are thigmokinesis or thigmotaxis functional when contact with soil particles occurs all the time. However, a touch or texture sense may be important to identify suitable feeding sites (Doncaster, Chapter 19) or contact by potential predators (Esser, 1963; Yeates, 1969). Other stimuli might warn of the presence of predators, but none is known.

### A. Intraspecific Communication

#### 1. Population Effects

Mutual stimulation by individuals of a species may evoke copulation, control the numbers and distribution of the species, or warn of attack. Larvae of *Meloidogyne hapla* and *M. javanica* attract each other (Bird, 1959) as do males of *Heterodera rostochiensis* (Green *et al.*, 1970). When stimuli act over a long period, it becomes difficult to distinguish between nutritional, drug, hormonal, pheremonal, and phytomonal effects. When the offspring develop faster than their parents, as did *Caenorhabditis briggsae* in experiments by Dougherty *et al.* (1959), it may be because the parents had modified the medium. Changes in environment can have the opposite effect and act so as to limit populations. The numbers of *Rhabditis belari* decreased, fewer males were produced, and more reproduction was parthenogenetic in old cultures or in those to which extracts from old cultures had been added (Nigon, 1952). Crowding induced different changes in *Caenorhabditis briggsae;* hermaphrodite forms became dauer larvae, the proportion of males increased, and amphimictic reproduction predominated (Yarwood and Hansen, 1969); but the end result was the same, a smaller population. An unknown substance extracted from old cultures of several *Deladenas* sp. caused the insect parasitic forms to develop in cultures of the mycophagous form (Bedding, personal communication). Lack of this substance at any stage initiated development of the mycophagous form, thus ensuring predominance of the form able to feed on the more readily available food, i.e., mycelium. When populations became great, the alternative insect food supply was used. The insects also dispersed the nematodes to new sources of food.

### 2. Feeding Stimulus

Fisher (1969) suggested that the leakage of contents after cells were punctured during feeding by *Aphelenchus avenae* stimulated more feeding and so increased fecundity in cultures at optimum density.

### 3. Attraction between Sexes

Sex attractants are known in two free-living nematodes, *Panagrolaimus rigidus* (Greet, 1964) and *Pelodera teres* (Jones, 1966), in which both sexes are mobile. The immobile females of many *Heterodera* species secrete chemicals that attracted their males (Green, 1966; Green and Plumb, 1970), but although males of *Meloidogyne* species aggregate around females in galls, Santos (1969) could not show that females produced an attractant for males of *M. arenaria. Heterodera* sex attractants are water soluble and pass through dialysis membranes. The male attractant secreted by females of *Heterodera schachtii* did not volatilize enough to form perceptible vapor gradients (Green, 1967) but transferred as a vapor to nearby drops of water (Greet *et al.*, 1968). Volatility would undoubtedly increase the dispersal of attractant when films of water were discontinuous, and diffusion would be much faster in the gaseous phase than in water films. At least six different male attractants exist in the genus *Heterodera.* Of 10 species tested, most secreted more than one attractant and most males responded to more than one (Green and Plumb, 1970).

## B. Interspecific Stimuli

### 1. Stimuli from Roots and Organisms

There are many examples of plant nematodes accumulating around roots, presumably in response to stimuli the roots produce (Table I). A notable feature is that hosts, nonhosts, and organisms associated with the roots may all attract nematodes. Viglierchio (1961) suggested that movement of *Meloidogyne hapla* larvae, attracted to rye in soil but repelled in sand, was dependent on a balance of attractant and repellent stimuli. The hatching agents and inhibitors of *Heterodera* and *Meloidogyne* eggs also attain a similar balance (Wallace, 1966; Shepherd and Clarke, Chapter 25). Although some bacteria from the rhizosphere of beet plants attracted *H. schachtii* larvae, others repelled them (Bergman and Van Duuren, 1959), and fungi, actinomycetes, and bacteria attracted or repelled *Aphelenchoides parietinus* (Katznelson and Henderson, 1963). The balance may be influenced by the surrounding medium, which, for example, had to contain sugar or cellobiose before *Brassica nigra* attracted *M. incognita* (Loewenberg *et al.*, 1960). Although host selection is rarely governed by attraction of the nematodes, the attraction of *M. hapla* and *H. schachtii* to nonhosts was weak and variable (Viglierchio, 1961). Barley roots did not attract *M. javanica* (Oteifa and Elgindi, 1961); *Desmodera,* tea, and turnip roots were less attractive to *Pratylenchus pratensis* than roots of *Crotalaria* or *Tephrosia* (Gadd and Loos, 1941) and resistant strains of alfalfa attracted fewer *Ditylenchus dipsaci* than did susceptible ones (Griffin, 1969).

That some stimuli originate from the roots and not from associated microorganisms was shown by the attraction of larvae of *H. rostochiensis, M. incognita,* and *P. penetrans* to roots under sterile conditions (Widdowson *et al.*, 1958; Peacock, 1959; Chen and Rich, 1963). Also, cutting the tops from seedlings made their roots less attractive to *P. penetrans* (Lavallee and Rohde, 1962).

The parts of roots most attractive to nematodes were the tip, the zone of elongation, and the zone of root hairs (Linford, 1939; Widdowson *et al.*, 1958; Wallace, 1966). *Trichodorus viruliferus* followed the tips of extending cord roots, the fast growing white anchoring roots of apple, and fed just behind the tip (Pitcher, 1967). However, excised root tips ceased to attract *P. penetrans* (Lavallee and Rohde, 1962) and repelled *M. hapla* (Wieser, 1956a), possibly because the roots were dying since when whole roots were removed from seedlings, the tips again lost their attraction, but the zones behind remained attractive (Widdowson *et al.*,

TABLE I

SOME STIMULI THAT CAUSE PLANT PARASITIC NEMATODES TO ACCUMULATE

| Nematode species | Living sources of stimuli[a] | | | Artificial stimuli[a] | | | | | |
|---|---|---|---|---|---|---|---|---|---|
| | Host plants | Poor or nonhosts | Micro-organisms | Cathode or anode | Reducing agents | $O_2$ or $KMnO_4$ | $CO_2$ | Amino acids | Heat |
| *Aphelenchoides* sp. | — | — | — | 1 | — | — | — | — | — |
| *Ditylenchus dipsaci* | 2–4 | 2–4 | — | 5 | 6 | 6, 7 | 4, 6, 7 | 5 | 8 |
| *Helicotylenchus multicinctus* | 9 | — | 9 | — | — | — | — | — | — |
| *Hemicycliophora paradoxa* | 10 | — | — | — | — | — | — | — | — |
| *H. similis* | 11 | 11 | — | — | — | — | — | — | — |
| *Heterodera avenae* (larvae) | 12 | — | — | — | — | — | — | — | — |
| *H. rostochiensis* (larvae) | 12–18 | — | — | — | — | — | — | — | 14, 15, 19 |
| *H. schachtii* (*larvae*) | 12, 17, 20–22 | 22 | 23 | 5 | 24 | 7 | 7 | — | — |
| *Meloidogyne* sp. (larvae) | 9 | — | 9 | — | — | — | — | — | — |
| *M. hapla* (larvae) | 3, 25, 26 | 22, 25 | — | — | 24 | — | 7 | — | — |
| *M. incognita* (larvae) | 27–29 | — | — | — | — | — | — | — | — |
| *M. incognita acrita* (larvae) | 22, 30, 31 | 22 | — | — | — | — | — | — | — |
| *M. javanica* (larvae) | 32–35 | — | — | — | 24 | 24 | 7, 32, 33 | 33, 34 | — |
| *Pratylenchus* sp. | — | — | — | 1 | — | — | — | — | — |
| *P. minyus* | — | — | — | — | — | — | 32 | — | — |
| *P. penetrans* | 36–38 | — | 37 | — | — | — | — | — | 8 |
| *P. pratensis* | 9, 39 | — | 9 | — | — | — | — | — | — |
| *Trichodorus viruliferus* | 40 | — | — | — | — | — | — | — | — |
| *Tylenchorhynchus claytoni* | — | — | — | — | — | — | — | — | 8 |

[a] References indicated by following numbers: 1. Caveness and Panzer (1960). 2. Blake (1962); 3. Griffin (1969); 4. Klingler (1963); 5. Jones (1960); 6. Klingler (1961); 7. Johnson and Viglierchio (1961); 8. El-Sherif and Mai (1969); 9. Linford (1939); 10. Luc (1961); 11. Khera and Zuckerman (1963); 12. Wallace (1958); 13. Kühn (1959); 14. Rode (1962); 15. Rode and Staar (1961); 16. Wallace (1960); 17. Weischer (1959); 18. Widdowson *et al.* (1958); 19. Rode (1965); 20. Baunacke (1922); 21. Kämpfe (1960); 22. Viglierchio (1961); 23. Bergman and Van Duuren (1959); 24. Bird (1959); 25. Lownsbery and Viglierchio (1961); 26. Wieser (1955); 27. Loewenberg *et al.* (1960); 28. Peacock (1959); 29. Peacock (1961); 30. Lownsbery and Viglierchio (1960); 31. Morikawa (1962); 32. Bird (1960); 33. Bird (1962); 34. Oteifa and Elgindi (1961); 35. Wallace (1966); 36. Chen and Rich (1963); 37. Edmunds and Mai (1967); 38. Lavallee and Rohde (1962); 39. Gadd and Loos (1941); 40. Pitcher (1967).

1958). By contrast, minor wounds, such as from nematodes penetrating the root, seemed to increase attractiveness to *M. incognita* and *M. javanica* (Peacock, 1959; Wallace, 1966). The attractiveness of sites of emerging lateral roots to larvae of *H. rostochiensis* (Widdowson *et al.*, 1958) and *H. schachtii* (Kämpfe, 1960), and of root galls to *M. javanica* (Bird, 1962), could have reflected the rupture of cells or the by-products of active cell growth. Roots extending rapidly after treatment with indoleacetic acid were no more attractive than untreated roots (Bird, 1962), but D-methoxyphenylacetic acid, a growth depressant, decreased root attractiveness so much that some parts were repellent (Wieser, 1956b).

### 2. Nature of Stimulants

Many stimuli from roots may attract nematodes. At least one was a water-soluble neutral organic compound of low molecular weight. Aqueous washings from roots attracted larvae of *Meloidogyne incognita acrita* (Morikawa, 1962), *H. rostochiensis,* and *H. schachtii* (Weischer, 1959); and an attractant for the last two was adsorbed on sintered glass (Wallace, 1958). An attractant for *Hemicycliophora paradoxa* from millet was adsorbed in soil and was active after drying and removal of the roots (Luc, 1961). Dialysis membranes that restrain large molecules did not stop roots from attracting nematodes nor did barriers of anionic and cationic exchange resins, which adsorb all acidic and basic materials including $CO_2$ (Blake, 1962). However, attraction of *M. incognita* larvae was stopped by activated charcoal, which adsorbs most organic materials but not $CO_2$ (Peacock, 1961).

Most of the known materials released by roots failed to attract, but gibberellic acid, tyrosine and glutamic acid attracted *Meloidogyne* larvae (Bird, 1959, 1962; Oteifa and Elgindi, 1961). Results were variable, suggesting that these compounds did not supply a complete stimulus or that there was an optimum concentration, as when Jones (1960) found *Ditylenchus dipsaci* was attracted by glutamic and aspartic acids at 1:100,000 but repelled by 1:1000. Some reducing agents, e.g., sodium dithionite, attracted nematodes. However, it seemed unlikely that it was their reducing property (redox potential) that was effective, because oxygen-enriched water or potassium permanganate, an oxidizing agent, also attracted them.

Carbon dioxide dissolved in water, partly as gas and partly as carboxyl ions, attracted *D. dipsaci,* larvae of *H. schachtii,* and *M. javanica* in sand (Johnson and Viglierchio, 1961) but failed to attract *M. hapla* or *M. javanica* in agar (Bird, 1959). Many nematodes were attracted by vapor gradients to a source of $CO_2$, and buffering the medium to prevent changes in pH along the gradient did not alter attraction (Klingler, 1961). Neither *D. dipsaci* nor larvae of *H. schachtii, M. javanica,* and *M. hapla* responded to pH changes (Bird, 1959, 1960). The other physical properties associated with $CO_2$ gradients, changes in oxygen concentration and redox potential, were simulated by gradients of oxygen, nitrogen, or hydrogen; but nematodes responded to none of these (Klingler, 1961; Bird, 1962).

Plant roots generate a negative electrical potential difference between the cell contents and external environment (Lundegårdh, 1942; Etherton and Higinbotham, 1960; Scott and Martin, 1962) of the same order of magnitude as that to which nematodes respond (Jones, 1960). The greatest potential drop would be across the cell walls but the charge alters from the root tip backward and could determine the root zone most attractive to a nematode once in contact with a root. The potential difference may be associated with salt uptake by the root but, although nonparasitic nematodes responded to some salts (Taniguchi, 1933), there have been few reports of these responses by parasitic nematodes other than the hatching of *Heterodera* eggs (Shepherd and Clarke, Chapter 25). Dilute aluminum chloride and cadmium chloride attracted *Tylenchorhynchus martini* (Ibrahim and Hollis, 1967) and ammonium nitrate attracted *H. rostochiensis* larvae (Rode, 1965) and *Rhabditis oxycerca* (Katznelson and Henderson, 1963).

Some nematodes are sensitive to heat gradients of the magnitude of those created by metabolism of living organisms. *Heterodera rostochiensis* larvae moved to their preferred temperature in a gradient of 0.2°C/cm (Rode and Staar, 1961; Rode, 1965). *Pratylenchus penetrans* and *D. dipsaci* accumulated at heat sources, e.g., heating wires or seeds, only 0.28°C greater than the temperature 4 cm away (El-Sherif and Mai, 1969), but preconditioning determined the preferred temperature to which *D. dipsaci* moved (Croll, 1967b).

## III. DISSEMINATION OF STIMULI

### A. Distances Nematodes Move to a Root

The reported distances nematodes moved to reach roots are mostly between 2 and 10 cm (Lavallee and Rohde, 1962; Pitcher, 1967; Viglierchio, 1961; Weischer, 1959), but Luc (1961) found that when millet plants were well established, soil up to 40 cm away was attractive to *Hemicycliophora paradoxa* (see Section III, C, 3). Many of these estimates were made under experimental conditions where the losses and diffusion of the chemicals were restricted. Small changes in the environment greatly alter the rate of loss and thus the effective range of a stimulant. The amount produced by the source, the rate of dispersal, and degradation of the stimulant chemical are dynamic; and since the concentration gradient and range depend on the balance of these factors, changes occur continuously.

## B. Loss of Stimuli (Disappearance by Degradation and Absorption)

Compared with aerial environments, soils are strongly buffered against change. The soil water and mineral matter have high specific heat, and salts buffer electrical stimuli. Clay and organic matter adsorb chemicals, and microbes degrade them; thus, stimuli which are long lasting in sterile or semisterile laboratory conditions may be short lived in soil. Only diffusable stimuli can be transmitted in soil where their evanescence is probably an advantage. Their effectiveness as signals of immediate conditions would diminish if the signals persisted long, because receptors would be fatigued and gradients might extend too far. Since the distance over which the stimuli act is short, roots attract only nematodes near them. Consequently, seedlings are unlikely to be rapidly overwhelmed by great numbers.

Heat is conducted largely by the water and solids of the soil. Conductivity in the soil atmosphere is small, and convection currents occur only in large crevices. The soil loses and gains heat slowly, for the soil moisture prevents rapid temperature change. Temperature gradients around individual sources of heat are short, but the heat produced by organisms during metabolism warms the soil so that gradients may extend from foci of biological activity in soil.

Chemicals occur as vapors or in solution in the soil water and may escape from the soil surface or be washed into deeper layers. The partition of substances between the vapor and aqueous phase depends on their vapor pressure and solubility in water. The few stimulants of which the properties are known are very soluble in water; thus, they would be mainly in solution in the soil water. The water in biologically active soil is likely to be saturated with $CO_2$, so that little can be lost except by diffusion to other regions or from the soil surface. Most organic materials in aqueous solution are probably soon degraded by microorganisms or made inactive by absorption or adsorption in the soil. For this reason, Kühn (1959) argued that gradients could not form in soil, but short stable gradients must extend across the volume in which the losses occur until the material secreted is balanced by that lost. The rate of biological degradation varies because the presence and numbers of an organism using a substrate depend on the prevalence of the substrate. Baunacke (1922) thought that young plants produced more active diffusate than older plants, but the difference might have been caused by an increase in the number of organisms degrading root secretions during the life of the plant.

## C. Dispersal of Stimuli

### 1. Diffusion of Stimuli

Chemicals in soil are probably dispersed mainly by diffusion but may be displaced by water flow. The same physical laws govern the formation of gradients of chemicals and heat; thus, the equation given below for chemical diffusion can be adapted for heat diffusion by changing the constant (Crank, 1956). Nematodes move fastest when soil structural moisture forces are small, in the range of 100–250 cm of water; at these suctions continuous air channels provide adequate oxygen (Wallace, 1963) and could contain vapor gradients. Volatile, water-soluble chemicals can diffuse in vapor or solution, but vapor gradients are longer because diffusion in air is usually about 10,000 times faster than in water. Many stimulants, for example, the *Heterodera* egg hatching factors (Shepherd and Clarke, Chapter 25; Green, unpublished) and millet root factor (Luc, 1961), were nonvolatile, but others, such as the attractant for *Heterodera schachtii* males, were slightly volatile (Greet *et al.*, 1968) so that diffusion occurs in both phases and an equilibrium gradient forms. This equilibrium gradient will be longer than an exclusively aqueous gradient but shorter than an exclusively vapor gradient. The profiles of vapor, aqueous, and equilibrium gradients are similar, although their dimensions differ (Crank, 1956).

In soil, ignoring major cracks, diffusion is probably equal in all directions over the short distances that stimuli seem to be effective. The concentrations of stimulants are usually small, so that the diffusion coefficient $d$ is the same in all parts of the gradient. Diffusion from a near spherical source of radius $a$ gives concentration $c$ at radius $r$ where $r$ is equal to or greater than $a$. At $r$ the changes of concentration with time $dc/dt$ equals $D(2dc/rdr + d^2c/dr^2)$ (Crank, 1956). When the gradient is stable $dc/dt = 0$, the loss of material equals the production at source and at some radius $b$, $c = 0$. When the rate of loss is $A$ then $c = [Ab^2/3D][0.5\ (1 - r^2/b^2) + (1 - b/r)]$. If the rate of loss can be related to the concentration, then $A = Rc$ where $R$ is a constant

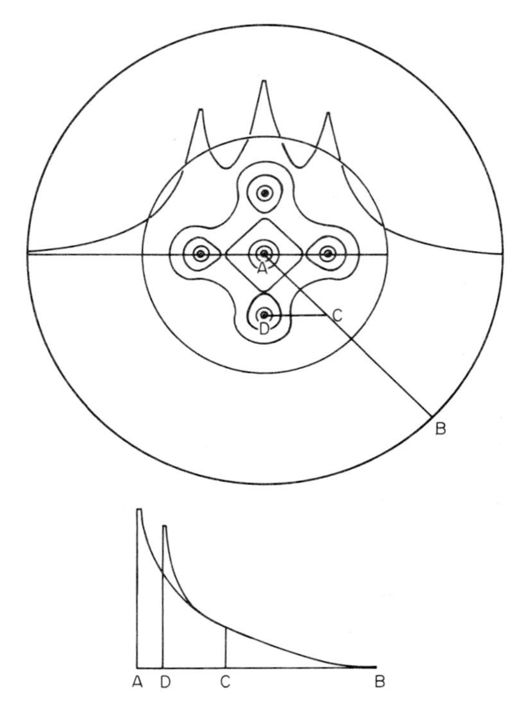


FIG. 1. The expected profiles of gradients formed by diffusion in an agar plate and in soil. In this plate diffusion is two dimensional, as in a cylinder, and in soil it is three dimensional, as in a sphere.

and $c = [(1 + Rb^2)/3D][0.5\ (1 - r^2/b^2) + (1 - b/r)]$ (Currie, private communication).

Thus, in soil gradients (Fig. 1), the concentration decreases rapidly near the source, but when diffusion is restricted to two dimensions, as in shallow layers of water or agar, diffusion is as from a source in the center of a flat cylinder (Crank, 1956). Then $c = [Ab^2/2D][0.5(r^2/b^2 - 1) - \log r/b]$ and the profile curves less (Fig. 1) as do the gradients of $CO_2$ measured by Klingler (1963). In narrow channels or strips, diffusion is unidimensional and the gradient approaches linearity.

## 2. Degeneration of Gradients

When secretion at the source stops, the stimulus would be expected to disappear gradually so that unless new sources appeared, nematodes would continue to be attracted toward the original source until the concentration is uniform or the gradient imperceptible.

### 3. Grouped Sources of Stimuli

Isolated sources giving discrete gradients are rare in nature, because except for seedlings, plants have many growing root tips and the gradients of sex attractants from females in a population must often merge. When some *H. rostochiensis* females stopped secreting attractant, others originally secreting less or situated farther away attracted and captured the males (Green *et al.*, 1970). Merging of gradients redirects diffusion, as does a physical barrier such as the glass plate of an observation chamber. Similarly, a large soil crevice redirects diffusion of a waterborne stimulant. The pattern of gradients around grouped sources was thought to be such that those from individual sources are discernible near and within the group but merge at a distance (Fig. 2). *Heterodera rostochiensis* males approaching a group of females seemed to be influenced and captured first by the steeper gradients near females located at the outer edges of the group although the concentration around the inner ones would probably be greater (Green *et al.*, 1970). The gradients around a group of females or established plant with many roots would

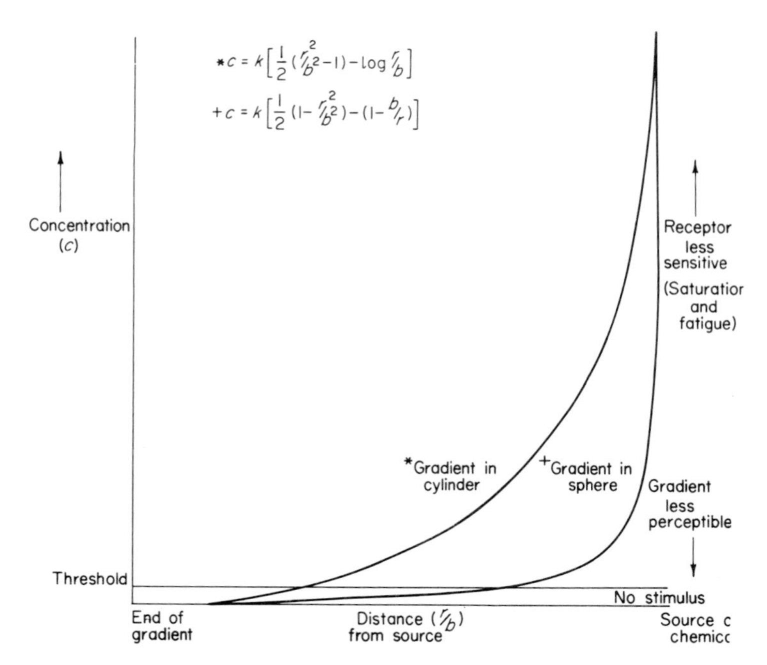


FIG. 2. Expected contours and profiles of gradients around a group of five females on agar. Successive contours represent equal increases in concentration of water soluble attractant.

be expected to extend farther in the soil because collectively more attractant is secreted.

### 4. Movement of Sources

Natural sources of stimulus are rarely stationary. When their rate of movement is of the same order of magnitude or less than the rate of diffusion, as with most root growth, the gradient would be shorter and steeper ahead than behind the source. This gradient pattern would aid such nematodes as *Trichodorus viruliferus* to follow the roots (Pitcher, 1967). Actively moving sources such as vermiform nematodes may move too often and too fast to establish more than transitory gradients over short distances, and these movments may disseminate secretions throughout the space they use so that gradients extend mainly from space into the surrounding soil. In effect the nematodes increase the diameter of the source. The movements of *Heterodera* males congregating on females, and other nematodes congregating on root tips, may disperse attractants and destroy the gradient in their immediate vicinity.

### 5. Soil Water Movement

Except in wet soil, water usually flows toward the absorbing zones of roots; thus, it would probably concentrate the attractant near the roots. Although such an effect would bias the gradients, Wallace (1960) found no measurable effect of moisture gradients in sand on the movements of *H. rostochiensis* larvae or on their attraction to roots. Rain percolates downward as an irregular front because of irregularities in soil structure. Secretions displaced downward might influence nematodes beneath plants, inducing them to move upward.

## IV. RESPONSES TO STIMULI

Stimuli may evoke responses through nerve connections between sensory and motor neurons. Alternatively, they may initiate hormone secretions. Nematodes can respond fully only when their environment and their condition allow it and adaptation in any part of the nervous system will vary the response. Preconditioning can affect the response as when *Ditylenchus dipsaci* preferred a temperature similar to that at which they were kept previously (Croll, 1967b).

Stimuli from organisms in the soil cannot carry directional information; thus, attraction and repulsion can occur only when nematodes respond to gradients. *Heterodera* males responded differently in different parts of a gradient of secretions from their females. Where they began to respond, their movment was unorientated although they slowly approached the female; nearer the female, they moved directly toward her; but closer still, the track became less direct (Green, 1966). If the receptors are similar to those of other animals, the receptor organ begins to respond as the concentration increases above the thresholds of the sensory receptor sites and then the receptor response increases with increasing concentration. The response is not necessarily proportional to the concentration, and adaptation may decrease the response after each increase. Too intense stimuli possibly saturate or fatigue the receptors so that they are insensitive, and nematodes orientate best at specific concentrations and gradients (Fig. 1).

### A. Changes in Rate of Movement, Arousal, and Orthokinesis

Male *Heterodera rostochiensis* became immobile unless external conditions changed frequently or until aroused and kept active by secretions from their females. Thus, in tubes of sand they dispersed widely only when female secretions were present (Green *et al.,* 1968; Evans, 1969). *Pratylenchus scribneri* larvae dispersed less than adults unless corn root exudates were present (Van Gundy, 1965). The hatching factors for *Heterodera* eggs make quiescent larvae move in ways that culminate in the stylet piercing the eggshell (Shepherd and Clarke, Chapter 25). The "arousal" response to a stimulus makes it unlikely that plant parasitic nematodes would aggregate by moving more slowly (orthokinesis) in response to the same stimulus. Although Kühn (1959) suggested that *H. rostochiensis* larvae aggregate orthokinetically, Weischer (1959) found they dispersed farther in sand with, rather than without, root diffusate. Various plant parasitic nematode larvae on roots were active (Blake, 1962; Linford, 1939; Pitcher, 1967; Widdowson *et al.,* 1958); whereas the animal parasite *Trichostrongylus colubriformis,* which probably received no arousal stimulus, became inactive (Wallace and Doncaster, 1964) and may have aggregated orthokinetically.

The rate nematodes move is determined by the rate waves are transmitted along their body, by their wavelength, and amplitude. Apparently the rate of transmission does not change, and it seems likely the propulsive waves are stimulated by an endogenous rhythm. The wavelength and amplitude seem to be determined mechanically; the greater the force generated, the greater the angle the wave makes with the body axis and the shorter the wavelength (Gray and Lissmann, 1964; Wallace and Doncaster, 1964). Stimuli therefore seem to initiate movement but not to control the rate of movement which is determined by other factors. Body waves in *Trichonema* larvae were similarly initiated by light but not controlled by it (Croll, 1966).

## B. Attraction

### 1. Klinokinesis

The simplest mechanism of attraction is where an increase in stimulation causes a decrease in the amount of turning and/or a decrease in stimulation increases the amount of turning; a direct relationship between stimulus and amount of turning causes repulsion. In either case the nematode must adapt to its basic rate of turning when at constant stimulation (Fraenkel and Gunn, 1960). Some tracks made by *Heterodera* males showed probable klinokinetic responses (Green, 1966), and on some time lapse ciné film sequences males aggregated quickly around a female although making no directed movements (Green, unpublished). The klinokinetic response of nematodes to chemical stimuli has not been tested because rapid changes in stimulus cannot be made without disturbing the medium, but *Trichonema* larvae do respond klinokinetically to changes of light intensity (Croll, 1965).

### 2. Orientated Responses (Taxis)

*Ditylenchus dipsaci* swimming in water possibly orientated to an electric potential gradient by adjusting the direction of the rhythmic motion of the head in response to variations in potential (Jones, 1960). When *D. dipsaci* orientated to sources of $CO_2$ while gliding on agar, the head swung widely as it does when swimming, so that the track became indistinct. The wide head movements allowed precise orientation; but as head movement may be independent of the body wave, movement is slow, so the track was deep (Klingler, 1963). *Heterodera* males orientating to females did not swing their head overwide but probed laterally. When orientating less directly, the head probed farther outward, usually only in a direction away from the source, and swung to the other side (Green, 1966).

The enforced alternate directions of the path make it impractical for the direction of movement to be determined by direct comparison of lateral stimuli. Relationship of the degree of turning or the distance before reversal of the direction of turn, during the wave form, to the degree of change of stimulus would probably result in asymmetrical or variable wave forms and is therefore unlikely since the wave form is fundamentally constant (Gray and Lissmann, 1964; Wallace and Doncaster, 1964).

Neither *Chromadorina viridis* nor *Panagrellus redivivus* seemed to perceive light or electrical stimuli intermittently; thus, Croll (1967a) concluded that continuous comparison of stimulation at paired receptors (tropotaxis) was the likely mechanism of orientation. The perpendicular positioning of the amphids to the wave made it unlikely they could effectively sense different stimuli so the comparison must be made between receptors at the head and tail.

### 3. Shock Responses

Shock responses, responses to a sudden change of stimulus as when an organism crosses a boundary between two contrasting environments, occurred when *Heterodera* males moved down gradient. They were characterized by continued head swinging and turning, with little forward movement (Green, 1966). Since the turning was random, the body was not orientated and the response combined orthokinesis and klinokinesis. Often the response began with a short backward movement of the whole nematode.

The *Wirrspuren* (confused tracks) with which *H. rostochiensis* larvae responded in temperature gradients (Rode and Staar, 1961; Rode, 1965) seem typical shock responses, as also do the sudden changes of direction when *Meloidogyne incognita* larvae approached roots (Morikawa, 1962). The behavior of *Ditylenchus dipsaci* when an electric potential gradient reversed was probably a shock response. Larvae stopped moving, squirmed, then swung their head in wide arcs until they faced toward the cathode (Jones, 1960).

The shock response suggests a possible mechanism of attraction, because after shock, movement stops and the head swinging occurs without propagating a propulsive wave in the body. A gradient may be defined as a series of zones of increasing concentration concentric with a point source, and attraction could occur if a shock response stopped forward movement when the nematode moved from one zone to another and perceived a decreased stimulus. Stopping without continued turning could not lead to attraction since it would be necessary for the head to move into a zone of increased concentration before forward movement could resume. The track would show a series of straight or nearly straight sections followed by sudden turns as the response occurred, leading indirectly up gradient as did tracks of *Heterodera* males (Green, 1966). In suitable conditions a response to a smaller stimulus on one side might prevent propagation of a propulsive wave during the lateral swing of the head to that side in the wave form. If this happened, the next head movement to the opposite side would cause an increased stimulus and propagate a propulsive wave. This could explain the characteristic orientation of *D. dipsaci* (Klingler, 1963) and *Heterodera* males (Green,

1966). The head would leave probing marks where it stopped and, when swinging, would smear the track unless it withdrew before turning. Movement would be well directed but slow because of the frequent stops. When moving directly up gradient, shock responses might occur on either side, giving the semblance of lateral probing, but when moving obliquely to a gradient, probing would be more frequently away from the source.

For such a mechanism of attraction to function the nematode must perceive an increase or decrease in stimulation rather than the level of stimulation. This could be achieved if the receptor adapts to the changing stimulus at a rate less than the increase in stimulation resulting from forward movement up the gradient. The receptor would only become fully adapted if the nematode stopped going up gradient or moved down gradient. Since the rate of adaptation would probably remain constant, the sensitivity of response would depend upon the gradient, the speed the nematode moves, and the direction of movement in relation to the direction of steepest ascent.

#### 4. Response in Soil

The soil imposes many restrictions on the behavior of nematodes since they are confined to tortuous passages of varying cross section, direction, and length (Jones *et al.*, 1969). Since diffusion is three dimensional, the stimulants rapidly become less concentrated with distance; the gradients are steep at first and thereafter attenuated; thus, the nematodes would probably respond klinokinetically which is a sensitive, though slow, response, little affected by the devious soil channels. Substances may diffuse through crumbs impenetrable to nematodes so that, even near the source, a direct response up gradient may be impossible. A trial and error shock response would lead the nematodes around the crumbs. Each choice situation would be a three-dimensional cone; thus, any turn from a decreased concentration would be likely to lead to an increase of concentration and restart movement.

## C. Retention at a Source of Stimulus

During movement back and forth along the root, *Meloidogyne* larvae move away, hesitate, and return (Linford, 1939). Similar responses can be seen in the tracks of *Ditylenchus dipsaci* near roots (Blake, 1962), at $CO_2$ sources (Klingler, 1963), and in the movements of male *Heterodera* sp. around females (Green, 1966). Typically, the nematodes wander a short distance away, stop, and turn until redirected toward the source of stimulus, producing a halo of radiating tracks similar to the "confused tracks" from shock responses during orientation. The slow response that allows them to move away before the shock response occurs probably indicates decreased sensory perception because receptors are saturated and fatigued.

The increased tendency for *Panagrellus redivivus* to circle at 16.5°C tends to keep it at this temperature although it moves continuously (Croll, 1969). Other stimuli, which may be tactile, chemical, or possibly electrical, must slow or stop movements and initiate feeding, cell penetration, or copulation (Doncaster, Chapter 19).

## V. DISCUSSION

Different behavioral patterns are needed in different environments and at different stages in life. Restraint of movement, unless stimulated, conserves energy. Quiescence is common in nematodes and probably has survival value. *Heterodera rostochiensis* males are active only when near mature females; *Heterodera* larvae are mostly confined to eggs until roots are near; the young stages of *Pratylenchus scribneri* are aroused by diffusates from roots on which they must feed to develop, whereas the adults, which must mate and would gain little from being inactive, do not respond. *Heterodera schachtii* males do not feed on and are not attracted to roots; their larvae are not attracted by females. So that actions will be properly timed, purposeful, and in the right context, stimuli usually evoke responses only in certain physiological or environmental conditions or in combination with other stimuli. Katznelson and Henderson (1963) suggested that most nematodes were attracted to biologically active regions of the soil. These usually occur around plant roots so that nematodes aggregate where there is food for them. Carbon dioxide and heat are ubiquitous products of metabolism and so act as very general stimuli. Since $CO_2$ is not degraded, the soil water must often be saturated with it, and receptors may become fatigued; but gradients of $CO_2$ occur in the air spaces and may be an attractant effective over long distances. Heat is rapidly adsorbed by the soil and forms shorter gradients, yet may restrain nematodes from wandering into soil where there are few other organisms. Some compounds exuded by roots provide specific stimuli but are easily degraded; thus, their gradients would be short and steep and may aid precise location from nearby. Linford (1939) noted that nematodes were less concentrated on rotting roots than on living roots, and Bird (1960) observed that root tips were more attractive than $CO_2$ sources. The preferred zone on a root might also be selected by a tactile sense responding to surface texture, surface electrical charge, or nonpolar chemicals.

Basic information is lacking on all aspects of the behavior of nematodes and almost nothing is known of the functioning of their sense organs and nervous systems.

## REFERENCES


Baunacke, W. E. (1922). *Arb. Biol. Reichsanst. Land- Forstwirt.* **11**, 185–288.

Bergman, B. H. H., and Van Duuren, A. J. (1959). *Meded. Inst. Ration. Suikerprod.* **29**, 27–52.

Bird, A. F. (1959). *Nematologica* **4**, 322–335.

Bird, A. F. (1960). *Nematologica* **5**, 217.

Bird, A. F. (1962). *Nematologica* **8**, 275–287.

Bird, A. F. (1966). *Nematologica* **12**, 359–361.

Blake, C. D. (1962). *Nematologica* **8**, 177–192.

Caveness, F. E., and Panzer, J. D. (1960). *Proc. Helminthol. Soc. Wash.* **27**, 73–74.

Chen, T., and Rich, A. E. (1963). *Plant Dis. Rep.* **47**, 504–507.

Crank, J. (1956). "The Mathematics of Diffusion," 347 pp. Oxford Univ. Press (Clarendon), London and New York.

Croll, N. A. (1965). *Parasitology* **55**, 579–582.

Croll, N. A. (1966). *Parasitology* **56**, 307–312.

Croll, N. A. (1967a). *Nematologica* **13**, 17–22.

Croll, N. A. (1967b). *Nematologica* **13**, 385–389.

Croll, N. A. (1969). *Nematologica* **15**, 389–394.

Dougherty, E. C., Hansen, E. L., Nicholas, W. L., Mollett, J. A., and Yarwood, E. A. (1959). *Ann. N. Y. Acad. Sci.* **77**, 176–217.

Edmunds, J. E., and Mai, W. F. (1967). *Phytopathology* **57**, 468–471.

El-Sherif, M., and Mai, W. F. (1969). *J. Nematol.* **1**, 43–48.

Esser, R. P. (1963). *Soil Crop Sci. Soc. Fla. Proc.* **23**, 121–138.

Etherton, B., and Higinbotham, N. (1960). *Science* **131**, 409–410.

Evans, K. (1969). *Nematologica* **15**, 433–435.

Fisher, J. M. (1969). *Nematologica* **15**, 22–28.

Fraenkel, G. S., and Gunn, D. L. (1960). "The Orientation of Animals," rev. from 1940 Ed., 376 pp. Oxford Univ. Press and Dover, New York.

Gadd, C. H., and Loos, C. A. (1941). *Ann. Appl. Biol.* **28**, 372–381.

Gray, H., and Lissmann, H. W. (1964). *J. Exp. Biol.* **41**, 135–154.

Green, C. D. (1966). *Ann. Appl. Biol.* **58**, 327–339.

Green, C. D. (1967). *Nematologica* **13**, 172–173.

Green, C. D., and Plumb, S. C. (1970). *Nematologica* **16**, 39–46.

Green, C. D., Greet, D. N., and Evans, K. (1968). *Rothamsted Exp. Sta. Rep.* 1967, pp. 146–147.

Green, C. D., Greet, D. N., and Jones, F. G. W. (1970). *Nematologica* **16**, 309–326.

Greet, D. N. (1964). *Nature* (*London*) **204**, 96–97.

Greet, D. N., Green, C. D., and Poulton, M. E. (1968). *Ann. Appl. Biol.* **61**, 511–519.

Griffin, G. D. (1969). *J. Nematol.* **1**, 9.

Ibrahim, I. K. A., and Hollis, J. P. (1967). *Phytopathology* **57**, 816.

Inglis, W. G. (1964). *Proc. Zool. Soc. London* **143,** 465–502.
Johnson, R. N., and Viglierchio, D. R. (1961). *Proc. Helminthol. Soc. Wash.* **28,** 171–174.
Jones, F. G. W. (1960). *Meded. Landbouwhogesch. Opzoekingssta. Staat Gent* **25,** 1009–1024.
Jones, F. G. W., Larbey, D. W., and Parrott, D. M. (1969). *Soil Biol. Biochem.* **1,** 153–165.
Jones, T. P. (1966). *Nematologica* **12,** 518–522.
Kämpfe, L. (1960). *Nematologica* **5,** 18–26.
Katznelson, H., and Henderson, V. E. (1963). *Nature (London)* **198,** 907–908.
Kerkut, G. A. (1967). *In* "Invertebrate Nervous Systems" (C. A. G. Wiersma, ed.), pp. 1–37. Univ. of Chicago Press, Chicago, Illinois.
Khera, S., and Zuckerman, B. M. (1963). *Nematologica* **9,** 1–6.
Klingler, J. (1961). *Nematologica* **6,** 69–84.
Klingler, J. (1963). *Nematologica* **9,** 185–199.
Kühn, H. (1959). *Nematologica* **4,** 165–171.
Lavallee, W. H., and Rohde, R. A. (1962). *Nematologica* **8,** 252–260.
Lee, D. L. (1965). "The Physiology of Nematodes," 154 pp. Oliver & Boyd, Edinburgh and London.
Lindvig, P. E., Greig, M. E., and Peterson, S. W. (1951). *Arch. Biochem. Biophys.* **30,** 241–250.
Linford, M. B. (1939). *Proc. Helminthol. Soc. Wash.* **6,** 11–18.
Loewenberg, J. R., Sullivan, T., and Schuster, M. L. (1960). *Phytopathology* **50,** 322–323.
Lownsbery, B. F., and Viglierchio, D. R. (1960). *Phytopathology* **50,** 178–179.
Lownsbery, B. F., and Viglierchio, D. R. (1961). *Phytopathology* **51,** 219–221.
Luc, M. (1961). *Nematologica* **6,** 95–106.
Lundegårdh, H. (1942). *Soil Sci.* **54,** 177–189.
Morikawa, O. (1962). *Nippon Oyo Dobutsu Konchu Gakkai-Shi* **6,** 34–38.
Nigon, V. (1952). *C. R. Acad. Sci.* **234,** 2568–2570.
Oteifa, B. A., and Elgindi, D. M. (1961). *Plant Dis. Rep.* **45,** 928–929.
Paramonov, A. A. (1954). *Tr. Gel'mintol. Lab. Akad. Nauk SSSR* **7,** 19–49. (Russian Translating Programme RTS 4255. National Lending Library, Boston Spa, Yorks, England, 1967.)
Peacock, F. C. (1959). *Nematologica* **4,** 43–55.
Peacock, F. C. (1961). *Nematologica* **6,** 85–86.
Pitcher, R. S. (1967). *Nematologica* **13,** 547–557.
Poinar, G. O. (1965). *Proc. Helminthol. Soc. Wash.* **32,** 148–151.
Ramisz, A. (1966). *Acta Parasitol. Pol.* **14,** 91–101.
Rode, H. (1962). *Nematologica* **7,** 74–82.
Rode, H. (1965). *Pedobiologia* **5,** 1–16.
Rode, H., and Staar, G. (1961). *Nematologica* **6,** 266–271.
Roggen, D. R., Raski, D. J., and Jones, N. O. (1966). *Science* **152,** 515–516.
Rohde, R. A. (1960). *Proc. Helminthol. Soc. Wash.* **27,** 121–123.
Santos, M. S. N. de A., (1969). *Rothamsted Exp. Sta. Rep.* 1968, Pt. 1, pp. 155–156.
Scott, B. I. H., and Martin, D. W. (1962). *Aust. J. Biol. Sci.* **15,** 83–100.
Taniguchi, R. (1933). *Proc. Imp. Acad. (Tokyo)* **9,** 432–435.
Van Gundy, S. D. (1965). *Nematologica* **11,** 19–32.
Viglierchio, D. R. (1961). *Phytopathology* **51,** 136–142.
Wallace, H. R. (1958). *Nematologica* **3,** 236–243.

Wallace, H. R. (1960). *Ann. Appl. Biol.* **48,** 107–120.
Wallace, H. R. (1963). "The Biology of Plant Parasitic Nematodes," 280 pp. Arnold, London.
Wallace, H. R. (1966). *Proc. Roy. Soc., London, Ser. B* **164,** 592–614.
Wallace, H. R., and Doncaster, C. C. (1964). *Parasitology* **54,** 313–326.
Weischer, B. (1959). *Nematologica* **4,** 172–186.
Widdowson, E., Doncaster, C. C., and Fenwick, D. W. (1958). *Nematologica* **3,** 308–314.
Wieser, W. (1955). *Proc. Helminthol. Soc. Wash.* **22,** 106–112.
Wieser, W. (1956a). *Proc. Helminthol. Soc. Wash.* **23,** 59–64.
Wieser, W. (1956b). *Science* **123,** 374–375.
Yarwood, E. A., and Hansen, E. L. (1969). *J. Nematol.* **1,** 184–189.
Yeates, G. W. (1969). *Nematologica* **15,** 1–9.
Yuen, P. H. (1967). *Can. J. Zool.* **45,** 1019–1033.
Yuen, P. H. (1968). *Nematologica* **14,** 554–564.

## CHAPTER 25

# Molting and Hatching Stimuli

AUDREY M. SHEPHERD AND A. J. CLARKE

*Rothamsted Experimental Station, Harpenden, Herts., England*

I. Molting . . . . . . . . . . . . . . . . . 267
A. Nematode Growth and the Function of Molting . . . . 267
B. Nematode Cuticle . . . . . . . . . . . . . 268
C. The Molting Process . . . . . . . . . . . . 269
II. Hatching . . . . . . . . . . . . . . . . 271
A. Introduction . . . . . . . . . . . . . . 271
B. Hatching of Some Animal Parasitic Nematodes . . . . . 271
C. Hatching of Plant Parasitic Nematodes . . . . . . . 272
References . . . . . . . . . . . . . . . 284

## I. MOLTING

### A. Nematode Growth and the Function of Molting

From embryo to adult, most nematodes molt four times, but a few species molt three or five times. Once the embryo is fully formed, cells stop proliferating except in the gonads and, in some species, in the intestine. Growth is thus largely from an increase in size of cells and not in their number. In some species cell size increases little between some of the stages but in others, for example, *Ascaris,* the increase is very great. Plant parasitic nematodes increase in length 3–10 times between first-stage larva and adult, *Ascaris* 400 times. It might be supposed that molting of the protective and only partially extensible cuticle would be necessary for size to increase, but Blake (1962) pointed out that *Ditylenchus dipsaci* larvae increase in size between molts. This has also been observed in *Aphelenchus avenae* (Fisher, 1970) and in several animal parasitic nematodes (Lee, 1965). In contrast to normal second-stage larvae, second-stage individuals of *Caenorhabditis briggsae* destined to become dauer larvae continue to grow without molting until they reach the size of a third-stage larva when they molt but retain the second-stage cuticle as a sheath (Yarwood and Hansen, 1969). Second-stage *Meloidogyne javanica* larvae become smaller during the first few days after entering the host root but then increase in size until the second molt occurs at about the fourteenth day. The second, third, and fourth molts all occur in quick succession without the larvae feeding; by about the nineteenth day the fourth molt is completed and the adult begins to feed. There is no growth during the molting phase but the adult grows rapidly, increasing its cross-sectional area up to four times in 7 days (Bird, 1959). *Ascaris* adults also grow enormously after the final molt, increasing their length about 20-fold (Watson, 1965). Stretching the existing cuticle could hardly account for increases such as these, especially since the cuticle becomes thicker not thinner as the nematode increases in size; thus, it must be assumed that new cuticle is laid down between as well as during molting. This led Crofton (1966) to suggest that the main function of molting in nematodes is not, as it is in arthropods, to accommodate new growth and expansion but to allow more complex structural changes and the laying down or modification of specialized organs such as the mouth stylet of plant parasitic forms.

### B. Nematode Cuticle

It is now generally accepted that the cuticle is secreted by the hypodermis and is not itself a living structure; it lacks the organelles associated with living tissues. But enzymes have been detected in various cuticular layers of *Ascaris* (Fairbairn, 1960a; Lee, 1961, 1962), *Nippostrongylus* (Lee, 1965), and *Meloidogyne* (Bird, 1958); these authors concluded that the nematode cuticle is a "metabolically active structure." However, Kan and Davey (1968) found no evidence of enzymes in the adult cuticle of *Phocanema decipiens*. Ultrastructural studies show that the cuticle is a complex, multilayered structure, which differs between species and often also between stages of the same species, especially in parasitic forms where changes in habitat occur between stages. The major constituent of cuticle is a collagenous-type protein (Bird, 1956, 1957; Clarke, 1968); but keratin, carbohydrates, lipid, and polyphenol-tanned proteins have also been found. The cuticle extends over the whole external surface of the body and, in a modified form, lines the anterior of the digestive tract as far as the esophago-intestinal junction, and the anal, genital, amphidial, phasmidial, and excretory openings. All cuticular material is sloughed at ecdysis.

## C. The Molting Process

### 1. Sequence of Events

The molting process includes the deposition of new cuticular material, the separation of the old cuticle from the new, and, finally, the shedding of the old cuticle (Rogers and Sommerville, 1957; Lee, 1965; Davey and Kan, 1967). In less specialized forms, the three phases follow each other rapidly. In some animal parasitic nematodes especially, but also in some plant parasitic and microbivorous forms, there is one stage in the life history during which the shedding of the old cuticle is delayed, or molting is not initiated until a specific stimulus received from the environment indicates that conditions are satisfactory.

What causes the nematode to begin to secrete a new cuticle is unknown; but Rogers and Sommerville (1957, 1963) suggested that molting is started by the nematode itself with secretions, possibly endocrine, that initiate the formation of the new cuticle. Early in molting there is usually a period of relative immobility called *lethargus,* during which the nematode appears darker, probably because much new protein is being produced by the hypodermis.

### 2. Stimulation of Exsheathing in Some Animal Parasitic Nematodes

That some species need an external stimulus to molt was first recognized and studied in animal parasites. For example, in the trichostrongyles, which are parasites of the alimentary tract, the final stage in the molting sequence, exsheathing, is delayed in the preparasitic third stage, and a stimulus is required from the host to trigger the production by the nematode itself of a fluid that causes exsheathment (Rogers and Sommerville, 1957, 1960, 1963; Sommerville, 1957; Silverman and Podger, 1964). The main stimulus from the host environment seems to be undissociated carbonic acid or dissolved carbon dioxide; but temperature, Eh, pH, and the presence of salts are also important. When all these factors are optimal, the stimulus acts after a short exposure; that is, the stimulus is the trigger and does not itself dissolve the cuticle. Larvae, once stimulated, release an exsheathing fluid into the space between the new and the old cuticle or sheath; this fluid digests, from inside, an area of the sheath near the excretory pore, ultimately releasing the larva. If exsheathing fluid is applied to unstimulated larvae, they too exsheath. Rogers and Sommerville (1957) found that the exsheathing fluid was heat labile and contained a dialyzable cofactor. Ligaturing different regions of the body indicated that exsheathing fluid was secreted by cells at the base of the esophagus. It is not the production of exsheathing fluid which is triggered by the stimulus but its release, possibly via the excretory system and through the excretory pore (Rogers and Sommerville, 1963). Apparently the component of the fluid that digests the old cuticle is leucine aminopeptidase (Rogers, 1965; Davey and Kan, 1967, 1968). In *Ascaris lumbricoides* and *Phocanema decipiens,* neurosecretory cells in the dorsal and ventral ganglia may be associated with the release of this enzyme (Davey, 1966; Davey and Kan, 1967, 1968), for their activity coincides with the disappearance of enzyme from the excretory gland and its appearance between the new and the old cuticle. The receptors for the stimulus are unknown but may be the amphids (Rogers, 1962; Rogers and Sommerville, 1963; Silverman and Podger, 1964).

### 3. Stimulation of Molting in Some Plant Parasitic Nematodes

Among the animal parasitic nematodes, the phase at which the molting cycle is blocked in the absence of a stimulus from the host is exsheathment. The earlier phases of protein secretion and separation of the cuticles occur normally, and the neurosecretory activity apparently associated with exsheathing is not concerned in the protein synthesis that produces the new cuticle (Davey and Kan, 1967).

Among the plant parasitic nematodes, an external stimulus for molting has been reported only in *Paratylenchus,* in which the block occurs much earlier in the cycle, for new cuticle is not laid down unless the stimulus is received. Rhoades and Linford (1959) showed that the nonfeeding, preadult stage of *Paratylenchus projectus* and *P. dianthus* (syn. *P. curvitatus*) can survive for long periods in moist soil without any host plants. In water, few of these larvae molted, but when placed in root diffusates of some host and nonhost plants, most larvae molted. Not all host plants stimulated molting. For example, *Trifolium pratense,* a host for both species, stimulated molting of *P. projectus* but not of *P. dianthus.* Rhoades and Linford also found that, in agar culture with host plants, fourth-stage larvae molted readily at first, but as the culture aged, fourth-stage larvae constituted an increasingly greater proportion of the total population. This suggests that, as with *Heterodera* hatching factors, young, actively growing roots produce more "active factor" than older ones.

As with the exsheathing stimulus in trichostrongyles, the molting stimulus for *P. nanus* in root diffusates seems to be a trigger mechanism (Fisher, 1966). Even when only briefly exposed and then transferred to water, larvae molted 1–2 weeks later. The stimulus is received or responded to in the anterior half of the nematode because, when ligatured, only this half of the larva molted. Boiling the exudate for 10 min removed the molting stimulus, but it was unaffected over a pH range from 4 to 8.

4. OTHER ASPECTS OF MOLTING

Wallace (1963) suggested that body size or shape may play a part in initiating molting. In *Rotylenchus buxophilus,* for example, the larval stages have distinct size ranges that do not overlap, thus, apparently each molt occurs when the larva has reached a certain size (Golden, 1956).

Fisher (1970) found that after hatching, larvae of *Aphelenchus avenae,* which feed on fungi and plant roots, molted three times at regular intervals when they had food. Each molt was initiated shortly after the preceding one was completed. A minimum amount of food had to be ingested by the larva if molting was to proceed, but if a larva took a large enough meal it could molt twice without feeding between molts.

## II. HATCHING

### A. Introduction

Just as molting usually proceeds without delay as soon as a larva reaches a certain stage of development, so too most species usually hatch as soon as the larva is fully embryonated or after it has molted once in the egg. In some environments, hatching, as molting, is interrupted and the life cycle of the nematode is fitted to that of the host, or eggs do not hatch during periods of adverse conditions in the host's environment.

### B. Hatching of Some Animal Parasitic Nematodes

The same factors that influence molting of trichostrongyles also influence the hatching of ascarid eggs. These are all components of the environment within the region of the alimentary tract where these nematodes are found. Thus $CO_2$ concentration is again important, as is the presence of reducing agents and certain salts, together with pH and temperature. The concentration of $CO_2$, either as undissociated carbonic acid or as gaseous $CO_2$, is critical; too large a concentration inhibits hatch (Lee, 1965).

Again, as with the molting stimulus, the hatching stimulus triggers the hatching process causing the larva inside the egg to secrete chitinase, lipase, and protease (Rogers, 1960). These enzymes, after passing through the modified innermost layer of the eggshell, dissolve an area of the outer layers through which the larva escapes. The permeability of the inner layer changes after stimulation. More trehalose, a component of the fluid bathing the larva inside the egg, escapes through the shells of stimulated eggs than from unstimulated ones (Fairbairn, 1960b).

## C. Hatching of Plant Parasitic Nematodes

### 1. Introduction

Shepherd (1962a) reviewed hatching in *Heterodera* spp. Little is known about hatching of other species. Baunacke (1922) first realized that *Heterodera* eggs are stimulated to hatch by root exudates from host plants. Hatching is also affected by several environmental factors, including temperature, soil moisture, soil aeration, pH, and organic and inorganic chemicals in the soil water. The chemicals may be derived from the soil itself or from plant organs or soil-inhabiting microorganisms. In addition, there are sometimes seasonal effects that are presumably a combination of some of the factors just mentioned or are related to diapause in the unhatched larvae.

### 2. Physical Factors Affecting Hatching

*a. Temperature.* Eggs of most nematodes will hatch over a wide range of temperatures. Table I shows the recorded *in vitro* optima. Some *Heterodera* spp. hatch best in fluctuating temperatures, simulating the

TABLE I
Optimum Temperature for Hatching of Some Plant Parasitic Nematodes

| Species | Optimum temp. (°C) | References |
|---|---|---|
| *Aphelenchus avenae* | 36 | Taylor (1962) |
| *Hemicycliophora arenaria* | 33 | McElroy and Van Gundy (1966) |
| *Heterodera avenae* | 10–18 | Winslow (1955), Fushtey and Johnson (1966) |
| *H. glycines* | 24 | Slack and Hamblen (1959) |
| *H. rostochiensis* | 20–25 | Lownsbery (1950), Fenwick (1951a), Oostenbrink (1967) |
| *H. schachtii* | 20–25 | Baunacke (1922), Wallace (1955a), Oostenbrink (1967) |
| *H. trifolii* | 17 | Oostenbrink (1967) |
| *Meloidogyne hapla* | 21–25 | Wuest and Bloom (1965), Bird and Wallace (1966) |
| *M. javanica* | 30 | Bird and Wallace (1966), Wallace (1966a) |
| *Tylenchulus semipenetrans* | 30 | O'Bannon (1968) |

diurnal fluctuations in field soils (Bishop, 1955; Wallace, 1955a; Juhl, 1968), but *Meloidogyne javanica* is unaffected by such fluctuations (Wallace, 1966a). Other effects of temperature are discussed in Section II, C, 2d.

*b. Soil Moisture.* Eggs of *Heterodera* and *Meloidogyne* survive a wide range of soil moistures, but in both genera the optimum moisture content for hatching corresponds to the point of inflection of the moisture characteristic of the particular soil type (Wallace, 1954, 1955b, 1966a,b, 1968b). The "field capacity" of the soil is a rough measure of this moisture content. At this suction pressure, the soil pores are almost emptied of water, except where particles are in contact, and the pore spaces are filled with air. At smaller and greater suctions, when the soil is saturated or very dry, there is little or no hatching. In *M. javanica,* inhibition of hatching at large suctions is reversible and probably has survival value (Baxter and Blake, 1969a).

*c. Soil Aeration.* All plant parasitic nematodes require oxygen for eggs to hatch. The oxygen tension in the soil depends on soil structure, moisture and depth, and also on the soil microorganisms present. The presence of small amounts of clay in soil greatly lessens the porosity, and consequently the volume of air, and decreases the hatch of *Heterodera* eggs; the greater the oxygen consumption of the soil, the smaller the hatch (Wallace, 1956a). The hatch of *Meloidogyne javanica* also increased with an increase in oxygen concentration over a range of 0.2–20% (Wallace, 1968a; Baxter and Blake, 1969b).

*d. Seasonal Effects.* The eggs of some species of *Heterodera* will not hatch *in vitro* at certain times of the year even when given all the usual requirements for hatching. This has been likened by some authors to diapause in insects (Oostenbrink, 1967; Shepherd and Cox, 1967) and is evident in *H. rostochiensis, H. cruciferae, H. avenae,* and *H. glycines* (Ross, 1962; Shepherd, 1962a; Oostenbrink, 1967; Shepherd and Cox, 1967).

The factors that initiate diapause in *Heterodera* spp. and the physiological processes involved are unknown, but fewer eggs of *H. rostochiensis* go into diapause when cysts are not subjected to the soil conditions of an European autumn but stored in the laboratory at 25°C (Cunningham, 1960; Shepherd and Cox, 1967). Diapause is broken a little sooner in *H. rostochiensis* and *H. cruciferae* when cysts are exposed to 30°C for 6 weeks (Shepherd and Cox, 1967), whereas *H. avenae* will hatch well only after several weeks at 5°–7°C (Cotten, 1962; Fushtey and Johnson, 1966).

*e. Hydrogen Ion Concentration.* The effect of pH has been examined

(Ellenby, 1946; Fenwick, 1951a; Robinson and Neal, 1953; Loewenberg *et al.,* 1960; Wallace, 1966a; Lehman, 1969), but results are inconclusive and often contradictory. This is partly because buffer solutions can themselves affect the eggs of some species and alter the chemistry of natural and artificial hatching agents. *In vitro, Heterodera* spp. hatch well from pH 3.0 to 8.0 and *Meloidogyne* spp. from pH 5.0 to 8.0.

### 3. Chemical Stimulation

*a. Effect of Exudates from Plants.* As far as is known, *Heterodera* is the only genus in which hatching is markedly influenced by plant root exudates. Among the other genera within the Heteroderidae, more eggs of *Hypsoperine ottersoni* hatched in root exudate of a host (canary grass) than in water, whereas the nonhost tomato did not stimulate hatching (Webber and Barker, 1968). The evidence for stimulation of hatching in *Meloidogyne* is rather scanty and often conflicting (Loewenberg *et al.,* 1960; Viglierchio and Lownsbery, 1960; Jones and Nirula, 1963; Ahmed and Khan, 1964a,b; Chidambaranathan and Rangaswami, 1964; Swarup and Pillai, 1964; Hamlen and Bloom, 1968). Where increased hatching was observed *in vitro,* it was small and barely significant. Wallace (1966a) found that significantly fewer eggs of *M. javanica* hatched in nonsterile soil than in sterile soil. Growing tomato seedlings in the nonsterile soil counteracted this inhibition of hatching, probably caused by microorganisms, and slightly (but not significantly) more larvae hatched than in sterile soil; adding tomato root exudate to the nonsterile soil did not prevent inhibition however.

*In vitro,* some larvae of most species of *Heterodera* hatch in water, but usually many more hatch in root exudates of host plants. However, this is a generalization and the proportion of eggs that hatch in water differs greatly between species and is often clearly related to other aspects of host–parasite relationships. Thus, in *H. rostochiensis,* which has a limited host range, usually only about 2% of eggs hatch in water. Theoretically at least, the presence of a source of food in the vicinity is assured in this way before most larvae emerge from the relative security of the egg and cyst. In *H. schachtii,* which has a wide host range, including many weeds, this degree of protection is less essential and the water hatch may exceed 30%. Just how effective this mechanism is in field soils is less certain. The hatch of *H. rostochiensis* during spring in the absence of a host crop is much greater than the 2% *in vitro* water hatch and is usually nearer 40%. Reports that soil water sometimes stimulates hatching of *H. rostochiensis* to some extent have attributed the effect to microorganisms (Ellenby, 1963; Ellenby and Smith, 1967; Giebel, 1963).

Hatching is usually stimulated by exudates from host plants, but some host plants produce inactive exudates and some nonhost plants produce active ones (see Shepherd, 1962a). Some species, for example, *H. avenae* and *H. glycines,* are not stimulated to hatch *in vitro* by any root exudates but hatch well in water. No hatching stimulant has been found, or any set of conditions, that will cause *H. goettingiana* to hatch *in vitro*. In the field, however, it behaves similarly to *H. rostochiensis.* Some eggs hatch under fallow or a nonhost crop, and many more hatch under a host crop (Shepherd, 1963). This emphasizes that results *in vitro* may not always reflect what happens in the field.

The most active root exudates are produced by plants during the period of maximum root growth (Winslow, 1955; Widdowson, 1958a,b). This is usually between 3 and 6 weeks from the beginning of root production (cf. Section I, C, 3). Aerial parts of plants may also produce substances that stimulate hatching although these are usually less potent than exudates from roots (Janzen and van der Tuin, 1956; Winslow and Ludwig, 1957; Golden, 1958).

Plotting cumulative hatch of larvae against log time gives a sigmoid curve (Fenwick, 1951b; Jones and Winslow, 1953) (Fig. 1). The curve for the number of larvae hatched plotted against the dilution of root exudate is hump-shaped (Fig. 2). There is an optimum concentration above and below which hatching is depressed; thus, two dilutions of diffusate can hatch the same number of larvae (Winslow and Ludwig, 1957; Hague, 1958). Therefore it is essential when assaying "natural" or

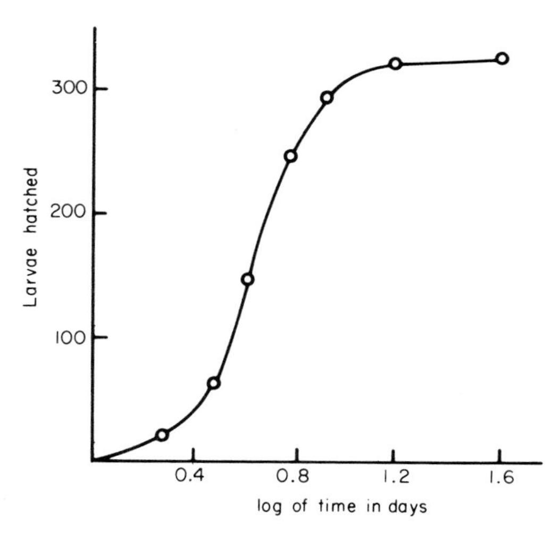


FIG. 1. Cumulative hatch of *Heterodera rostochiensis* larvae plotted against log time (from Fenwick, 1951b).

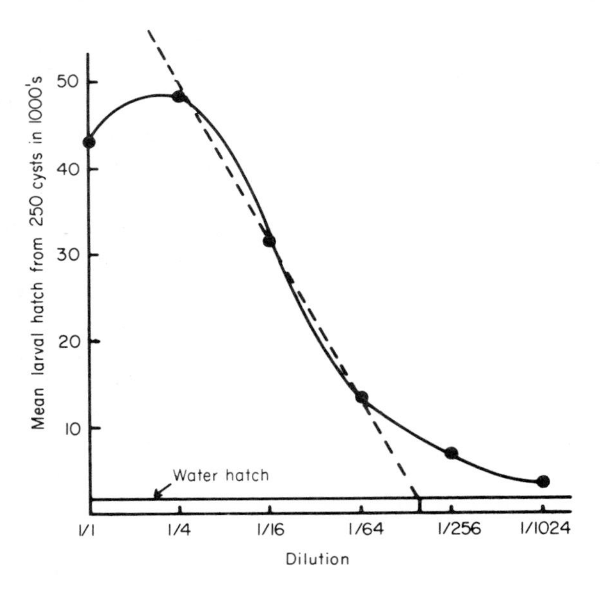


FIG. 2. Typical hatch–dilution curve for *Heterodera rostochiensis,* with the best straight line indicating the approximate threshold dilution. (After Shepherd, 1962.)

"artificial" compounds for hatching activity to test a series of dilutions. As dilution is increased from the optimum, the curve is approximately a straight line, which, when continued downward to where it intercepts the line for the hatch in water, gives the best estimate of the dilution at which activity reaches zero.

*b. Inhibition of Hatching.* The hatching of *Heterodera* spp. can be inhibited by too little oxygen, too much carbon dioxide, and solutions with high osmotic pressures, e.g., concentrations of $0.2\,M$ or more (Wallace, 1963). Dilute solutions of chemicals (e.g., 1–10 m$M$) may also inhibit hatching. It is difficult to distinguish inhibition from toxicity, and often the distinction is not made. The degree of inhibition is easier to assess when the nematode species hatches well in water. Clarke and Shepherd (1964) listed organic and inorganic compounds that inhibited the hatching of *H. schachtii.* Inhibition was expressed as

$$\%\ \text{inhibition} = 100 - (100\,H_s/H_w)$$

where $H_w$ is the hatch in water and $H_s$ the hatch in the test substance. Several simple compounds inhibited hatching almost completely, e.g., oxalic acid and *p*-cresol. In hatching tests on seven *Heterodera* spp. many inorganic ions were inhibitory (Clarke and Shepherd, 1966b). In tests on *H. glycines,* Lehman (1969) found that the cations $Mg^{2+}$

and $NH_4^+$ were more inhibitory than $Ca^{2+}$, and among anions $NO_3^-$ was more inhibitory than $Cl^-$ and $SO_4^{2-}$.

Organic materials in the soil may also inhibit hatching. Sembdner and Schreiber (1960) found that when some plant root diffusates were mixed with potato root diffusate, hatching of *H. rostochiensis* was substantially inhibited. Root diffusate of *Tagetes* spp., however, did not inhibit hatching of *H. rostochiensis* (Hesling *et al.,* 1961; Omidvar, 1961; Wilski, 1963) despite the naturally occurring nematicide in the roots. James (1966, 1968) showed that the fungus causing brown root rot of tomatoes produced substances that inhibited the hatching of *H. rostochiensis.* Cellulose soil amendments are reported to inhibit the emergence of larvae of *H. tabacum* from cysts (Miller *et al.,* 1968).

Naturally produced organic and inorganic substances in the soil may stimulate or inhibit hatching, and the number of larvae released will depend on the balance between them. Inhibitors may be more important than hatching agents for nematodes that hatch freely in water, e.g., *Meloidogyne* spp. and *H. avenae* [see also Section II, C, 3a, Wallace (1966a)].

Besides inhibitors from sources outside *Heterodera* cysts or *Meloidogyne* egg masses, there is evidence that inhibitors may be liberated inside the cysts or within the egg masses as hatching proceeds (see Shepherd, 1962a; Ishibashi, 1967). Kaul (1962) suggested that the phenolic pigments of *H. rostochiensis* cyst walls may be hatch inhibitors.

*c. The Isolation and Identification of Natural Hatching Factors.* Although attempts to identify the hatching factors for *Heterodera* spp. began in the 1930's, the nature and structure of these compounds is still unknown. The very small concentration of the factors in root diffusates, the difficulties of purifying the crude starting materials, and the time-consuming bioassay methods have greatly impeded progress. An attempt to replace the bioassay for the *H. rostochiensis* hatching factor by a colorimetric method based on the reaction with alkaline picric acid or 3,5-dinitrobenzoic acid (Ellenby, 1958) was unsuccessful because color formation was not specific (Schreiber *et al.,* 1961).

The hatching factors for the beet, cabbage, and hop cyst nematodes have been less well studied than the factor(s) for *H. rostochiensis* (see Shepherd, 1962a). Potato root diffusate was the usual source of raw material, but diffusates from tomato (Calam *et al.,* 1949a) and *Solanum nigrum* (Russell *et al.,* 1949) and the ethanolic extracts of potato plants (Janzen and van der Tuin, 1956) have also been used. The hatching factor in diffusates or extracts was concentrated by adsorption on charcoal or ion exchange resins and evaporation *in vacuo.* Calam *et al.*

(1949b) and Marrian *et al.* (1949) used one procedure to purify the factor and Clarke and Widdowson (1966) another. The products isolated were resins or gums, but none was demonstrably pure. Both products hatched many eggs, even in very dilute solutions. Clarke and Widdowson (1966) obtained a yield of 18 $\mu$g/liter potato root diffusate which hatched larvae at a dilution of 0.05 $\mu$g/ml (i.e., 5 parts in $10^8$). The resin obtained by Marrian *et al.* (1949) was "fully active at dilutions of 1 in $10^7$." A crystalline material isolated from tomato root diffusate (Hartwell *et al.*, 1960) was much less effective.

The hatching factor is dialyzable, stable at pH 1–9 at room temperature, but rapidly inactivated at higher pH values (Calam *et al.*, 1949a). Marrian *et al.* (1949) reported that the factor was acidic (equivalent weight 250–290) and contained a lactone group and one or two hydroxyl groups. A possible formula of $C_{18}H_{24}O_8$ was suggested. However, until the factor is isolated in a pure state, conclusions based on the analytical or degradative data must be accepted with reserve. New procedures such as thin layer chromatography and electrophoresis are more effective in purifying the hatching factor than methods previously used. The introduction of these procedures and new methods of quantitative assay should remove the last obstacles to the isolation of the hatching factor and the determination of its structure.

*d. Artificial Hatching Agents.* Artificial hatching agents for *Heterodera* spp. were sought soon after it was found that exudates from roots stimulated hatching (see Shepherd, 1962a), in the hope of using them to diminish the numbers of cyst nematode larvae in soil. Rademacher (1930) and Rademacher and Schmidt (1933) compared hatching of *H. schachtii* by more than 300 compounds, and many more have since been tested on this and other *Heterodera* species. Compounds may be toxic, inhibit hatching, be inactive, or hatch different numbers of eggs. The response may also depend on the concentration. Larvae seem to emerge and survive in many synthetic and inorganic compounds at concentrations of about 1–10 m*M*. Some of these compounds hatch species *in vitro* for which no active root diffusate is known. For example, eggs of *H. glycines* will not hatch *in vitro* in any host or nonhost root diffusate yet tested, but they hatch well in 3 m*M* zinc chloride and in 3 m*M* flavianic acid (Clarke and Shepherd, 1966b, 1967). Clarke and Shepherd (1966b, 1968) compared hatching agents in terms of a hatch rating

$$100(H_s - H_w/H_d - H_w)$$

where $H_s$ is the hatch in the test substance at optimum concentration, $H_w$ the hatch in water, and $H_d$ the hatch in root diffusate, all determined simultaneously. At optimum concentrations the most effective hatching agents hatch at least as many eggs per cyst as the appropriate root diffusate, sometimes many more, but the natural hatching factors act at smaller concentrations. For example, as many eggs of *H. rostochiensis* hatched in 0.3 m*M* (i.e., 79 μg/ml) picrolonic acid (hatch rating 102) as in potato root diffusate, but the same number hatched in an aqueous solution containing only 8 μg/ml of partially purified hatching factor from potato root diffusate.

Because of the selective response of *Heterodera* spp. to both natural and artificial hatching agents, these can sometimes be used to distinguish between species. Solutions of artificial hatching agents are of known composition and less variable than root diffusates, and they are useful tools for studying the mechanism of hatching. Their use, either alone or with a nematicide, has been suggested for controlling field populations of *Heterodera*, but their effectiveness in soil has yet to be demonstrated.

Many synthetic compounds hatch *H. schachtii* well and some of them hatch other *Heterodera* spp. (Shepherd, 1962b; Clarke and Shepherd, 1964, 1966a,b,c, 1967, 1968; Viglierchio and Yu, 1966). These substances are effective at millimolar concentrations, and some are quite specific. Clarke and Shepherd tested the ability of over 300 organic compounds to hatch *H. schachtii* and *H. rostochiensis*. Active compounds came from widely different chemical groups, e.g., acids, dyes, and heterocyclic compounds. Many that hatched *H. schachtii* did not hatch *H. rostochiensis*. Table II shows the hatch ratings for both species of some compounds that hatch *H. schachtii* very well. The only organic compounds that hatched *H. rostochiensis* as well as potato root diffusate were anhydro-

TABLE II
THE HATCH RATING FOR *H. rostochiensis* OF SOME COMPOUNDS THAT WERE VERY ACTIVE HATCHING AGENTS FOR *H. schachtii*[a,b]

| Compound | m*M* concn. | *H. schachtii* | *H. rostochiensis* |
|---|---|---|---|
| Nicotinic acid | 10 | 318 | 4 |
| *o*-Bromophenolindophenol | 3 | 298 | 0 |
| *o*-Chlorophenolindo-2,6-dichlorophenol | 2 | 273 | 1 |
| Tolylene blue | 3 | 273 | 47 |
| *o*-Aminobenzoic acid | 10 | 250 | 2 |
| *m*-Chlorophenolindophenol | 1 | 243 | (−)[c] |
| Picric acid | 2 | 238 | 4 |
| Ethylenethiuram monosulfide | 0.4 | 211 | 1 |

[a] *N.B.*: Hatch rating of 100 ≡ hatch in root diffusate.
[b] Modified from Clarke and Shepherd (1968).
[c] Hatch less than in water.

tetronic acid (Calam *et al.*, 1949b) (hatch rating 90) and picrolonic acid (Clarke and Shepherd, 1966c, 1968) (hatch rating 102). Some simple analogs of these compounds hatched *H. schachtii* but not *H. rostochiensis*.

Suitably placed polarizable atoms seem to be an essential feature in the structure of good hatching agents. A distance of about 4 Å between polarizable atoms was associated with those for *H. schachtii* and about 6.7 Å with those for *H. rostochiensis* (Clarke and Shepherd, 1968). Compounds that hatched both species contained pairs of polarizable atoms at approximately both these distances apart.

Flavianic acid hatched *H. cruciferae, H. glycines, H. schachtii,* and *H. trifolii* well but hatched only few eggs of *H. goettingiana* and hardly any of *H. carotae* and *H. rostochiensis* (Clarke and Shepherd, 1967). Janzen (1968) found that more *H. rostochiensis* larvae emerged from dried cysts placed in flavianic acid solutions than when cysts were first soaked in water, as is customary in hatching tests with this species.

Dithiocarbamates and some other organic compounds widely used as fungicides, insecticides, and soil sterilants, or their breakdown products, stimulate eggs of some species to hatch. Solutions of sodium ethylene-bis-dithiocarbamate (nabam) hatched eggs of *H. schachtii* (Steele, 1961, 1963; Clarke and Shepherd, 1966a), *H. tabacum,* and *Meloidogyne* spp. (Miller and Stoddard, 1958; Miller, 1967). Miller (1967) found that several organic fungicides besides dithiocarbamates stimulated the emergence of *H. tabacum* larvae and commented:

> Fungicides that stimulate emergence of nematode larvae might well be injurious if not combined with appropriate nematocides when used on living plants. For example, nabam soil drenches applied around tomato and stephanotis vines growing in soil infested with *M. hapla* increased galling of roots so much that the plants died, whereas plants in infested soil not treated with nabam were merely stunted.

Clarke and Shepherd (1966a) showed that nabam solutions break down to give several compounds, some inhibiting and others stimulating hatching of *H. schachtii.* The compound mainly responsible for the hatch-stimulating property of nabam solutions is ethylenethiuram monosulfide (hatch rating 211). Nabam solutions containing manganese sulfate were inactivated because the manganese ion catalyzed decomposition to the hatch-inhibiting carbon disulfide.

Sodium *N*-methyl dithiocarbamate (metham-sodium, Vapam), which readily decomposed in aqueous solution, killed the eggs of *H. schachtii,* preventing hatching (Clarke and Shepherd, 1966a). Methyl isothiocyanate, the main decomposition product, killed *H. schachtii* eggs at concentrations above 0.1 m*M*. Miller and Lukens (1966) found that Vapam stimulated the hatching of *H. tabacum* but killed the larvae which emerged, and they attributed both effects to methyl isothiocyanate.

Rademacher (1930) found that some inorganic compounds hatched *H. schachtii*, and Wallace (1956b) compared the effect of some ions on this species. He reported that $Ca^{2+}$, $Mg^{2+}$, $K^{+}$, and $Na^{+}$ all caused more larvae to hatch than water. Clarke and Shepherd (1966b) tested the effect of many different ions on the hatching of nine *Heterodera* species and confirmed that $Ca^{2+}$ and $Mg^{2+}$ hatched *H. schachtii* a little but found that $K^{+}$ and $Na^{+}$ did not; several other ions were as active as sugar beet root diffusate. The hatching activity of a particular cation sometimes varied with the anion present, but the differences were not consistent for all *Heterodera* spp. Some inorganic ions inhibited and some promoted hatching according to the species of *Heterodera*, the ion, and its concentration. Some anions and some cations stimulated hatching. The ions most abundant in the soil were not the most active.

As with the organic compounds, some inorganic ions hatched several *Heterodera* species and some only few. For example, $Zn^{2+}$ hatched many eggs of eight out of the nine species; others, such as $Sr^{2+}$ and $Li^{+}$, only two. Closely related ions sometimes had very different effects; $Zn^{2+}$ hatched *H. glycines* well but $Cd^{2+}$ inhibited hatching of this species. Only vanadate ions hatched *H. rostochiensis* as effectively as potato root diffusate. Except for $Cu^{2+}$, orthovanadate, and $[UO_2]^{2+}$, the ions moderately or very active with any one of the other species were also active with *H. schachtii;* but many that hatched *H. schachtii* did not hatch other species.

The hatching properties of mixtures of ions is less well understood. Lehman (1969) tested the response of *H. glycines* to pH and to combinations of the cations $Mg^{2+}$, $NH_4^+$ and $Ca^{2+}$ with the anions $NO_3^-$, $SO_4^{2-}$, and $Cl^-$ in 5 m$M$ phthalate buffer at pH 5.6 and also $CaCl_2$ and $NH_4NO_3$ over a range of pH. The influence of pH was secondary to that of the inorganic ion present at pH values greater than 4; $CaSO_4$ stimulated hatching more than the buffer solution or water. Loewenberg *et al.* (1960) found that suitable combinations of inorganic ions increased the emergence and favored the survival of *Meloidogyne incognita* larvae.

Rademacher (1930) reported that calcium hypochlorite hatched *H. schachtii* eggs, and it was recommended for controlling field populations (see Shepherd, 1962a). Strong solutions kill larvae and also dissolve cyst walls, eggshells, and cuticles, but Smedley (1936) found that low concentrations caused *H. rostochiensis* larvae to emerge. Shepherd (1963) used calcium hypochlorite to hatch eggs of *H. goettingiana;* Fox and Kerekes (1969) used sodium hypochlorite to hatch eggs of *H. carotae,*

*H. glycines, H. weissi,* and Osborne's cyst nematode. Hatching appeared to be passive, the eggshell dissolving in the reagent.

4. Hatching Behavior of Nematodes

Observing the behavior of nematodes before and during hatching by cine and time-lapse photographic techniques has provided information on the physiology of the hatching process. The eggshell consists of two or sometimes three layers, the innermost being lipid or a lipid-type material, the next a mixture of chitin with a collagen-type protein, and the outer, when present, protein. The larva must penetrate these layers either by chemical or physical means (or both) before hatching can take place. Larvae move vigorously within the egg before hatching, which may help to disrupt the inner lipid layer. Bird (1958) observed that the inner layer of the eggshell of *Meloidogyne javanica* was destroyed by the movement of the larvae before hatching.

Time-lapse film (Doncaster and Shepherd, 1967) showed that larvae in eggs of *Heterodera rostochiensis* moved little before and for several days after stimulation with potato root exudate. Then slow and intermittent movements began, and these gradually increased in vigor and duration. The length of time before movement began, and also before hatching, differed considerably between individuals. Emergence was always preceded by some minutes or hours of exploratory stylet thrusting at either or both ends of the egg, with the head pressed hard against the shell. Stylet thrusting finally became purposeful and precise, and the larva made a line of perforations so close that each broke back into the previous one, making a straight slit across the end of the egg. The bodily thrust of the larva then carried it out through the slit as soon as this was long enough. The three pharyngeal glands became active soon after stimulation, in fact, before movement was detected; and by the time the larva emerged the gland ducts were swollen with secretions. Emission of gland secretions was not seen however, and the shell did not stretch under the pressure from the head of the larva.

Larvae often moved actively after stimulation, and some made exploratory stylet thrusts but failed to begin purposeful stylet thrusting. Such larvae eventually became quiescent again and failed to hatch. This suggests that rupture of the inner shell layer by larval movement either does not occur in *Heterodera* or is less important than in some other nematodes and may partly explain why some of the eggs in cysts remain unhatched at the end of a season, even when a host crop is grown. Clearly then, *Heterodera* larvae have three behavior patterns in hatching. These are movement, exploratory stylet thrusting, and purposeful stylet thrusting; unless the last is reached, hatching does not take place.

When eggs of *H. rostochiensis* were stimulated with an artificial hatching agent, sodium metavanadate, the same basic behavior patterns occurred as in root diffusate, but movement inside the egg started earlier and was more vigorous. Exploratory stylet thrusting started sooner, but purposeful stylet thrusting was less well coordinated. The eggshell became softened, and the larva was able to push its way out without completing the slit in the eggshell. The dorsal pharyngeal glands were less active than when stimulated with root diffusate, but the subventral gland ducts were slightly distended. Most eggs hatched much sooner than in root diffusate under the conditions in the observation chamber.

In *Ditylenchus destructor,* stylet thrusting occurs just before hatching, the stylet often penetrating the end of the shell which, in contrast to *Heterodera,* becomes softened and stretches easily as the larva moves and eventually bursts out (Anderson and Darling, 1964). Whether a slit is cut in the shell by the larva is uncertain. Taylor (1962) observed behavior in *Aphelenchus avenae* similar to that of *Ditylenchus* and noted that the metacorpus of the larva pulsated at an early stage, before shell softening occurred, although no expulsion of secretions was observed. In *Criconemoides xenoplax,* the stylet apparently plays no part in hatching but the eggshell just gives way to pressure from the larva (Seshadri, 1964).

## 5. The Hatching Mechanism in *Heterodera*

There are several ways in which hatching agents might function. For example, they could exercise a physical effect on the egg or larva by changing the permeability of a barrier and allowing the passage of water and solutes (Dropkin *et al.,* 1958). This might happen if hatching agents were taken up by the binding material in some part of the egg or larva thus changing the structure and function of that material (Clarke and Shepherd, 1966b). Alternatively, hatching agents might act chemically by affecting the metabolism of the larva inside the egg.

There is both direct and indirect evidence of the existence of permeability barriers isolating the unhatched larva from its environment and controlling the flow of water and solutes. Wihrheim *et al.* (1968) used $^{14}$C-labeled phenylalanine to study penetration into cysts of *H. tabacum.* The cyst wall offered little hindrance to the passage of the labeled material, but the newly hatched larva had a definite barrier. Evidence for a barrier in the eggshell was inconclusive. Ellenby (1965) showed with interference microscopy that *H. rostochiensis* cyst walls are readily permeable to buffer solutions.

Ellenby (1968) also demonstrated that the eggshell of *H. rostochiensis* is freely permeable to water when wet but becomes increasingly impermeable as it dries. The inhibition of hatching of *Meloidogyne* spp. and of *H. rostochiensis* by solutions of high osmotic pressure also suggests a relationship between permeability and hatching (Dropkin *et al.*, 1958). The eggshell of *M. javanica* becomes more permeable to water before the eggs hatch (Wallace, 1968b,c). In eggs of *Ascaris lumbricoides*, the vitelline membrane becomes more permeable before hatching occurs (Fairbairn and Passey, 1957; Fairbairn, 1961) and the activated *A. lumbricoides* larva produces enzymes that weaken the eggshell. Enzymic attack on the eggshell has been observed in the hatching of *M. javanica* (Bird, 1968) but not of *Heterodera* spp. However, Doncaster and Shepherd (1967) noted that the pharyngeal glands of the unhatched larva of *H. rostochiensis* became active in potato root diffusate and also, although less so, in solutions of the artificial hatching agent sodium metavanadate, but no emission of fluids from the glands was seen. The sodium metavanadate solution softened the egg shell before hatching, but the potato root diffusate did not.

Although it is sometimes suggested that the natural hatching factors have a metabolic function, there is not as yet evidence to support this belief. The differences observed by Doncaster and Shepherd (1967) in the action of sodium metavanadate and potato root diffusate on larvae in eggs of *H. rostochiensis* are slighter than the similarities; the larva goes through the same basic routine in both. There is no evidence that these two hatching agents operate different mechanisms, which makes it unlikely that the natural factor is a key metabolite. Other evidence appears to support this view. For example, the respiration rates of hatched larvae of *H. rostochiensis*, and of larvae active in the egg before hatching, do not differ greatly (Sembdner *et al.*, 1961). Added glucose increased the respiration rate of cysts suspended in potato root diffusate and also increased the rate of oxygen uptake of hatched larvae in water and in potato root diffusate. The respiration rate of free larvae in potato root diffusate did not differ significantly from that in water. It was concluded that there was no direct relation between oxygen uptake and the rate of hatching.

## REFERENCES

Ahmed, A., and Khan, A. H. (1964a). *Indian Phytopathol.* **17,** 98–101.
Ahmed, A., and Khan, A. M. (1964b). *Indian Phytopathol.* **17,** 102–109.
Anderson, R. V., and Darling, H. M. (1964). *Nematologica* **10,** 131–135.
Baunacke, W. E. (1922). *Arb. Biol. Abt. Land- Forstwirt. Berlin* **11,** 185–288.
Baxter, R. I., and Blake, C. D. (1969a). *Ann. Appl. Biol.* **63,** 183–190.
Baxter, R. I., and Blake, C. D. (1969b). *Ann. Appl. Biol.* **63,** 191–203.

Bird, A. F. (1956). *Exp. Parasitol.* **5,** 350–358.
Bird, A. F. (1957). *Exp. Parasitol.* **6,** 383–403.
Bird, A. F. (1958). *Nematologica* **3,** 205–212.
Bird, A. F. (1959). *Nematologica* **4,** 31–42.
Bird, A. F. (1968). *J. Parasitol.* **54,** 475–489.
Bird, A. F., and Wallace, H. R. (1966). *Nematologica* **11,** 581–589.
Bishop, D. (1955). *Ann. Appl. Biol.* **43,** 525–532.
Blake, C. D. (1962). *Ann. Appl. Biol.* **50,** 713–722.
Calam, C. T., Raistrick, H., and Todd, A. R. (1949a). *Biochem. J.* **45,** 513–519.
Calam, C. T., Todd, A. R., and Waring, W. S. (1949b). *Biochem. J.* **45,** 520–524.
Chidambaranathan, A., and Rangaswami, G. (1964). *Curr. Sci.* **33,** 724–725.
Clarke, A. J. (1968). *Biochem. J.* **108,** 221–224.
Clarke, A. J., and Shepherd, A. M. (1964). *Nematologica* **10,** 431–453.
Clarke, A. J., and Shepherd, A. M. (1966a). *Ann. Appl. Biol.* **57,** 241–255.
Clarke, A. J., and Shepherd, A. M. (1966b). *Ann. Appl. Biol.* **58,** 497–508.
Clarke, A. J., and Shepherd, A M. (1966c). *Nature (London)* **211,** 546.
Clarke, A. J., and Shepherd, A. M. (1967). *Nature (London)* **213,** 419–420.
Clarke, A. J., and Shepherd, A. M. (1968). *Ann. Appl. Biol.* **61,** 139–149.
Clarke, A. J., and Widdowson, E. (1966). *Biochem. J.* **98,** 862–868.
Cotten, J. (1962). *Nature (London)* **195,** 308.
Crofton, H. D. (1966). "Nematodes," 160 pp. Hutchinson Univ. Library, London.
Cunningham, P. C. (1960). *Sci. Proc. Roy. Dublin Soc., Ser. B* **1,** 1–4.
Davey, K. G. (1966). *Amer. Zool.* **6,** 243–249.
Davey, K. G., and Kan, S. P. (1967). *Nature (London)* **214,** 737–738.
Davey, K. G., and Kan, S. P. (1968). *Can. J. Zool.* **46,** 893–898.
Doncaster, C. C., and Shepherd, A. M. (1967). *Nematologica* **13,** 476–478.
Dropkin, V. H., Martin, G. C., and Johnson, R. W. (1958). *Nematologica* **3,** 115–126.
Ellenby, C. (1946). *Nature (London)* **157,** 451–452.
Ellenby, C. (1958). *J. Helminthol.* **32,** 219–226.
Ellenby, C. (1963). *Nature (London)* **198,** 1110.
Ellenby, C. (1965). *Nematologica* **11,** 297–299.
Ellenby, C. (1968). *Proc. Roy. Soc. Ser. B* **169,** 203–213.
Ellenby, C., and Smith, L. (1967). *Ann. Appl. Biol.* **59,** 283–288.
Fairbairn, D. (1960a). *In* "Nematology: Fundamentals and Recent Advances with Emphasis on Plant Parasitic and Soil Forms" (J. N. Sasser and W. R. Jenkins, eds.), pp. 267–296. Univ. of North Carolina Press, Chapel Hill, North Carolina.
Fairbairn, D. (1960b). *In* "Host Influence on Parasite Physiology" (L. A. Stauber, ed.), pp. 50–64. Rutgers Univ. Press, New Brunswick, New Jersey.
Fairbairn, D. (1961). *Can. J. Zool.* **39,** 153–162.
Fairbairn, D., and Passey, R. F. (1957). *Exp. Parasitol.* **6,** 566–574.
Fenwick, D. W. (1951a). *J. Helminthol.* **25,** 37–48.
Fenwick, D. W. (1951b). *J. Helminthol.* **25,** 49–56.
Fisher, J. M. (1966). *Aust. J. Biol. Sci.* **19,** 1073–1079.
Fisher, J. M. (1970). *Aust. J. Biol. Sci.* **23,** 411–419.
Fox, J. A., and Kerekes, M. G. (1969). *J. Nematol.* **1,** 8–9.
Fushtey, S. G., and Johnson, P. W. (1966). *Nematologica* **12,** 313–320.
Giebel, J. (1963). *Biul. Inst. Ochr. Rosl.* **21,** 157–160.
Golden, A. M. (1956). *Md. Agr. Exp. Sta., Bull.* **A-85,** 1–28.
Golden, A. M. (1958). *Plant Dis. Rep.* **42,** 188–193.
Hague, N. G. (1958). *Nematologica* **3,** 149–153.

Hamlen, R. A., and Bloom, J. R. (1968). *Phytopathology* **58,** 515–518.
Hartwell, W. V., Dahlstrom, R. V., and Neal, A. L. (1960). *Phytopathology* **50,** 612–615.
Hesling, J. J., Pawelska, K., and Shepherd, A. M. (1961). *Nematologica* **6,** 207–213.
Ishibashi, N. (1967). *Noji Shikenjo Kenkyo Hokoku* **11,** 177–219.
James, G. L. (1966). *Nature (London)* **212,** 1466.
James, G. L. (1968). *Ann. Appl. Biol.* **61,** 503–510.
Janzen, G. J. (1968). *Nematologica* **14,** 601–602.
Janzen, G. J., and van der Tuin, F. (1956). *Nematologica* **1,** 126–137.
Jones, F. G. W., and Nirula, K. K. (1963). *Plant Pathol.* **12,** 148–154.
Jones, F. G. W., and Winslow, R. D. (1953). *Nature (London)* **171,** 478–479.
Juhl, M. (1968). *Tidsskr. Planteavl* **72,** 42–63.
Kan, S. P., and Davey, K. G. (1968). *Can. J. Zool.* **46,** 235–241.
Kaul, R. (1962). *Nematologica* **8,** 288–292.
Lee, D. L. (1961). *Nature (London)* **192,** 282–283.
Lee, D. L. (1962). *Parasitology* **52,** 241–260.
Lee, D. L. (1965). "The Physiology of Nematodes," 154 pp. Oliver & Boyd, Edinburgh and London.
Lehman, P. S. (1969). *J. Nematol.* **1,** 14–15.
Loewenberg, J. R., Sullivan, T., and Schuster, M. L. (1960). *Phytopathology* **50,** 215–217.
Lownsbery, B. F. (1950). *Phytopathology* **40,** 18.
McElroy, F., and Van Gundy, S. D. (1966). *Phytopathology* **56,** 889.
Marrian, D. H., Russell, P. B., Todd, A. R., and Waring, W. S. (1949). *Biochem. J.* **45,** 524–528.
Miller, P. M. (1967). *Plant Dis. Rep.* **51,** 202–206.
Miller, P. M., and Lukens, R. J. (1966). *Phytopathology* **56,** 967–970.
Miller, P. M., and Stoddard, E. M. (1958). *Science* **128,** 1429–1430.
Miller, P. M., Taylor, G. S., and Wihrheim, S. E. (1968). *Plant Dis. Rep.* **52,** 441–445.
O'Bannon, J. H. (1968). *Nematologica* **14,** 12–13.
Omidvar, A. M. (1961). *Nematologica* **6,** 123–129.
Oostenbrink, M. (1967). *Meded. Rijksfac. Landbouwwetensch. Gent* **32,** 503–539.
Rademacher, B. (1930). *Arch. Pflanzenbau* **3,** 750–787.
Rademacher, B., and Schmidt, O. (1933). *Arch. Pflanzenbau* **10,** 237–296.
Rhoades, H. L., and Linford, M. B. (1959). *Science* **130,** 1476–1477.
Robinson, T., and Neal, A. L. (1956). *Phytopathology* **46,** 665–667.
Rogers, W. P. (1960). *Proc. Roy. Soc., Ser. B* **152,** 367–386.
Rogers, W. P. (1962). "The Nature of Parasitism," 287 pp. Academic Press, New York.
Rogers, W. P. (1965). *Comp. Biochem. Physiol.* **14,** 311–321.
Rogers, W. P., and Sommerville, R. I. (1957). *Nature (London)* **179,** 619–621.
Rogers, W. P., and Sommerville, R. I. (1960). *Parasitology* **50,** 329–348.
Rogers, W. P., and Sommerville, R. I. (1963). *Advan. Parasitol.* **1,** 109–177.
Ross, J. P. (1962). *Phytopathology* **53,** 608–609.
Russell, P. B., Todd, A. R., and Waring, W. S. (1949). *Biochem. J.* **45,** 528–530.
Schreiber, K., Osske, G., and Sembdner, G. (1961). *Nematologica* **6,** 17–24.
Sembdner, G., and Schreiber, K. (1960). *Nematologica Suppl.* II, 127–140.
Sembdner, G., Osske, G., and Schreiber, K. (1961). *Biol. Zentralbl.* **80,** 551–561.
Seshadri, A. R. (1964). *Nematologica* **10,** 540–562.

Shepherd, A. M. (1962a). "The Emergence of Larvae from Cysts in the Genus *Heterodera*," Tech. Commun. No. 32, 90 pp. Commonw. Bur. of Helminthol.. St. Albans, Herts., England. Commonwealth Agricultural Bureaux, Farnham Royal, Bucks, England.
Shepherd, A. M. (1962b). *Nature (London)* **196**, 391–392.
Shepherd, A. M. (1963). *Nematologica* **9**, 143–151.
Shepherd, A. M., and Cox, P. M. (1967). *Ann. Appl. Biol.* **60**, 143–150.
Silverman, P. H., and Podger, K. R. (1964). *Exp. Parasitol.* **15**, 314–324.
Slack, D. A., and Hamblen, M. L. (1959). *Phytopathology* **49**, 319–320.
Smedley, E. M. (1936). *J. Helminthol.* **14**, 11–20.
Sommerville, R. I. (1957). *Exp. Parasitol.* **6**, 18–30.
Steele, A. E. (1961). *J. Amer. Soc. Sugar Beet Technol.* **11**, 528–532.
Steele, A. E. (1963). *J. Amer. Soc. Sugar Beet Technol.* **12**, 296–298.
Swarup, G., and Pillai, M. J. (1964). *Indian Phytopathol.* **17**, 88–97.
Taylor, D. P. (1962). *Proc. Helminthol. Soc. Wash.* **29**, 52–54.
Viglierchio, D. R., and Lownsbery, B. F. (1960). *Nematologica* **5**, 153–157.
Viglierchio, D. R., and Yu, P. K. (1966). *J. Amer. Soc. Sugar Beet Technol.* **13**, 698–715.
Wallace, H. R. (1954). *Nature (London)* **173**, 502–503.
Wallace, H. R. (1955a). *J. Helminthol.* **29**, 3–16.
Wallace, H. R. (1955b). *Ann. Appl. Biol.* **43**, 477–484.
Wallace, H. R. (1956a). *Ann. Appl. Biol.* **44**, 57–66.
Wallace, H. R. (1956b). *Ann. Appl. Biol.* **44**, 274–282.
Wallace, H. R. (1963). "The Biology of Plant Parasitic Nematodes," 280 pp. Arnold, London.
Wallace, H. R. (1966a). *Proc. Roy. Soc., Ser. B* **164**, 592–614.
Wallace, H. R. (1966b). *Nematologica* **12**, 57–69.
Wallace, H. R. (1968a). *Nematologica* **14**, 223–230.
Wallace, H. R. (1968b). *Nematologica* **14**, 231–242.
Wallace, H. R. (1968c). *Parasitology* **58**, 377–391.
Watson, B. D. (1965). *Quart. J. Microsc. Sci.* **106**, 83–91.
Webber, A. J., Jr., and Barker, K. R. (1968). *Proc. Helminthol. Soc. Wash.* **35**, 34–37.
Widdowson, E. (1958a). *Nematologica* **3**, 6–14.
Widdowson, E. (1958b). *Nematologica* **3**, 173–178.
Wihrheim, S. E., Miller, P. M., and Dimond, A. E. (1968). *Phytopathology* **58**, 843–847.
Wilski, A. (1963). *Biul. Inst. Ochr. Rosl.* **19**, 231–237.
Winslow, R. D. (1955). *Ann. Appl. Biol.* **43**, 19–36.
Winslow, R. D., and Ludwig, R. A. (1957). *Can. J. Bot.* **35**, 619–634.
Wuest, P. J., and Bloom, J. R. (1965). *Phytopathology* **55**, 885–888.
Yarwood, E. A., and Hansen, E. L. (1969). *J. Nematol.* **1**, 184–189.

# CHAPTER 26

# Mode of Action of Nematicides

C. E. CASTRO AND I. J. THOMASON

*Department of Nematology, University of California, Riverside, California*

I. The State of Knowledge . . . . . . . . . . . . . 289
II. Gross Effects . . . . . . . . . . . . . . 290
A. Time–Death Studies . . . . . . . . . . . . 290
B. Influence of the Environment . . . . . . . . . . 291
C. Resistance . . . . . . . . . . . . . . 292
III. Permeation Characteristics . . . . . . . . . . . 292
IV. Model Systems . . . . . . . . . . . . . . 294
V. Hypotheses . . . . . . . . . . . . . . . 295
References . . . . . . . . . . . . . . . 296

## I. THE STATE OF KNOWLEDGE

Knowledge of the physiology of plant parasitic nematodes must still be considered to be in its infancy (Lee, 1965). Although there is no reason to suspect an unusual array of biochemical patterns in these animals, the information at hand is limiting in its capacity to sustain a sophisticated interpretation of intoxication. Equally limiting is the paucity of knowledge of the fate of alkyl halides in living systems. However, the beginnings of an understanding of the wide scope of biodehalogenation is slowly emerging (Castro and Bartnicki, 1968).

We shall limit our considerations to the alkyl halide nematicides since this class of toxicant, though not particularly selective, dominates the commercial nematicide field, and alkyl halide nematicides have been studied more than their nonhalogenated counterparts. At this stage of development any projection of modes of action of nematicides at a molecular level must be speculative. In fact our later suggestions rest entirely upon studies of model systems. New information about the gross biological effects of intoxication by these entities and some grasp of their permeation characteristics have been obtained and are outlined below.

## II. GROSS EFFECTS

### A. Time–Death Studies

Most studies on the effect of exposure time on the toxicity of alkyl halide nematicides to plant parasitic nematodes have been carried out in soil (McBeth and Bergeson, 1955; Johnson and Lear, 1966). Under these conditions a longer interval (up to 2 weeks) between dosing the soil and planting is required for satisfactory root-knot nematode control with DBCP (1,2-dibromo-3-chloropropane) than with 1,3-D (1,3-dichloropropane) or EDB (ethylene dibromide). One conclusion drawn was that DBCP is slower in its "toxic action" than 1,3-D or EDB. Studies of this nature tend to confuse behavior of the toxicant in the soil matrix with its ability to kill once it is in contact with the nematode. Furthermore, if eggs or egg masses are present in the soil, they may influence the results obtained. 1,2-Dibromo-3-chloropropane is a relatively poor ovicide. When exposed *in vitro* for 96 hr, 200 times the dosage of DBCP was required to kill *Meloidogyne javanica* eggs in egg masses than was required to kill second-stage larvae (Mojé and Thomason, 1963). *In vitro* studies have shown that a 96-hr exposure period is sufficient to assess the toxicity of DBCP to *M. javanica*. 1,2-Dibromo-3-chloropropane reaches its maximum internal concentration in 4 hr when *Aphelenchus avenae* is exposed to a solution of $0.53 \times 10^{-2}$ *M* DBCP (Marks *et al.*, 1968). The fact that maximum internal concentration is reached in 4 hr and that toxicity *in vitro* is adequately assessed in a 96-hr exposure suggest that some factor other than inherently slow toxic action is responsible for the slow performance of DBCP in soil.

More critical assessments of the effects of exposure time can be made if nematodes are exposed to alkyl halide nematicides for specific periods of time and then removed from the solution of the toxicant (Thomason *et al.*, 1968; Evans and Thomason, 1969). Most adult females of *A. avenae* exposed to $0.53 \times 10^{-2}$ *M* EDB for 8 hr can recover motility and reproduce. Exposure for 24 hr is lethal. On the other hand, most juvenile stage larvae do not survive exposure for longer than 4 hr. This agrees with Chitwood's (1952) observation that immature forms of nematodes are more readily killed by nematicides. In addition, larval stages in the molt are more easily killed by EDB than nonmolting stages (Evans and

TABLE I
RELATIVE SENSITIVITY OF NEMATODES TO EDB ($0.53 \times 10^{-2}$ *M*) AS INFLUENCED BY EXPOSURE TIME[a]

| Exposure time (hr) | Percent kill | | |
|---|---|---|---|
| | *Aphelenchus avenae* | *Meloidogyne javanica* | *Tylenchulus semipenetrans* |
| 0.5 | 17 | 4 | 95 |
| 1 | 26 | 17 | 94 |
| 2 | 22 | 37 | 93 |
| 4 | 0 | 74 | 95 |
| 8 | 14 | 99 | 97 |
| 24 | 78 | 100 | 100 |

[a] Mixed population of *A. avenae,* second stage larvae of *M. javanica* and *T. semipenetrans*. Number of animals exposed per test unit is 600 ± 20.

Thomason, 1969). As one would suspect, all genera of nematodes do not react in the same way. Second-stage larvae of the citrus nematode, *Tylenchulus semipenetrans,* will not survive a 0.5-hr exposure to $0.53 \times 10^{-2}$ *M* EDB, whereas second-stage *M. javanica* will survive up to 2–4 hr exposure (Table I).

Exposure time appears to be extremely important above certain minimal dosage levels, and our limited experience indicates that this will vary considerably between nematode species and alkyl halide nematicide.

## B. Influence of the Environment

### 1. TEMPERATURE

Temperature has a profound influence on the efficacy of some alkyl halide nematicides. Because most studies (Goring and Youngson, 1957; Lear and Johnson, 1963) have been conducted in soil, the direct effect of temperature on the nematode–nematicide interaction has not been determined. In soil, EDB is not effective below 10°C, whereas 1,3-D gives satisfactory control of nematodes down to 5°C and lower. Thorough mixing of EDB with the soil did not improve control at 10°C suggesting that it was not lack of contact with the nematode that was responsible (Lear and Johnson, 1963). The permeation of EDB into *A. avenae* was shown by Marks *et al.* (1968) to be similar at 5° and 25°C. Furthermore, molting forms of *A. avenae,* which are normally very sensitive, survived an 8-hr exposure to $0.53 \times 10^{-2}$ *M* EDB at 5°C. The ability of *A. avenae* to survive exposure to EDB decreases as the temperature increases from

5° to 35°C, thus suggesting that active metabolism is important in the toxicity of EDB. This is further supported by the observation that toxicity of EDB is sharply reduced in an anaerobic environment. (Evans and Thomason, unpublished results).

#### 2. pH

pH has not been reported to have a major effect on the toxicity of alkyl halide nematicides, although these substances can hydrolyze in basic or even neutral media (Castro and Belser, 1966).

An example of the marked influence that pH can have on a nematicidal chemical is found in butyric acid (Sayre *et al.*, 1965). At pH 4 a $1 \times 10^{-2}$ *M* butyric acid solution killed 90% of the second-stage larvae of *M. javanica* when exposed for 24 hr. At pH 5.3 butyric acid was only partially effective, and at pH 7.0 it was inactive. At pH 7.0 butyric acid is predominantly ionized and is presumably not capable of penetrating the nematode readily. Thus, if a particular substance contains an acidic or basic moiety, pH may enhance its toxicity by altering its permeating capacity.

### C. Resistance

No case of acquired resistance by a nematode to the halogenated hydrocarbon nematicides has been reported. The rather low level of selection pressure (one application per year, short persistence in soil) would not favor the development of resistance. In our laboratory, repeated exposure of *A. avenae* to highly toxic dosages of EDB did result in the selection of resistant strains.

## III. PERMEATION CHARACTERISTICS

The general pattern of permeation of nematodes by water, alkyl halide nematicides, and other molecules has been charted. Although a great many species have not been examined, the results to date indicate that general observable characteristics of the process are typical. For the details of the method and the mathematics of the treatment the reader is referred to Marks *et al.* (1968).

For our purposes here it will suffice to briefly note the broad aspects of permeation and draw some conclusions that frame any consideration of intoxication by alkyl halides.

The overall rate of permeation of nematodes is given by Eq. (1):

$$\text{overall rate in} = (\text{rate in}) - (\text{rate out}) \quad (1)$$

wherein rate in $= k_1$ (No) and rate out $= k_{-1}$ (Ni); $k_1$ and $k_{-1}$ are pseudo-first-order rate constants for the in and out processes, respectively, and (No) and (Ni) are the concentrations of permeating substance without and within the animals. The equilibrium constant for permeation ($k_{eq}$) is defined as

$$K = k_1/k_{-1} \quad (2)$$

The evaluation of some of these kinetic parameters from the integrated rate equations are given in Table II.

Table III portrays a comparison of the constants for *A. avenae* produced on a culture of *Rhizoctonia solani* grown on wheat kernels and upon *R. solani* growing on potato-dextrose agar in petri plates.

The following conclusions are relevant to intoxication:

(1) Permeation is a dynamic process in which the nematodes rapidly come to equilibrium with their aqueous environment. This dynamism is emphasized by considering that the average residence time of a water molecule in *A. avenae* is approximately 0.9 sec.

(2) The equilibrium concentration of nematicides in the animal exceeds that in the external solutions. Thus, the equilibrium concentration of ethylene dibromide in *A. avenae* exposed to a $0.53 \times 10^{-2}\,M$ solution is $1.34 \times 10^{-2}\,M$, and the animals remain alive for hours under these conditions (cf. Section II of this chapter).

(3) Live nematodes are selective toward substances permeating them. Moreover, each substance enters at its own rate independent of others. For example, ethylene dibromide does not alter the rate with which water

TABLE II
COMPARISON OF THE $K$, $k_1$, AND $k_{-1}$ VALUES FOR DIFFERENT CHEMICALS[a] WITH DIFFERENT SPECIES OF NEMATODES

| | EDB[b] | | | HTO[b,c] | | | DBCP[b] | | |
|---|---|---|---|---|---|---|---|---|---|
| Nematode | $K$ | $k_1$ | $k_{-1}$ | $K$ | $k_1$ | $k_{-1}$ | $K$ | $k_1$ | $k_{-1}$ |
| *A. avenae* | 2.8 | 0.40 | 0.13 | 1 | 0.15 | 0.15 | 5.4 | 0.065 | 0.012 |
| *Caenorhabditis* sp. | 6 | 0.22 | 0.04 | 1 | 0.09 | 0.09 | | | |
| *T. semipenetrans* | 9 | 0.54 | 0.06 | | | | | | |
| *A. tritici* | 20 | 0.64 | 0.03 | 1 | 0.08 | 0.08 | | | |

[a] All values represent an average of at least three determinations. Reproducibility was within 5%.

[b] The units for $K$ and $k_{-1}$ in all cases are (min$^{-1}$). The equilibrium constants are dimensionless.

[c] Here HTO stands for water.

TABLE III
COMPARISON OF PERMEATION PARAMETERS FOR *A. avenae* GROWN UPON "WHEAT KERNELS" OR "AGAR PLATES"

| Agar | Wheat | Substrate |
|---|---|---|
| $k_1$ 0.4 | 1.0 | |
| $k_{-1}$ 0.13 | 0.06 | EDB |
| $K$ 2.8 | 16 | |
| $k_1$ 0.15 | 0.22 | |
| $k_{-1}$ 0.15 | 0.22 | HTO |
| $K$ 1 | 1 | |

enters or exits and it has no effect upon the slow entrance of sodium acetate. Thus, these nematicides do not appear to function by altering the permeation capacities of the nematodes.

(4) The nematode cuticle is not a barrier to penetration by nematicides. They readily enter and not simply by ingestion. If ingestion were the major mode of entry, the rates for water and solute uptake would be the same. This is not true for any substance thus far examined.

(5) Although a wide enough array of substances has not been examined to warrant a very detailed account of the nature of the selectivity, it is clear that water, a polar molecule, enters more rapidly than the nematicides. They in turn permeate much more swiftly than simple salts, glucose, or glycine. Thus, the movement of water must be the dominant event in osmoregulation.

(6) The habitat of nematodes can greatly alter their permeation rates. Thus, the *A. avenae* reared on *Rhizoctonia*-wheat (cf. Table III) have a greater capacity for the nematicide than the animals reared on *Rhizoctonia*-PDA. These results give an indication of the important influence environment is likely to have upon permeation.

## IV. MODEL SYSTEMS

The nematicidal activity of alkyl halides has been compared with their susceptibility toward bimolecular nucleophilic attack (Mojé, 1960). A very loose correlation exists and has led to the suggestion that alkyl halides do in fact alkylate nucleophilic sites on proteins, for example, on the —SH groups of cysteine, and death ensues. This suggestion accords with known means of inhibition of enzymic activity by iodoacetate and iodoacetamide. The model systems corresponding to this view are reactions of the form $RX + Q = QR + X^-$, wherein $Q$ = hydroxide, iodide, mercaptide, etc. These are slow reactions with the alkyl halide nematicides. Thus, second-order rate constants for the solvolysis of *cis*- and *trans*-1,3-dichloropropene are of the order of 0.1–0.3 liter/mole/hr. Hence, under less basic physiological conditions a large number of elastic collisions of these molecules with potentially reactive sites not leading to reaction will occur.

On the other hand, an exceedingly rapid reaction of alkyl halides with Fe(II) porphyrins has been observed (Castro, 1964). Thus, *cis*-1,3-dichloropropene oxidizes $^{II}Fe$ porphyrins with a rate constant of $\sim 4 \times 10^4$ liter$^2$/mole$^2$/min. The broad spectrum biocide DDT is even more rapid ($k_3 \sim 3 \times 10^7$ liter$^2$/mole$^2$/min). These rapid transformations also occur with EDB and DBCP and parallel the general reactivity of alkyl halides to low valent transition metal species (Castro and Kray, 1963, 1966; Kray and Castro, 1964). In addition to $^{II}Fe$ porphyrins, the low valent complexes of $B_{12}$ are quite reactive toward alkyl halides (Schrauzer, 1968; Wood *et al.*, 1968). Thus, given the extremely rapid reaction of low valent metal ions with alkyl halides, and the even more rapid reactions with biometallics, proteins and coenzymes containing low valent metal species as their prosthetic groups must be most susceptible to reaction with organic halides. Put another way, the rates of reaction of biometallics with alkyl halides are so rapid that reaction will most certainly occur if the halide diffuses to the site of the metals. Indeed it is likely that low valent metal centers are the most reactive site in biological systems toward simple alkyl halides.

## V. HYPOTHESES

Undoubtedly there are a multiplicity of sites for the interaction of any biocide with a living system. The question to be broadly put is how many and what are the most sensitive and damaging in nematodes? At low levels of nematicides (1–200 ppm) nematodes appear to die quietly. They lie straight. Indeed death is preceded by a quiescence that can be reversed if the animals are removed from the toxicant. On the other hand, higher doses of the alkyl halides result in a contortion of the animal in what visually appears to be a violent ending. We believe these observations can be explained by involving two modes of intoxication. The first quiescent stage and quiet death at low concentrations could be the result of direct oxidative flooding of iron centers in the respiratory sequence. The organism works increasingly less effectively to respire and gradually dies. This view is supported by our observation that potassium cyanide, which is well known to complex the iron of hemoproteins, results in the same kind of slow, straight-bodied death. The slowness of this latter process is the result of the inability of salts to rapidly permeate the animals. The high concentration induced contorted death with the alkyl halides may be ascribed to the less delicate, almost brute forces, alkylation of a great many sites in the animal. The stress from this internal fusillade results in the severe contortions observed.

## REFERENCES

Castro, C. E. (1964). *J. Amer. Chem. Soc.* **86**, 2310–2311.
Castro, C. E., and Bartnicki, E. W. (1968). *Biochemistry* **7**, 3213–3218.
Castro, C. E., and Belser, N. O. (1966). *J. Agr. Food Chem.* **14**, 69–70.
Castro, C. E., and Kray, W. C. (1963). *J. Amer. Chem. Soc.* **85**, 2768–2773.
Castro, C. E., and Kray, W. C. (1966). *J. Amer. Chem. Soc.* **88**, 4447–4455.
Chitwood, B. G. (1952). *Advan. Chem. Ser.* **7**, 91–99.
Evans, A. F., and Thomason, I. J. (1969). *J. Nematol.* **1**, 287. (Abstr.)
Goring, C., and Youngson, C. (1957). "Factors Influencing Diffusion and Nematode Control by Soil Fumigants." Dow Chem. Co., Midland, Michigan.
Johnson, D. E., and Lear, B. (1966). *Plant Dis. Rep.* **50**, 915–916.
Kray, W. C., and Castro, C. E. (1964). *J. Amer. Chem. Soc.* **86**, 4603–4608.
Lear, B., and Johnson, D. E. (1963). *Soil Sci.* **95**, 322–325.
Lee, D. E. (1965). "The Physiology of Nematodes," 154 pp. Freeman, San Francisco, California.
McBeth, C. W., and Bergeson, G. B. (1955). *Plant Dis. Rep.* **39**, 223–225.
Marks, C. F., Thomason, I. J., and Castro, C. E. (1968). *Exp. Parasitol.* **22**, 321–337.
Mojé, W. (1960). *Advan. Pest Contr. Res.* **3**, 181–217.
Mojé, W., and Thomason, I. J. (1963). *Phytopathology* **53**, 428–431.
Sayre, R. M., Patrick, Z. A., and Thorpe, H. J. (1965). *Nematologica* **11**, 263–268.
Schrauzer, G. W. (1968). *Accounts Chem. Res.* **1**, 97–103.
Thomason, I. J., Castro, C. E., and Marks, C. F. (1968). *Abstr. Pap. 1st. Int. Congr. Plant Pathol., London* p. 201.
Wood, J. M., Kennedy, F. S., and Wolfe, R. S. (1968). *Biochemistry* **7**, 1707–1713.

# CHAPTER 27

# Senescence, Quiescence, and Cryptobiosis

A. F. Cooper, Jr. and S. D. Van Gundy

*Department of Nematology, University of California, Riverside, California*

I. Introduction . . . . . . . . . . . . . . . . 297
II. Senescence . . . . . . . . . . . . . . . . 299
III. Quiescence . . . . . . . . . . . . . . . . 304
A. Anhydrobiosis . . . . . . . . . . . . . . . . 304
B. Anoxybiosis . . . . . . . . . . . . . . . . 307
C. Cryobiosis . . . . . . . . . . . . . . . . 309
IV. Cryptobiosis . . . . . . . . . . . . . . . . 310
V. Summary . . . . . . . . . . . . . . . . 314
References . . . . . . . . . . . . . . . . 315

## I. INTRODUCTION

The words *senescence, quiescence,* and *cryptobiosis* describe quantitatively decreasing levels (time rates of change) of metabolic activity which may occur during the aging of an organism. Which mode is operative depends upon the metabolic status and adaptability of the individual and upon the kinds and intensities of environmental stress.

Senescence ("normal" aging) is characterized by an average pattern of metabolic activities whereby the life cycle can be completed in a given environment (see Fig. 1).

When the environmental balance shifts unfavorably, some organisms experience a metabolic "slowdown" called quiescence: movement slows or ceases, and the normal sequence of life cycle functions is delayed.

With further increase or continuation of environmental stress a few biological systems enter cryptobiosis which is a "shutdown" of metabolism to levels not detectable by currently available biochemical methods or physical instrumentation (Clegg, 1967; Keilin, 1959; Kostük, 1965).

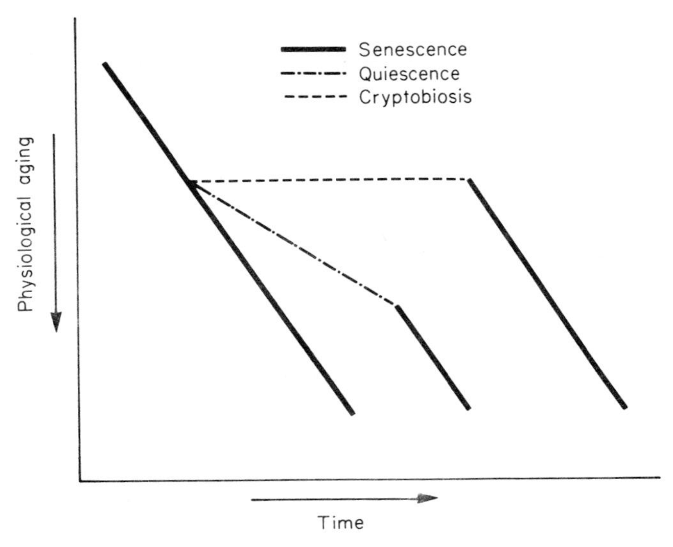


FIG. 1. The relationship of senescence, quiescence, and/or cryptobiosis to physiological aging.

Structural integrity is maintained (or only slowly declines) and the potential for resumption of quiescence, motility, and even the normal patterns of senescence is retained, often for extended periods of time.

Appreciable numbers of the soil-inhabiting species of the Nematoda are capable of quiescence and/or cryptobiosis which permits survival in environments not consistently favorable for growth and reproduction and delays the aging process, a capability shared with few of the lower metazoans (coelenterates and flatworms) or the vastly more complex higher metazoans (mollusks and arthropods). This, along with their adaptability to axenic (sterile) or monoxenic (with one host organism) culture (Evans, 1970; Rothstein and Nicholas, 1969), rapid reproduction rates in laboratory culture (populations often peak in 2–3 weeks), five well-defined life cycle stages, hermaphroditic or parthenogenetic reproduction (easy establishment of clones), body wall transparency (favoring nondestructive observation), and—except for the reproductive system—the fact that all the cells are postmitotic, makes certain of the nematodes nearly ideal for studies of senescence, quiescence, and cryptobiosis.

Cryptobiosis in nematodes was first observed by Needham (1743) who revived *Anguina tritici* from galled wheat by suspending them in water. Since that time, cryptobiosis or some degree of quiescence (dormancy) has been demonstrated in several nematode species (Bosher,

1960; Cooper, 1969; Ellenby, 1968b; Endo, 1962; Evans, 1968; Klingler and Lengweiler-Rey, 1969), some remaining viable for periods of up to 20–30 years without normal metabolism (Steiner and Albin, 1946). Often the qualitative and quantitative differences between quiescence and cryptobiosis are difficult to demonstrate. However, both are depressed metabolic states reversible within a short time after environmental stress is relieved (Bhatt, 1967; Bhatt and Rohde, 1970; Cooper, 1969; Kostük, 1965), in contrast with nematode or insect molting or diapause which are states of heightened metabolic activity that require completion of regenerative cycles before normal senescence resumes.

This review deals with the biological and physiological characteristics of the Nematoda which may be of significance in studies of biological aging. In general, the discussion is confined to soil-inhabiting nematodes, but supporting evidence is drawn from studies of animal parasitic forms and other kinds of organisms.

## II. SENESCENCE

Information about senescence in nematodes is found primarily in studies on survival. These have been reviewed in detail by Van Gundy (1965) and Wallace (1964). Related information can be obtained in Chandler and Read (1961), Fairbairn (1960), Hyman (1951), Rogers (1962), Rogers and Sommerville (1963), Thorne (1961), Wallace (1961), and Winslow (1960). All studies on survival have been complicated by starvation and have usually been confined to a specific larval stage; thus, little information on longevity and mortality of nematodes is yet available. The noninfective larvae of some plant and animal parasitic nematodes are known to survive for extended periods without exogenous energy input (Fielding, 1951). Larvae in eggs and the newly hatched larvae of many species are unable to feed, so must exist on endogenous reserves accumulated by the adult female (Giovannola, 1936; Poinar and Leutenegger, 1968; Rogers, 1948; Van Gundy, 1958, 1959, 1965).

Although many investigators (Fairbairn, 1960; Rogers, 1962; Van Gundy, 1965; Wallace, 1966) have suggested that lipids are essential for survival of nonfeeding nematodes, there is little quantitative data on the amount utilized. It has been shown by histochemical methods that both lipid and glycogen reserves are built up in earlier stages (Chowdhury *et al.*, 1958; Giovannola, 1936; Jones, 1955) and that decreases in nematode activity and infectivity are related to their depletion. Physical activity levels and rate of lipid utilization have been correlated with aging or starvation in third-stage larvae of *Ancylostoma caninum* (Clark, 1969), *Ascaridia galli* (Elliott, 1954), *Haemonchus contortus* (Rogers, 1939, 1940), *Neoaplectana carpocapsae* (Poinar and Leutenegger, 1968), and *Nippostrongylus brasiliensis* (Haley and Clifford, 1960; Wilson, 1965). Wilson (1965) reported lipids comprised 15–20% of the tissue solids in infective larvae of *N. brasiliensis* and were catabolized at a rate of 0.9% per day over a period of 8 days at 25°C. He demonstrated, however, that more nonlipids than lipids were consumed. No glycogen was present, and total carbohydrate did not decrease. These results indicated that endogenous lipids are the principal but not the only source of energy reserves in the survival of infective larvae. The eggs of *Ascaris lumbricoides* had a 27% decrease in saponifiable lipids at 30°C (Passey and Fairbairn, 1955) but did not appear to utilize any of their stored carbohydrates. Barrett (1969) reported that infective larvae of *Strongyloides ratti* do not feed but rely entirely on endogenous food reserves deposited during the first and second larval stages. He stated that lipids comprised 25% (neutral lipids were 82% and phospholipids 18% of this) of the dry weight of the infective larvae. After 12 days at 20°C, only about 9% remained. Lipid catabolism was linear for the first 4 days, then it decreased; only the neutral lipids decreased, while the phospholipids were unchanged during aging. There was some glycogen catabolism, from 3 to 1.7% of the dry weight after 12 days. This is in agreement with the results of Sivapalan and Jenkins (1966) who reported that only neutral lipids were catabolized in nonfeeding *Panagrellus redivivus*. It might be expected that only the neutral lipids would be utilized since phospholipids have, in general, a structural rather than storage reserve function. Van Gundy *et al.* (1967) reported that lipids comprise about 30% of the dry weight of *Tylenchulus semipenetrans* and *Meloidogyne javanica*, a figure similar to that quoted by Tracey (1958) for three species of *Ditylenchus* and by Krusberg (1967) for *Pratylenchus penetrans*, *Aphelenchoides ritzemabosi*, and *Tylenchorhynchus claytoni*. They reported 5.2% weight loss per day for the two plant parasitic nematodes, 1.8% lipid and 3.4% nonlipid. Although they did not specifically demonstrate protein utilization, they speculated that this may explain some of the loss of dry weight. With fresh unstarved fourth-stage larvae and adult *Aphelenchus avenae* and *Caenorhabditis* sp., the neutral lipid content for each species was 33 and 36% of the dry weight, respectively. When starved under well-aerated conditions at 27°C, these nematodes utilized lipids linearly for the first 10 days, after which the rate of catabolism decreased (Cooper, 1969). Over the first 10 days there was a 37% decrease in total dry weight for *A. avenae* and a 43% decrease in *Caenorhabditis* sp., with the neutral lipids accounting for 23 and 27% of this decrease, respectively. Comparison of the glycogen catabolism of these two nematodes indicated aerobically nonfeeding *A. avenae* maintained a constant amount (about 8% of the dry weight) of glycogen while *Caenorhabditis* sp. had utilized almost all of its internal glycogen (3% of the dry weight) within 72 hr.

The respiration rates of free-living larvae (Schwabe, 1957) and third-stage larvae (Rogers, 1948) of *Nippostrongylus muris* decreased with time (up to 12 days). Van Gundy *et al.* (1967) noted similar decreases in the respiration rate in second-stage larvae of *Meloidogyne javanica* over 8 days. Barrett (1969) reported that as infective larvae of *S. ratti* aged, there was an exponential decrease in the rate of respiration and also a corresponding drop in the rate of substrate utilization, motility, and infectivity. The respiratory quotient (RQ) of the infective larvae remained the same regardless of age, indicating that a similar substrate was being catabolized. A similar decrease in respiration and substrate utilization was measured in *Caenorhabditis* sp.; there was, however a shift in RQ (from 0.96 to 0.43) after the nematode had been starved for 20–24 hr (Cooper, 1969). Infectivity and motility in second-stage larvae of *T. semipenetrans* was not correlated with decreased oxygen uptake with time (Van Gundy *et al.*, 1967). In these studies there was a rapid initial reduction (24–28 hr) in respiration rates, after which it remained constant for 6 days. However, substrate utilization did decrease with time. With *A. avenae* there was neither an initial drop in oxygen consumption nor a decreased lipid utilization with time up to 10 days (Cooper, 1969). There was, however, a decrease in neutral lipid usage after 10 days. Wilson (1965) suggested that decreasing rates of oxygen uptake and substrate utilization during survival may be an indication of a physiological adaptation which enables the nematode (larvae or eggs) to conserve stored energy reserves.

Van Gundy *et al.* (1967) reported that the enzymes esterase and acid phosphatase, both of which function in the hydrolysis of lipids and phospholipids, had lower activity in aged than in fresh larvae of *M. javanica.*

The only clear-cut study on aging conducted with soil-inhabiting nematodes has been done by Gershon (1970) and Erlanger and Gershon (1970). In these studies *Caenorhabditis briggsae, Caenorhabditis elegans,* and *Turbatrix aceti* were employed under axenic conditions through the full life cycle. With this system the results were complicated neither by other living organisms nor by starvation.

Gershon (1970) reported that adults of *T. aceti* did age and that the life span at 50% mortality was 25 days, with some individuals surviving to 45 days. He showed (Gershon, personal communication) similar results for *C. elegans* and *C. briggsae*. Erlanger and Gershon (1970) studied the activity of representative enzymes: acetylcholinesterase, α-amylase, malic dehydrogenase, acid ribonuclease, and acid phosphatase. It was found that the activity of acetylcholinesterase, α-amylase, and malic dehydrogenase decreased, while that of the acid hydrolase decreased at first but then rose again in the older ages. Gershon (personal communication) found that the activity of isocitrate lyase in *T. aceti* decreased with the age of the adult nematodes, and there was an increase in a protein which could be precipitated with the antibody (from rabbit) formed with isocitrate lyase as the antigen, thus suggesting an increase in nonfunctional protein within the cells with age.

Studies on the survival of plant parasitic nematodes by Van Gundy and co-workers (Van Gundy and Thomason, 1962; Van Gundy *et al.*, 1967; Van Gundy, unpublished results) also indicate that senescence does occur in nematodes (Fig. 2). Although these studies were conducted on starving nematodes, data on endogenous food reserves (lipids) indicated that lipid depletion was not the causal factor of early mortality in the population even though it may have been at later stages. In soil, two of these nematodes (*M. javanica* and *T. semipenetrans*) had entered quiescence, which enabled the nematodes to survive without any apparent physiological aging. When compared to those *in vitro*, 64-day-old larvae of *M. javanica* had a physiological age of 4 days, and 128-day-old *T.*

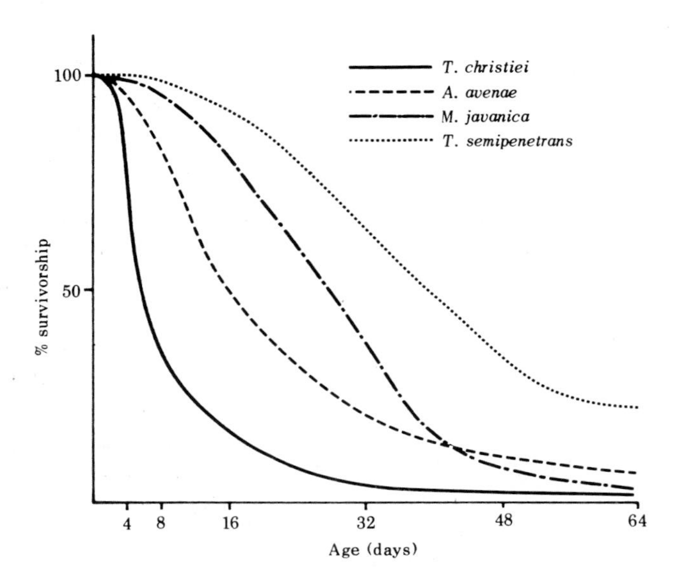


FIG. 2. Survival of nonfeeding plant parasitic nematodes aged at 27°C.

*semipenetrans* larvae had a physiological age of 16 days. It was postulated that decreased oxygen tension was responsible for reduced physiological aging. In still other studies (Van Gundy *et al.*, 1967), it appeared that aging of *M. javanica* larvae actually began in the egg and continued throughout the life of the nematode unless interrupted by unfavorable environmental factors. Under these conditions, developing larvae, hatched immediately from the egg, retained a higher level of infectivity and motility than larvae stored for some time in the egg before hatching. Nonfeeding *A. avenae* appears to age physiologically and survives in an aerobic environment for about 40–50 days on its endogenous food reserves (Cooper, 1969). When placed under prolonged oxygen stress, however, they survive for more than 90 days without apparent physiological aging. The most interesting feature of this study was that aging of second-, third-, and fourth-stage larvae and adults (eggs not tested) could be reversibly stopped and resumed at will.

Several theories on aging have been proposed, but most are difficult to test experimentally. One describes aging as a genetically programmed phenomenon. Another attributes aging to random somatic mutations which accumulate in postmeiotic cells (Curtis, 1966). As more knowledge of nematode genetics is gained, it may be possible to test these theories using soil-inhabiting nematodes. A recent theory of aging in human cells was proposed by Tappel (1968). According to Tappel, the aging process(es) develops from a sequence of biochemical reactions that occur in the following order: (1) peroxidation of polyunsaturated lipids, resulting in the formation of free radical intermediate products (direct reaction of oxygen and polyunsaturated lipids to form unstable peroxides); (2) interaction of free radicals with cell enzymes and membranes, resulting in the disruption of their structures and functions through polymerization and protein chain-scission reactions; and (3) specific damage to lysosome organelles which contain enzymes destructive to cellular components, which lead to both tissue destruction and the formation of age pigments. These processes may be accelerated in the absence of antioxidants such as vitamin E, ascorbic acid, and methionine.

Nematodes contain polyunsaturated lipids (Krusberg, 1967; Sivapalan and Jenkins, 1966) and possibly age-pigment granules composed of lipofuchsin which appear to develop from lysosomelike granules (Lee, 1969). The usual complement of intercellular organelles, common in the metazoa, are also seen in nematodes (Johnson *et al.*, 1970a,b; Lee, 1969; Miller, 1967); thus, it appears likely that the same processes of aging can occur in postmitotic cells of nematodes as those found in other organisms, including humans. Nematodes provide a "relatively" simple biological system in which aging process(es) may be studied in the future.

## III. QUIESCENCE

Comparatively few places in the world have conditions which allow soil-inhabiting nematodes to reproduce and develop throughout the year. Usually there is a cold or dry season which stops development and/or reduces metabolism (Van Gundy, 1965). This leads to quiescence, usually accompanied by a reduced metabolism which increases as conditions rise above the threshold for normal activity, thus enabling nematodes to survive environments that are not consistently favorable.

Quiescence can be induced by a wide variety of environmental stresses. Generally, quiescent states have been classified in terms of the type of environmental stress involved. The quiescent period may be limited by metabolic exhaustion and extremes of environmental conditions.

### A. Anhydrobiosis

The stem nematode, *Ditylenchus dipsaci,* possesses outstanding ability to survive anhydrobiosis. Specimens have been revived after storage in dry plant material for 23 years (Fielding, 1951). Isolated fourth-stage larvae of *D. dipsaci,* at 50% relative humidity, survive for at least one month (Wallace, 1962). Fourth-stage larvae of *Aphelenchus avenae* can survive desiccation for over a year on a dry potato dextrose agar plus fungus mat, while free fourth-stage larvae and adults can survive for only minutes (Cooper, 1969; Evans, 1968; Mankau and Mankau, 1964). Third-stage larvae of *Caenorhabditis* sp. cannot survive desiccation either singly or in a dried bacterial culture for more than a few minutes (Cooper, unpublished data). Ellenby (1968a), working with "nematode wool" (aggregates of *D. dipsaci*), reported nematodes in the center of the aggregations survived longer than those on the outside of the mass. They also dried sooner. He also reported evidence that in the isolated larva there was a gradual decrease in the permeability of the drying cuticle and that survival may be related to the rate at which drying occurs.

*Aphelenchus avenae,* exposed to desiccation, has been reported to swarm, and it was believed that congregation into masses of nematodes facilitated survival during anhydrobiosis (Evans, 1968; Townshend, 1964). These reports also lend support to the hypothesis that slow drying favors longest survival during anhydrobiosis (Ellenby, 1968a; Endo, 1962; Fielding, 1951; Franklin, 1937; Wallace, 1962). This may partly explain the longer survival of *Heterodera* sp. eggs in cysts than as free eggs and why *A. tritici* or *D. dipsaci* survive desiccation better in plant material than as free larvae. In addition to slowing the rate of drying they may also protect these organisms from extreme desiccation.

Little is known about the mechanisms of survival during anhydrobiosis, and in general the relationships between moisture and temperature are difficult to assess. Larvae of *Ditylenchus dipsaci* and *Trichostrongylus colubriformis* have survived in 50% relative humidity at 24°C for 32 days (Wallace, 1962; Wallace and Doncaster, 1964), and *Trichostrongylus axei* at 29% relative humidity and 38°C for 104 days (Stewart and Douglas, 1938).

Cysts of the golden nematode, *Heterodera rostochiensis,* are formed from the body wall of the mature female swollen with eggs. The cyst wall, composed of tanned protein (Ellenby, 1946), is resistant to decay in soil and eggs remain viable in dry-stored cysts for 5 years (Ellenby, 1955), but free larvae will survive drying for only 4 min (Kämpfe, 1959). Endo (1962), working with *H. glycines,* has shown that, upon drying, eggs in cysts in soil, eggs in cysts not in soil, free eggs, and free larvae will remain viable for months, weeks, days, and minutes, respectively.

Why some nematodes survive anhydrobiosis better than others is not known. Van Gundy (1965) suggested that distribution of water and proteinaceous constituents in the cells may be such that water withdrawal does not cause the disruption of their intimate structure. He further speculated that desiccation resistance in *D. dipsaci* was related to high lipid content. However, Krusberg (1967) found no correlation between lipid content and desiccation resistance in five soil-inhabiting nematodes: *D. triformis, D. dipsaci,* and *Aphelenchoides ritzemabosi* survived desiccation, whereas *Pratylenchus penetrans* and *Tylenchorhynchus claytoni* were killed by desiccation. However, Baxter and Blake (1969) found that 21.3% *P. thornei* survived drying to 80% relative humidity, and of the survivors 59% survived 20 weeks in air-dried soil. Wallace and Greet (1964) suggested *Tylenchorhynchus icarus* resists desiccation and may survive in dry soil surface layers for several days. Cooper (1969) found very little quantitative difference in the lipid content of *A. avenae* and *Caenorhabditis* sp.; however, *A. avenae* could survive desiccation, while *Caenorhabditis* could not.

One metabolic study has been conducted on nematodes recovering from a state of anhydrobiosis. Bhatt and Rohde (1970), working with *A. tritici,* reported no oxygen uptake was detected in desiccated nematodes and that only when placed in water did they "recover" and resume oxygen uptake. They found that nematodes desiccated for a longer period of time require more time to resume maximum oxygen consumption. Individuals dried for 1 year required 6–10 hr to obtain maximum oxygen uptake, while those desiccated for 10 years needed 70–80 hr. The reasons for this are unknown, but it appears that gradual physiological alterations occur during anhydrobiosis which could be dependent upon: (1) a greater degree of dehydration, (2) gradual decrease in enzymic activity and/or loss of enzymes, (3) utilization of intermediate metabolites that must be resynthesized, and (4) a buildup of inhibitory products that must be either metabolized and/or excreted.

When discussing environmental factors which may induce quiescence in soil-inhabiting nematodes, it is very difficult to separate the effects of high osmotic pressures from those of dehydration. There are no convincing data which clearly demonstrate induction of quiescence by high osmotic pressures which does not also involve internal dehydration. For this reason, osmobiosis will be considered a special form of anhydrobiosis in this review.

Saprobic nematodes from soil high in decaying matter tolerate wide ranges of osmotic potential (Stephenson, 1942; Sachs, 1950; Osche, 1952). Marine forms may have considerable tolerance to salinity (Croll and Viglierchio, 1969; Winslow, 1960). The work of Wallace and Greet (1964), Blake (1961), and Roggen (1966) suggests that the phytoparasitic nematodes *D. dipsaci, Tylenchorhynchus icarus,* and *Xiphinema index* can tolerate osmotic pressures up to 10 atm, a level considerably higher than in most agricultural soils.

Bhatt and Rohde (1970), studying the influence of high osmotic pressures on oxygen uptake in *D. dipsaci, A. tritici, A. agrostis,* and *Pratylenchus penetrans,* reported an increase in oxygen consumption as osmotic potentials were increased to 22.5 atm. They also reported that when these nematodes were placed in a hypotonic solution (distilled water), there was a gradual decrease in oxygen uptake with time; however, these results were confounded by the effect of starvation.

Viglierchio *et al.* (1969), comparing the physiological responses of several free-living and plant parasitic nematodes to osmotic stress, noted wide differences: Some were capable of osmoregulation through the movement of ions while others were not. They found that fourth-stage *D. dipsaci* larvae that can tolerate considerable desiccation had high osmotic pressure in their tissues and tolerated high osmotic pressures rather than adapting rapidly to changes in osmotic pressure. *Rhabditis* spp. with poikilosmotic tissues apparently has a mechanism for ion uptake which successfully adapts the nematode to short-term osmotic fluctuations but fails to do so during prolonged exposure to high osmotic pressures. This phenomenon may be related to their inability to tolerate desiccation and may also be the case with many soil-inhabiting forms since those best able to survive desiccation are also able to "tolerate" high osmotic pressure.

This "capacity to tolerate" high osmotic potential, and presumably desiccation, may explain why *D. dipsaci* and *Aphelenchoides* spp., which normally live in environments of rapid fluctuating moisture, can endure accelerated desiccation, while others can survive anhydrobiosis, provided desiccation is not too rapid.

### B. Anoxybiosis

Although the specific oxygen requirements of these nematodes are little understood, it has been suggested that a continuing and abundant supply of oxygen is necessary for growth and reproduction (Cooper *et al.*, 1970; Dropkin, 1966; Van Gundy and Stolzy, 1961, 1963a,b; Wallace, 1964, 1966). Many soil-inhabiting nematodes live in environments that can be, for varying periods of time, low in oxygen content (Letey *et al.*, 1962; Van Gundy *et al.*, 1968). Many can be considered facultative anaerobes, living a predominantly aerobic existence, with oxidative metabolism the main energy-producing process. However, they can survive microaerobic and anaerobic periods using fermentative metabolism (Cooper, 1969). The capacity of a nematode for both oxidative and fermentative metabolism certainly enhances its chance of survival under unfavorable lowered oxygen concentrations. This has been demonstrated in a comparative study on the influence of aerobic (5–21% $O_2$), microaerobic (less than 5% $O_2$), and anaerobic (0% $O_2$) environments of the neutral lipid (oxidative) and glycogen (fermentative) catabolism of starving *A. avenae* and a *Caenorhabditis* sp. (Cooper, 1969). *Caenorhabditis* sp. survived anoxybiosis for less than 80 hr, while *A. avenae* survived for 30 days and longer. With these nematodes, neutral storage lipids were the main energy source under aerobic conditions. There was no neutral lipid catabolism under anaerobic conditions. Analyses of aerobic, microaerobic, and anaerobic nonfeeding *Caenorhabditis* showed successively greater net usage of storage glycogen (initial content equals 3% of the dry weight). With *A. avenae,* however, there was no net glycogen (about 8% of the dry weight) utilization under aerobic conditions except under microaerobic or anaerobic conditions. *Caenorhabditis* utilized all of its detectable internal glycogen in less than 72 hr, while *A. avenae* still had some glycogen remaining after 20 days, which may explain why some soil-inhabiting nematodes survive anoxybiosis better than others. At 5% oxygen, neutral lipid utilization by *A. avenae* was ~50% of that of lipid catabolism at 21% $O_2$ and there was no usage below 4% oxygen. A similar decrease in the oxygen uptake was observed: At 5% oxygen (compared to 21% oxygen controls) there was a 55% reduction in respiration and below 5% oxygen there was no oxygen uptake.

Survival of animal parasitic (Davey, 1938; von Brand and Simpson, 1945), plant parasitic (Van Gundy and Stolzy, 1963a,b; Van Gundy *et al.*, 1962), and free-living (Bryant *et al.*, 1964) nematodes was extended several days in the presence of oxygen. Van Gundy (1965) suggested the higher energy yield of oxidative (lipid) catabolism may favor survival compared to the lower energy yield of fermentative (glycogen) catabolism, and this also correlates well with the motility levels associated with the presence and absence of oxygen. However, the inability of certain nematodes to survive prolonged anoxybiosis may not result from the quantitative differences in energy production between oxidative or fermentative catabolism but may be caused by other factors, such as depletion of storage glycogen, enzyme deactivation, or the inability to excrete toxic metabolites.

Soil-inhabiting nematodes are primarily aerobic organisms, and many of their physiological and behavioral functions [for example, egg laying (Le Jambre and Whitlock, 1967; Van Gundy and Stolzy, 1961, 1963a), egg hatch (Van Gundy and Stolzy, 1963a), development, movement, and molt (Cooper, 1969; Van Gundy and Stolzy, 1961, 1963a,b)] are dependent on an abundant supply of oxygen. The reasons for this are probably twofold: (1) These physiological and behavioral functions require more energy than can be produced by anaerobic glycolysis, thus requiring oxidative metabolism; (2) toxic or inhibitory metabolic end products may accumulate as a result of anaerobic fermentation; such could be the case in egg hatch where there may be no method for excretion of those end products.

Even though soil-inhabiting nematodes can be considered as facultative anaerobes, at least three forms studied maintain oxidative metabolism over a wide range of oxygen concentrations. *Aphelenchus avenae, Caenorhabditis* sp., and *Caenorhabditis briggsae* have the ability to regulate oxygen consumption and metabolize aerobically even when exposed to oxygen concentrations ranging from 4 to 21% (Cooper, 1969; Bryant *et al.*, 1964).

Van Gundy (1965) suggested that relative capability to achieve anoxybiosis may be an important factor governing the population maximum reached by particular species in a given habitat. This appears to be the case with most nematode populations where this has been studied (Cooper *et al.*, 1970; McElroy, 1967; Van Gundy and Stolzy, 1961, 1963a,b). It also appears that anoxybiosis can enhance the survival of individual nematodes, at least with *A. avenae* (Cooper, 1969) and *Meloidogyne javanica* (Van Gundy *et al.*, 1967). Aerobically starved *A. avenae* died in about 30–35 days at 27°C, while those maintained in anoxybiosis survived up to 60 days without loss of viability. This may be true for many soil-inhabiting nematodes.

### C. Cryobiosis

Generally, nematodes can be grouped into those which are freeze-susceptible and those which are freeze-resistant. Those nematodes which are considered freeze-susceptible can be further catgorized into cold-sensitive and cold-tolerant. Cold-tolerant nematodes are those which survive chilling injury induced between 10°– 0°C. Nematodes are considered freeze-resistant if they are not injured by freezing (Sayre, 1964; Van Gundy, 1965).

Based on motility and survival, when storage temperatures were reduced from 15° to 3°C, *Trichodorus christiei* and *M. javanica,* although freeze-susceptible, were considered sensitive, while *Pratylenchus scribneri* and *Tylenchulus semipenetrans* were tolerant (Thomason *et al.*, 1964; Van Gundy and Thomason, 1962). No information was given for temperatures below 3°C. An example of a freeze-resistant nematode is *A. avenae* (Cooper, unpublished data). The free-living nematode *Caenorhabditis* sp. was considered to be freeze-susceptible since it could not survive freezing (Cooper, unpublished data). Animal-parasitic nematodes such as *Strongylus edentatus* (Hyman, 1951), *S. vulgaris* (Hyman, 1951), and *Nematodirus filicollis* (Poole, 1956) can be classified as freeze-resistant because they have survived in a frozen condition for up to one year, while *Angiostrongylus cantonensis* is freeze-susceptible since it could not survive freezing (Alicata, 1967).

Eggs of most species studied appear to be more tolerant to freezing temperatures than are larvae (Croften, 1963; Hyman, 1951; Sayre, 1964; Wallace, 1964). In certain nematodes, larval stages are more resistant to freezing than other larval stages or adults (Van Gundy, 1965). The fourth-stage larvae of *Paratylenchus projectus* will resist freezing at —4°C, while third-stage larvae and adults will not (Rhoades and Linford, 1959). Third-stage larvae of *Ostertagia circumcincta* survived —6°C for 2 weeks, but second-stage larvae died after 24 hr (Furman, 1944).

The reason(s) some nematodes (and larval stages of certain nematodes) can survive freezing is unknown. Generally, the lowering of tissue temperature results in a corresponding decrease in the metabolism

(Overgaard Nielsen, 1949; von Brand, 1960). Van Gundy *et al.* (1967) found that the body contents of *M. javanica* and *T. semipenetrans* were conserved at low temperatures. Cooper (1969) found that there was a decrease in the oxygen uptake in *A. avenae* and *Caenorhabditis* sp. when the temperature decreased from 27° to 20°C. Santmyer (1956) reported a linear decrease in the oxygen consumption for *Panagrellus redivivus* between 35° and 20°C, below which there was a more gradual decrease in oxygen uptake to 5°C. However, as the temperature became lower, metabolic imbalances and/or partial loss of active ion transport may have resulted (Van Gundy, 1965). At temperatures near absolute zero, all enzymic activity and metabolism probably ceases. There is some evidence in plants and animals that the ratio of unsaturated to saturated lipids in the nuclear membranes may explain why some organisms can escape chilling 0°–10°C injury while others do not (Lyons *et al.*, 1964). Sayre (1964) speculated that freeze-resistant nematodes accumulate metabolic products, thus increasing the cytoplasmic viscosity and solute concentration and decreasing the amount of free water. Thus when super-cooled these nematodes are more tolerant to ice formation, probably because smaller ice crystals are formed which do not damage the cells (Smith, 1961). Nematodes in which this occurs are generally considered to be freeze-resistant.

The influence of low temperatures on hatching, reproduction, movement, development, etc., of soil-inhabiting nematodes and several animal parasitic forms has been reviewed (Crofton, 1963; von Brand, 1960; Wallace, 1964).

## IV. CRYPTOBIOSIS

Certain organisms can enter such extreme dormancy that no functions of a living system appear to be present. These organisms are not dead however; the condition is a reversible one. Keilin (1959) proposed the term *cryptobiosis* to describe this form of dormancy, and he defined it as "the state of an organism when it shows no visible signs of life and when its metabolic activity becomes hardly measurable or comes reversibly to a standstill."

Many nematodes enter a state of cryptobiosis when they lose contact with water or when exposed to other unfavorable environmental conditions accompanied by suspension (or great reduction) of all life processes for considerable periods of time. "Prolonged" anhydrobiosis, anoxybiosis, or cryobiosis may induce cryptobiosis in some nematodes. Cryptobiosis appears to be restricted to soil-inhabiting and plant parasitic species (Steiner, 1940; Van Gundy, 1965).

It is over 200 years since Needham (1743) first observed that the white fibrous substances from the "wheat galls" became worms or "eels" upon the addition of water. This was the first demonstration of nematode cryptobiosis. Since then a number of nematodes have been reported to be capable of surviving prolonged cryptobiosis (French and Barraclough, 1962; Harrison and Hooper, 1963; Hirschmann, 1962; Rhoades and Linford, 1959; von Brand and Simpson, 1945). *Anguina tritici,* the wheat gall nematode, was revived after 28 years dry storage (Fielding, 1951).

Nematodes can survive cryptobiosis for long periods of time; for example, fourth-stage larvae of *D. dipsaci* lyophilized for 5 years at —80°C (Bosher and McKeen, 1954), a *Plectus* species after 125 hr at —190°C (Hyman, 1951), second-stage larvae of *A. tritici* and fourth-stage larvae of *D. dipsaci* after 23 years in dry plant tissue (Fielding, 1951), and *Actinolaimus hintoni* and *Dorylaimus keilini* after 10 years in dried mud (Lee, 1961). The longest period of cryptobiosis recorded is 39 years for *Tylenchus polyhypnus* (Steiner and Albin, 1946). These nematodes must, however, remain dry in order to stay in a state of cryptobiosis.

Very few metabolic studies have been conducted on nematodes in a state of cryptobiosis. While working with *A. avenae* (Cooper, 1969) it was discovered that a cryptobiotic state could be induced with prolonged (6 days and longer) anoxybiosis. When second-, third-, fourth-stage larvae and adults were maintained under an anaerobic ($N_2$) condition for longer than 120 hr (at 27°C), they entered into cryptobiosis. While in this state there was no measurable metabolism (either glycogen or neutral lipid metabolism). Even after being returned to an aerobic environment, it was several hours before oxygen uptake, neutral lipid catabolism, or glycogen anabolism could be detected. These nematodes survived 60 days of cryptobiosis with $>85\%$ recovery. During this entire period there was no measurable metabolic activity. However, it appeared that larvae survived the treatment better than adults.

Kostük (1965), utilizing histochemical methods, reported no detectable lipid, protein, or glycogen metabolism in *A. tritici* in a cryptobiotic state produced by anhydrobiosis. He concluded that the fermentation of simple sugars (aerobic glycolysis) takes place during cryptobiosis. In addition, he suggested these same reactions may also function in *Ditylenchus allii, Aphelenchoides besseyi,* and *Heterodera major* during cryptobiosis.

Clegg (1967), while studying cryptobiosis in the eggs of brine shrimp, *Artemia salina,* reported no decrease in the concentration of trehalose, glycerol, and glycogen during storage in a dry state for periods up to 6 years. He did not study the decrease in lipid concentrations in these organisms. However, the work of Cooper (1969) (anoxybiosis), Kostük (1965) (anhydrobiosis), and Krusberg (1967) (anhydrobiosis) suggests that lipid content (and lipid catabolism) plays no major role in determining whether an organism can survive in a cryptobiotic state.

The metabolic reactions and pathways occurring while an organism is in cryptobiosis are unknown, if indeed there is any metabolic activity. It is not understood whether cryptobiosis in nematodes, as induced by anhydrobiosis, affects the organism metabolically in a manner similar to that induced by anoxybiosis or cryobiosis. It is not difficult to visualize the complete inhibition of enzymic activity near absolute zero, as has been demonstrated by the survival of *Plectus* sp. exposed to —253°C for 26 hr and —270°C for 8 hr (Hyman, 1951). The metabolic activity occurring during cryptobiosis as induced by anhydrobiosis is more difficult to assess. Many soil-inhabiting nematodes have been stored for long periods in a desiccated state at temperatures between 10° and 30°C (Baines *et al.*, 1959; Bergeson, 1959; Fielding, 1951; Raski *et al.*, 1965). The degree of dehydration is sometimes difficult to determine (Ellenby, 1968a,b). But assuming there was complete dehydration, the question arises as to whether or not enzymic activity (even at temperatures of molecular activity) can occur while the nematode is in a solid state; more than likely it cannot. It is difficult, however, to visualize the complete lack of metabolic (enzymic) activity in cryptobiotic nematodes as a result of a prolonged lack of oxygen in an aqueous medium (Cooper, 1969) at temperatures between 25° and 30°C. It has been suggested (Van Gundy, 1965) that the length of time a particular nematode can survive in a cryptobiotic state is limited by the rate of metabolic exhaustion. There are likely to be states of cryptobiosis however in which there is no metabolism occurring, and the exhaustion of food reserves is of no importance in determining the length of time a nematode can endure. It is difficult to explain what is occurring in a nematode like *A. avenae* in which cryptobiosis can be a result of anhydrobiosis, cryobiosis, or anoxybiosis (Cooper, 1969; Evans, 1968; Mankau and Mankau, 1964; Townshend, 1964). Whether or not it is the same "type" of cryptobiosis in all cases is not known.

This unique ability of *A. avenae* (and possibly other soil-inhabiting nematodes) to enter a cryptobiotic state in response to several different types of environmental factors makes it a valuable organism for the study of the physiological, biochemical, and morphological changes occurring during cryptobiosis (and for the study of biological senescence). This nematode can be cultured in very large numbers (Evans, 1970) and can be readily manipulated experimentally, with at least the third- and fourth-stage larvae and adults capable of surviving cryptobiosis (Cooper, 1969; Evans, 1968).

A very interesting physiological difference was observed with *A. avenae* when contrasting the transition anoxybiosis with cryptobiosis. While this nematode was still metabolically active under anoxybiosis, measurable quantities of ethanol were being excreted (catabolized from storage glycogen); however, as the nematodes entered into the state of cryptobiosis, glycogen catabolism and ethanol production apparently ceased. Anaerobic cryptobiosis generally occurred when the ethanol concentration in the incubation medium reached 8 m*M* (Cooper, 1969).

There is at least one example where biological differences can be observed in a nematode, depending on how cryptobiosis was induced. With *A. avenae* (Evans, 1968; Mankau and Mankau, 1964), cryptobiosis as induced by desiccation results in the nematode's becoming tightly coiled, while cryptobiosis, as induced by the lack of oxygen, results in the nematode's becoming straight (Cooper, 1969). These differences are not understood; however, the coiling may reduce the rate of water loss, while anoxybiosis causes the nematode to relax. The authors (Cooper and Van Gundy, unpublished data) now have evidence that the lack of oxygen (up to 7 days) has no effect on the survival of desiccated nematodes, but it is unknown what the effect of desiccation would be on nematodes that are already in a state of cryptobiosis as a result of a lack of oxygen.

Nematodes can survive in a cryptobiotic state for a long period of time, and their resistance to nematicides and high temperatures appears to increase (Kostük, 1965). Evans and Thomason (1970) noted that *A. avenae* was almost unaffected by the nematicide ethylene dibromide (EDB) while in a state of cryptobiosis (induced by anoxybiosis) when compared to nematodes maintained in an aerobic environment. Cryptobiosis makes the control of some plant parasitic nematodes difficult and facilitates the transportation of parasites within infested plants and soil, bulbs, and root crops. Cryptobiosis may explain the failure of flooding to control certain soil-inhabiting nematodes (Bird and Jenkins, 1965). Prolonged anoxybiosis (caused by the lowering of oxygen tensions in the soil because of flooding) may induce cryptobiosis, whereby the nematodes remain in a very inactive state, probably increasing their chances of survival. When the soil is returned to a condition for planting, the nematodes recover and resume crop damage.

Although the above-mentioned experimental condition may not occur in nature, many soil-inhabiting nematodes naturally live in habitats subjected to extremes of cold, drying, and low oxygen concentrations (Van Gundy *et al.*, 1967; Wallace, 1964). Cryptobiosis, therefore, may have importance in the survival of the species in which it occurs and has the effect of extending a comparatively short survival span of a few months in some nematodes to as long as 10–30 years.

As pointed out by others (Buck, 1965; Keilin, 1959), the biological significance of cryptobiosis does not reside only in the question of whether or not all life processes can come to a complete but reversible halt. Of equal interest are the underlying biochemical, physiological, and morphological circumstances which enable such metabolic regulation and thereby permit such unusual tolerance to conditions that rapidly destroy the vast majority of living organisms.

## V. SUMMARY

Consideration of the "state of the art" of research in quiescence, cryptobiosis, and senescence is complicated by the fact that most researchers have other fields of interest and by the extreme difficulty of working experimentally with nematodes in a state of cryptobiosis before the problems of detecting greatly reduced metabolism have been overcome. It is evident that there has been a great deal of study on the life cycles of numerous soil-inhabiting nematodes, but very little information is available on the life expectancy of an individual species. Some forms are able to withstand desiccation, lack of oxygen, or unusually low temperatures, while other species cannot. Some nematodes survive unfavorable conditions best at a specific stage of development, while others can survive them during most any stage of their life cycle. Soil-inhabiting species which can survive one kind of unfavorable environmental extreme are often able to survive other kinds as well. Generally, those that do not endure one form cannot survive another. For nonfeeding nematodes prolonged survival, under favorable conditions, is dependent on the conservation of energy (mainly lipids), while prolonged survival under unfavorable conditions is not understood.

The significance of anhydrobiosis, cryobiosis, anoxybiosis, and cryptobiosis, then, is that many nematodes are able to inhabit environments not consistently favorable and that a comparatively short life cycle can be extended for varying periods of time up to 30 years.

Briefly, we have attempted to develop the following important concepts:

(1) Many soil-inhabiting nematodes have a physiological stage (or stages) which enables them to endure anhydrobiosis, cryobiosis, or anoxybiosis.

(2) If unfavorable environmental conditions are such that metabolic

functions cannot proceed, several species go into a state of cryptobiosis during which all metabolic functions are greatly reduced or reversibly stopped.

(3) The underlying biochemical, physiological, and morphological circumstances of cryptobiosis in nematodes have important biological implications in areas of metabolic regulation.

(4) Most soil-inhabiting forms can be considered facultative anaerobes, thus having the capacity for oxidative and/or fermentative catabolism.

(5) Senescence does appear to occur.

(6) While in a state of quiescence or cryptobiosis, physiological aging is slowed or stopped.

(7) Soil-inhabiting nematodes may prove useful organisms for the general study of biological senescence.

## REFERENCES


Alicata, J. E. (1967). *J. Parasitol.* **53,** 1064–1066.

Baines, R. C., Van Gundy, S. D., and Sher, S. A. (1959). *Calif. Agr.* **13,** 16–18.

Barrett, J. (1969). *Parasitology* **59,** 343–347.

Baxter, R. I., and Blake, C. D. (1969). *Ann. Appl. Biol.* **63,** 191–203.

Bergeson, G. B. (1959). *Nematologica* **4,** 344–354.

Bhatt, B. D. (1967). Ph.D. Thesis, 69 pp. Univ. of Massachusetts, Amherst, Massachusetts.

Bhatt, B. D., and Rohde, R. A. (1970). *J. Nematol.* **2,** 277–285.

Bird, G. W., and Jenkins, W. R. (1965). *Plant Dis. Rep.* **49,** 517–518.

Blake, C. D. (1961). *Nature (London)* **192,** 144–145.

Bosher, J. E. (1960). *Proc. Helminthol. Soc. Wash.* **27,** 127–128.

Bosher, J. E., and McKeen, W. E. (1954). *Proc. Helminthol. Soc. Wash.* **21,** 113–117.

Bryant, C., Nicholas, W. L., and Jantunen, R. (1964). *Nematologica* **10,** 409–418.

Buck, J. (1965). *J. Insect Physiol.* **11,** 1503–1516.

Chandler, A. C., and Read, C. P. (1961). "Introduction to Parasitology," 10 Ed., 756 pp. Wiley, New York.

Chowdhury, A. B., Ray, H. N., and Bhaduri, N. V. (1958). *Bull. Calcutta Sch. Trop. Med.* **6,** 59.

Clark, F. E. (1969). *Exp. Parasitol.* **24,** 1–8.

Clegg, J. S. (1967). *Comp. Biochem. Physiol.* **20,** 801–809.

Cooper, A. F. (1969). Ph.D. Thesis, 108 pp. Univ. of California, Riverside, California.

Cooper, A. F., Van Gundy, S. D., and Stolzy, L. H. (1970). *J. Nematol.* **2,** 182–188.

Crofton, H. D. (1963). *Commonw. Bur. Helminthol. Tech. Commun.* **35,** 104 pp.

Croll, N. A., and Viglierchio, D. R. (1969). *Proc. Helminthol. Soc. Wash.* **36,** 1–9.

Curtis, H. J. (1966). "Biological Mechanisms of Aging," 275 pp. Springfield, Illinois.

Davey, D. G. (1938). *J. Exp. Biol.* **15,** 217–224.

Dropkin, V. H. (1966). *Ann. N.Y. Acad. Sci.* **139,** 39–52.

Ellenby, C. (1946). *Nature (London)* **157**, 302–303.
Ellenby, C. (1955). *Ann. Appl. Biol.* **43**, 1–11.
Ellenby, C. (1968a). *Experientia* **24**, 84–85.
Ellenby, C. (1968b). *Proc. Roy. Soc., Ser. B* **169**, 203–213.
Elliott, A. (1954). *Exp. Parasitol.* **3**, 307–320.
Endo, B. Y. (1962). *Phytopathology* **52**, 80–88.
Erlanger, M., and Gershon, D. (1970). *Exp. Gerontol.* **4**, 131–138.
Evans, A. A. F. (1968). Ph.D. Thesis, Univ. of Adelaide, South Australia. 176 pp.
Evans, A. A. F. (1970). *J Nematol.* **2**, 99–100.
Evans, A. A. F., and Thomason, I. J. (1970). *J. Nematol.* **1**, 287.
Fairbairn, D. (1960). *In* "Nematology: Fundamentals and Recent Advances with Emphasis in Plant Parasitic and Soil Forms" (J. N. Sasser and W. R. Jenkins, eds.), pp. 267–296. Univ. of North Carolina Press, Chapel Hill, North Carolina.
Fielding, M. J. (1951). *Proc. Helminthol. Soc. Wash.* **18**, 110–112.
Franklin, M. T. (1937). *J. Helminthol.* **15**, 69–74.
French, N., and Barraclough, R. M. (1962). *Nematologica* **7**, 309–316.
Furman, D. P. (1944). *Amer. J. Vet. Res.* **5**, 79–86.
Gershon, D. (1970). *Exp. Gerontol.* **4** 125–130.
Giovannola, A. (1936). *J. Parasitol.* **22**, 207–218.
Haley, A. J., and Clifford, C. M. (1960). *J. Parasitol.* **46**, 579–582.
Harrison, B. D., and Hooper, D. J. (1963). *Nematologica* **9**, 158–160.
Hirschmann, H. (1962). *Proc. Helminthol. Soc. Wash.* **29**, 30–43.
Hyman, L. (1951). "The Invertebrates: Acanthocephala, Aschelminthes and Entoprocta. The Pseudocoelomate Bilateria," Vol. III, 572 pp. McGraw-Hill, New York.
Johnson, P. W., Van Gundy, S. D., and Thomson, W. W. (1970a). *J. Nematol.* **2**, 42–58.
Johnson, P. W., Van Gundy, S. D., and Thomson, W. W. (1970b). *J. Nematol.* **2**, 59–79.
Jones, C. A. (1955). *J. Parasit.* **41**, Suppl., 48. (Abstr.)
Kämpfe, L. (1959). *Verh. Deut. Zool.* **25**, 278–386.
Keilin, D. (1959). *Proc. Roy. Soc. London Ser. B* **150**, 149–191.
Klingler, V. J., and Lengweiler-Rey, V. (1969). *Z. Pflanzenkr. (Pflanzenpathol.) Pflanzenschutz* **76**, 193–208.
Kostük, N. A. (1965). *Tr. Gel'mintol. Lab. Akad. Nauk SSSR* **16**, 55–57.
Krusberg, L. R. (1967). *Comp. Biochem. Physiol.* **21**, 83–90.
Lee, C. C. (1969). *Exp. Parasitol.* **24**, 336–347.
Lee, D. L. (1961). *Parasitology* **51**, 237–240.
Le Jambre, L. F., and Whitlock, J. H. (1967). *J. Parasitol.* **53**, 887.
Letey, J., Stolzy, L. H., Valoras, N., and Szuszkiewicz, T. E. (1962). *Agron. J.* **54**, 316–319.
Lyons, J. M., Wheaton, T. A., and Pratt, H. K. (1964). *Plant Physiol.* **39**, 262–268.
McElroy, F. D. (1967). Ph.D. Thesis, 92 pp. Univ. of California, Riverside, California.
Mankau, R., and Mankau, S. K. (1963). *In* "Soil Organisms" (J. Doeksen and J. van der Drift, eds.), pp. 271–280. North-Holland Publ., Amsterdam.
Miller, J. H. (1967). *J. Parasitol.* **53**, 94–99.
Needham, T. (1743). *Phil. Trans.* **42**, 634–641.
Osche, G. (1952). *Z. Morphol. Oekol. Tiere* **41**, 54–77.
Overgaard Nielsen, C. (1949). *Natura Jutlandica* **2**, 1–131.
Passey, R. F., and Fairbairn, D. (1955). *Can. J. Biochem. Physiol.* **33**, 1033–1046.

Poinar, G. O., and Leutenegger, R. (1968). *J. Parasitol.* **54,** 340–350.
Poole, J. B. (1956). *Can. J. Comp. Med.* **20,** 169–172.
Raski, D. J., Hewitt, W. B., Taylor, C. E., and Taylor, R. H. (1965). *Nematologica* **11,** 44–45.
Rhoades, H. L., and Linford, M. B. (1959). *Science* **130,** 1476–1477.
Rogers, W. P. (1939). *J. Helminthol.* **17,** 195–202.
Rogers, W. P. (1940). *J. Helminthol.* **18,** 183–192.
Rogers, W. P. (1948). *Parasitology* **39,** 105–109.
Rogers, W. P. (1962). "The Nature of Parasitism," 287 pp. Academic Press, New York.
Rogers, W. P., and Sommerville, R. I. (1963). *Advan. Parasitol.* **1,** 109–177.
Roggen, D. R. (1966). *Nematologica* **12,** 287–296.
Rothstein, M., and Nicholas, W. L. (1969). *In* "Chemical Zoology" (M. Florkin and B. T. Scheer, eds.), Vol. 3, pp. 289–328. Academic Press, New York.
Sachs, H. (1950). *Zool. Jahrb.* (*Syst.*) **79,** 209–272.
Santmyer, P. H. (1956). *Proc. Helminthol. Soc. Wash.* **23,** 30–36.
Sayre, R. M. (1964). *Nematologica* **10,** 168–179.
Schwabe, C. W. (1957). *Amer. J. Hyg.* **65,** 325–337.
Sivapalan, P., and Jenkins, W. R. (1966). *Proc. Helminthol. Soc. Wash.* **33,** 149–157.
Smith, A. U. (1961). "Biological Effects of Freezing and Super Cooling," 462 pp. Williams & Wilkins, Baltimore, Maryland.
Steiner, G. (1940). *Proc. Int. Congr. Microbiol., New York* pp. 434–435.
Steiner, G., and Albin, F. E. (1946). *J. Wash. Acad. Sci.* **36,** 97–99.
Stephenson, W. (1942). *Parasitology* **34,** 253–265.
Stewart, M. A., and Douglas, J. R. (1938). *Parasitology* **30,** 477–490.
Tappel, A. L. (1968). *Geriatrics* **23,** 97–105.
Thomason, I. J., Van Gundy, S. D., and Kirkpatrick, J. D. (1964). *Phytopathology* **54,** 192–195.
Thorne, G. (1961). "Principles of Nematology," 553 pp. McGraw-Hill, New York.
Townshend, J. L. (1964). *Can. J. Microbiol.* **10,** 727–737.
Tracey, M. V. (1958). *Nematologica* **3,** 179–183.
Van Gundy, S. D. (1958). *Nematologica* **3,** 283–294.
Van Gundy, S. D. (1959). *Proc. Helminthol. Soc. Wash.* **26,** 67–72.
Van Gundy, S. D. (1965). *Annu. Rev. Phytopathol.* **3,** 43–68.
Van Gundy, S. D., and Stolzy, L. H. (1961). *Science* **134,** 665–666.
Van Gundy, S. D., and Stolzy, L. H. (1963a). *Nematologica* **9,** 605–612.
Van Gundy, S. D., and Stolzy, L. H. (1963b). *Nature* (*London*) **200,** 1187–1189.
Van Gundy, S. D., and Thomason, I. J. (1962). *Phytopathology* **52,** 366–367.
Van Gundy, S. D., Stolzy, L. H., Szuszkiewicz, T. E., and Rackham, R. L. (1962). *Phytopathology* **52,** 628–632.
Van Gundy, S. D., Bird, A. F., and Wallace, H. R. (1967). *Phytopathology* **57,** 559–571.
Van Gundy, S. D., McElroy, F. D., Cooper, A. F., and Stolzy, L. H. (1968). *Soil Sci.* **106,** 270–274.
Viglierchio, D. R., Croll, N. A., and Gortz, J. H. (1969). *Nematologica* **15,** 15–21.
von Brand, T. (1960). *In* "Nematology: Fundamentals and Recent Advances with Emphasis in Plant Parasitic and Soil Forms" (J. N. Sasser and W. R. Jenkins, eds.), pp. 233–266. Univ. of North Carolina Press, Chapel Hill, North Carolina.
von Brand, T., and Simpson, W. F. (1945). *Proc. Soc. Exp. Biol. Med.* **60,** 368–371.
Wallace, H. R. (1961). *Helminthol. Abstr.* **30,** 1–22.

Wallace, H. R. (1962). *Nematologica* **7,** 91–101.
Wallace, H. R. (1964). "The Biology of Plant Parasitic Nematodes," 280 pp. Arnold, London.
Wallace, H. R. (1966). *Nematologica* **12,** 57–69.
Wallace, H. R., and Doncaster, C. C. (1964). *Parasitology* **54,** 313–326.
Wallace, H. R., and Greet, D. N. (1964). *Parasitology* **54,** 129–144.
Wilson, P. A. G. (1965). *Exp. Parasitol.* **16,** 190–194.
Winslow, R. D. (1960). *In* "Nematology: Fundamentals and Recent Advances with Emphasis on Plant Parasitic and Soil Forms" (J. N. Sasser and W. R. Jenkins, eds.), pp. 341–415. Univ. of North Carolina Press, Chapel Hill, North Carolina.

# Author Index

Numbers in italics refer to the pages on which the complete references are listed.


## A

Acedo, J., 79, *89,* 176, *180*
Adams, R. E., 128, *136,* 148, *157*
Ahmed, A., 274, *284*
Albin, F. E., 299, 311, *317*
Alhassan, S. A., 190, *207*
Alicata, J. E., 309, *315*
Allen, M. W., 24, *32,* 36, 45, *48, 49,* 56, 59, *69,* 185, 187, 190, 196, 197, 199, 200, *207, 210*
Allen, T. C., Jr., 187, 196, *209*
Amici, A., 193, *207*
Andersen, S., 27, *32,* 55, 64, *69*
Anderson, R. V., 141, 143, 144, 149, *156,* 167, *180,* 283, *284*
Andreasson, B., 63, *71*
Arndt, C. H., 138, 155, *156*
Asai, K., 112, *116*
Athow, K. L., 197, 198, *207*
Atkinson, G. F., 120, *135*
Ayala, A., 187, 190, 196, 197, 199, 200, *207*

## B

Baines, R. C., 57, 58, 59, 60, *69,* 312, *315*
Balasubramanian, M., 95, 114, *115*
Bally, W., 57, *69*
Banage, W. B., 237, *245*
Barker, K. R., 81, 83, *89,* 162, 168, *180,* 188, *208,* 274, *287*
Barr, G., 146, *156*
Barraclough, R. M., 311, *316*
Barrett, J., 300, 301, *315*
Barrons, K. C., 112, *115*
Bartnicki, E. W., 289, *296*
Bateman, D. F., 82, 83, *89*
Batten, C. K., 125, 127, *136*
Baunacke, W. E., 251, 255, *264,* 272, *284*
Baxter, R. I., 173, *180,* 238, 243, *245,* 273, *284,* 305, *315*
Bedi, A. S., 55, 64, *69*
Behrens, J., 195, *207*
Belser, N. O., 292, *296*
Benton, A. W., 220, *233*
Bělař, K., 12, 23, 24, *32*
Benedict, W. G., 124, *135*
Bennet-Clark, T. A., 144, *156*
Benton, A. W., 76, 86, 87, *89*
Bergeson, G. B., 123, 128, 133, *135, 136,* 171, *182,* 197, 198, *207,* 290, *296,* 312, *315*
Bergman, B. H. H., 99, 109, *115,* 250, 251, *264*
Bernhard, W., 107, *116*
Betto, E., 203, *207*
Bhaduri, N. V., 299, *315*
Bhatt, B. D., 238, 239, 240, 241, *245, 246,* 299, 305, 306, *315*
Bingefors, S., 55, *71,* 168, 178, *180*
Birchfield, W., 57, *69*
Bird, A. F., 36, 39, 41, 42, 44, 45, 46, 47, *48, 49,* 58, 60, 62, *69,* 75, 79, 81, 82, 84, 85, 86, 87, *89, 90,* 94, 95, 97, 105, 107, 109, *115,* 129, *135,* 141, 142, 150, 151, *156,* 163, 171, 177, 179, *180,* 215, 216, 217, 218, 219, 220, 222, 231, *233, 234,* 248, 251, 252, 253, 263, *264,* 268, 272, 282, 284, *285,* 300, 301, 302, 303, 309, 310, 313, *317*
Bird, G. W., 179, *180,* 190, *207,* 313, *315*
Bishop, D., 273, *285*
Blake, C. D., 172, 173, 175, *180,* 238, 243, *245,* 251, 252, 259, 262, *264,* 267, 273, *284, 285,* 305, 306, *315*
Blanchard, F. A., 170, *180*
Blickenstaff, M. L., 165, 168, *182*
Bloom, J. R., 133, *135,* 272, 274, *286, 287*
Blumer, S., 194, *207*
Boell, E. J., 244, *246*
Bolander, W. J., 123, *135*

Boone, W. R., 171, *181*
Bos, L., 195, *207*
Bosher, J. E., 54, *70,* 162, 169, 174, *182,* 298, 311, *315*
Boswell, T. E., 163, 166, *180*
Boubals, D., 206, *207*
Boveri, T., 2, *32, 33*
Bovien, P., 52, *69*
Bowman, P., 133, *135*
Bowman, T., 200, *209*
Braun, A. J., 188, *210*
Breece, J. R., 186, 193, *207*
Brim, C. A., 55, *70*
Brodie, B. B., 122, 126, *135*
Brun, J.-L., 6, *34,* 63, *69*
Brunt, A. A., 205, *207*
Bryant, C., 79, *89,* 308, *315*
Brzeski, M. W., 161, 163, 170, 171, *180, 184*
Buck, J., 314, *315*
Buecher, E. J., Jr., 169, *180*
Bütschli, O., 1, *33*
Bumbu, I. V., 81, 83, *89*
Burris, R. H., 244, 245, *246*
Byars, L. P., 160, 163, 164, 169, 171, 177, *180*


## C


Cadman, C. H., 92, *115,* 186, 193, 196, 198, 200, *207, 208*
Calam, C. T., 277, 278, 280, *285*
Cameron, J. W., 57, 58, 59, 60, *69*
Campbell, W. A., 124, *135*
Casida, L. E., Jr., 161, *180*
Castillo, J. M., 173, 176, *180,* 232, *233* 242, 243, *246*
Castro, C. E., 289, 290, 291, 292, 295, *296*
Caubel, G., 189, 194, *207*
Caullery, M., 17, *33*
Caveness, F. E., 251, *264*
Chambers, J., 200, 205, *209, 210*
Chancey, E. L., 87, *90*
Chandler, A. C., 299, *315*
Chandler, P. A., 128, *136*
Chang, L. M., 86, *89,* 178, *180*
Chapman, R. A., 188, *208*
Chapman, V., 20, *34*
Chen, T. A., 94, *115,* 138, 147, 155, *156,* 161, 162, 163, 173, 177, 178, *181, 182, 190, 207, 208,* 250, 251, *264*
Chernvak, E. K., 166, 168, *181, 183*
Chiaravalle, P. D., 55, *69,* 240, *246*
Chiba, K., 170, *181*
Chidambaranathan, A., 274, *285*
Chin, D. A., 153, *156,* 178, *181*
Chitwood, B. G., 53, *69,* 92, *115,* 290, *296*
Christie, J. R., 17, 25, *33,* 92, 93, 94, 97, 105, 108, 112, *115,* 138, 155, *156,* 163, *181,* 185, 188, 190, *207*
Chow, H. H., 220, *233*
Chowdhury, A. B., 299, *315*
Clark, F. E., 300, *315*
Clark, S. A., 151, *156*
Clarke, A. J., 214, 215, 216, 217, 218, 222, 232, *233,* 268, 276, 278, 279, 280, 281, 283, *285*
Clegg, J. S., 297, 311, *315*
Clifford, C. M., 300, *316*
Cohn, E., 57, *69,* 84, *89,* 188, 189, 191, 192, 194, *207*
Colbran, R. C., 56, *69*
Cole, R. J., 227, 228, 230, *233*
Collis-George, N., 239, *246*
Comas, M., 17, *33*
Comenga Gerpe, M., 222, *233*
Converse, R. H., 188, *210*
Cooke, D. A., 189, 190, *207, 208*
Cooper, A. F., Jr., 237, 238, *246,* 299, 300, 301, 303, 304, 305, 307, 308, 309, 310, 311, 312, 313, *315, 317*
Cooper, J. I., 192, *207*
Cooper, W. E., 122, 126, *135*
Cotten, J., 19, 20, *33,* 55, *69,* 273, *285*
Coursen, B. W., 121, *135*
Cox, P. M., 214, 215, 216, 218, 222, 232, *233,* 273, *287*
Crank, J., 255, 256, *264*
Cremer, M. C., 187, *207*
Crittenden, H. W., 92, *115,* 125, 126, 129, 132, *135, 136*
Crofton, H. D., 143, 144, *156,* 236, *246,* 268, *285,* 309, 310, *315*
Croll, N. A., 161, 167, 168, 178, *181, 184,* 253, 258, 260, 261, 263, *264,* 306, *315, 317*
Cropley, R., 194, *207*

Crosse, J. E., 129, *135, 136,* 162, 169, 172, 174, 175, *181*
Crossman, L., 163, *181*
Cryan, W. S., 164, *181*
Culbreth, W., 170, *183*
Cunningham, P. C., 273, *285*
Curtis, H. J., 303, *315*
Cutler, H. G., 231, *233*

D

Daems, W. T., 39, *49*
Dahlstrom, R. V., 278, *286*
Dalmasso, A., 24, *33,* 92, *115,* 189, 194, *207*
Dantuma, G., 55, 62, *70*
Darling, H. M., 27, 28, *34,* 53, 54, *70,* 166, 167, 168, *180, 181,* 189, *208,* 283, *284*
Das, S., 161, 162, *181,* 197, 198, 199, 200, *207*
Dasgupta, D. R., 57, *69*
Daulton, R. A. C., 56, 60, *69*
Davey, D. G., 308, *315*
Davey, K. G., 268, 269, 270, *285, 286*
Davide, R. G., 25, *33*
Davis, D. W., 171, *181*
Davis, R. A., 105, *115,* 131, *135*
Debrot, E. A., 189, 198, 201, *207*
DeBruyn Ouboter, M. P., 52, 62, *69*
deGuiran, G., 112, *115*
de Maeseneer, J., 86, *89*
DeMott, H. E., 242, *246*
Den Ouden, H., 26, *33,* 122, *136,* 170, *181*
Deubert, K. H., 79, 86, *89, 90,* 124, 132, *135,* 164, 173, 175, *181, 182,* 243, *246*
Dias, H. F., 193, 198, 199, *207*
Dickinson, S., 140, *156*
Dickson, D. W., 220, *233*
Dieter, A., 85, *89*
Diller, V. M., 170, *180*
Dimond, A. E., 283, *287*
DiSanzo, C. P., 188, *208*
Dolliver, J. S., 162, 165, 168, *181*
Doncaster, C. C., 36, *49,* 139, 140, 141, 142, 143, 144, 149, 150, *156,* 166, *184,* 250, 251, 252, 259, 260, *266,* 282, 284, *285,* 305, *318*
Douglas, J. R., 305, *317*
Dougherty, E. C., 23, 28, *34,* 160, 169, *181,* 217, 218, *234,* 249, *264*
Dropkin, V. H., 56, *69,* 81, 82, 83, *89,* 92, 97, 103, 105, 107, 111, 112, 113, 114, *115,* 166, 171, 180, *181,* 245, *246,* 283, 284, *285,* 307, *315*
DuCharme, E. P., 57, *69,* 162, 163, 170, 173, 177, 179, *181*
Duggan, J. J., 55, *69*
Duke, P. L., 62, *70*
Dunn, E., 126, *135*
Dunnett, J. M., 55, 64, *69*
Dunning, R. A., 190, *208*
Dutky, S. R., 228, 230, *233*
Dwinell, L. D., 131, *135*

E

Edmunds, J. E., 124, 132, 133, *135,* 175, 178, *181,* 251, *264*
Elgindi, D. M., 250, 251, 252, *265*
Ellenby, C., 26, *33,* 36, 45, *49,* 79, 80, *89, 90,* 214, 218, 232, *233, 234,* 274, 277, 283, *285,* 299, 304, 305, 312, *316*
Elliott, A., 300, *316*
Ells, H. A., 78, 88, *89*
El-Sherif, M., 221, *233,* 251, 253, *264*
Endo, B. Y., 99, 105, 107, 109, 111, 112, 113, 114, *115, 117,* 155, *156,* 299, 304, 305, *316*
Entner, N., 242, *246*
Epps, J. M., 56, 60, 62, *69,* 92, *117*
Epstein, A. H., 188, *208*
Eriksson, K. B., 27, *33,* 54, 64, 65, *69,* 168, 178, *180, 181*
Erlanger, M., 301, 302, *316*
Esser, R. P., 248, *264*
Estes, L. W., 92, *117*
Estey, R. H., 153, *156,* 178, *181*
Etherton, B, 253, *264*
Evans, A. A. F., 140, 153, *156,* 290, *296,* 298, 299, 304, 312, 313, *316*
Evans, K., 259, *264*

F

Fadeel, A. A., 238, *246*
Fairbairn, D., 236, *246,* 268, 272, 284, *285,* 299, 300, *316*
Fassuliotis, G., 22, *33*
Faulkner, L. R., 123, 133, *135,* 166, 167, 168, *181*
Feder, W. A., 163, 166, 168, 173, *181, 183*

Feldmesser, J., 95, *115,* 163, 170, 171, 173, *181*
Fenwick, D. W., 163, 166, *181, 184,* 250, 251, 252, 259, *266,* 272, 274, 275, *285*
Ferris, J. M., 173, *183*
Ferris, V. R., 173, *183*
Fielding, M. J., 92, *115,* 299, 304, 311, 312, *316*
Fisher, J. M., 139, 140, 153, 155, *156,* 188, *208,* 249, *264,* 267, 270, 271, *285*
Fisher, K. D., 168, *183*
Flegg, J. J. M., 92, *115,* 188, 189, 190, 191, *208, 209*
Flor, H. H., 64, *69*
Flores, H., 188, *208*
Fortin, J. A., 154, *157,* 175, *183*
Fox, J. A., 28, *33, 34,* 66, *70,* 281, *285*
Fraenkel, G. S., 261, *264*
Franklin, M. T., 92, *115,* 155, *156,* 304, *316*
French, N., 311, *316*
Fritzsche, R., 186, 192, 194, 196, *208, 210*
Frost, R. R., 195, *208*
Fujiwara, A., 170, *181*
Fullmer, H. M., 101, *115*
Fulton, J. P., 186, 193, *208*
Furman, D. P., 309, *316*
Fushtey, S. G., 272, 273, *285*

## G

Gadd, C. H., 250, 251, *264*
Garcia-Hidalgo, F., 204, *209*
Gershon, D., 301, 302, *316*
Gibbins, L. N., 63, *69,* 221, *234*
Gibbs, A. J., 187, 195, 196, *208*
Giebel, J., 61, *71,* 86, *90,* 113, *117,* 274, *285*
Gillard, A., 56, *69*
Gimenez-Martin, G., 101, *116*
Giovannola, A., 299, *316*
Gipson, I., 99, 111, *116*
Glick, D., 244, *246*
Godfrey, G. H., 93, *116*
Goffart, H., 75, 83, 84, *89*
Goheen, A. C., 185, 186, 191, 193, 196, 205, 206, *208, 209*
Gold, A. H., 200, *210*
Golden, A. M., 56, 60, 62, *69,* 271, 275, *285*
Gonzales-Fernandez, A., 101, *116*
Gonzalez, C., 242, *246*
Goode, M. J., 122, *135*
Goodey, T., 24, *33,* 52, *69,* 138, *156,* 187, *208*
Gooding, G. W., Jr., 206, *208*
Goodman, R. M., 161, 163, 179, *180, 182*
Goplen, B. P., 56, *69*
Goring, C., 291, *296*
Gortz, J. H., 306, *317*
Gotz, P., 226, *234*
Graham, C. W., 188, *208*
Grandison, G. S., 63, *69,* 221, *234*
Gray, H., 259, 260, *264*
Green, C. D., 28, *33,* 139, 140, 141, 142, 150, *156,* 189, *208,* 248, 249, 255, 257, 259, 260, 261, 262, *264*
Green, P. B., 143, *156*
Greet, D. N., 236, 240, 241, *246,* 248, 249, 255, 257, 259, *264,* 305. 306, *318*
Greig, M. E., 248, *265*
Griffin, G. D., 187, 188, 189, 195, *208, 211,* 250, 251, *264*
Grogan, R. G., 186, 196, 197, 199, *210*
Guile, C. T., 54, 60, 62, *69*
Gunn, D. L., 261, *264*
Gysels, H., 76, 85, *89,* 220, *234*

## H

Hague, N. G., 275, *285*
Haley, A. J., 300, *316*
Hamblen, M. L., 272, *287*
Hamlen, R. A., 274, *286*
Hanks, R. W., 162, 170, *181*
Hanna, M. R., 128, *135*
Hansen, E. L., 164, 169, *180, 181, 182, 183,* 217, 218, *234,* 249, *264, 266,* 268, *287*
Hansen, O. H., 217, 218, *234*
Harris, J. E., 143, *156*
Harrison, B. D., 186, 187, 188, 189, 191, 192, 193, 194, 195, 196, 197, 198, 199, 206, *207, 208,* 311, *316*
Hart, W. H., 186, 193, *207*
Hartwell, W. V., 278, *286*
Hartwig, E. E., 92, *117*
Hastings, R. J., 54, *70,* 162, 169, 174, *182*
Hawn, E. J., 128, *135*
Hechler, H. C., 17, 22, 24, 32, *33,* 143, 144, 153, *156*

Heiling, A., 75, 83, 84, *89*
Helgeson, J. P., 113, 114, *115*
Henderson, V. E., 250, 253, 263, *265*
Hendrix, F. F., 124, *135*
Hertwig, P., 3, 12, 23, *33*
Hesling, J. J., 52, 54, *70*, 277, *286*
Hewitt, W. B., 185, 186, 191, 193, 194, 196, 197, 200, 204, 205, 206, *208*, *209*, 312, *317*
Hieb, W. F., 230, *234*
Higinbotham, N., 253, *264*
Hildebrandt, A. C., 162, 165, 168, *181*
Hirano, K., 132, *135*
Hirschmann, H., 4, 5, 14, 18, 19, 20, 21, 28, *33*, *34*, 311, *316*
Hirumi, H., 138, 147, 155, *156*, 163, *182*, 190, *208*
Högger, C. H., 167, 179, *182*
Hoff, J. K., 190, *208*
Hollings, M., 194, *208*
Hollis, J. P., 39, *49*, 190, *207*, 253, *264*
Holter, H., 244, *246*
Holtzmann, O., 112, *116*
Honda, H., 23, *33*
Hooper, D. J., 53, 63, *71*, 155, *156*, 220, *234*, 311, *316*
Horner, C. E., 189, *208*
Houston, B. R., 126, 133, *136*
Howard, H. W., 59, 60, *70*
Howell, R. K., 60, 61, *70*
Huang, C. S., 93, 97, 103, 114, *116*, 163, 165, 168, 171, *182*
Huijsman, C. A., 55, 59, 64, *70*, *71*
Huisingh, D., 220, *233*
Hull, R., 189, *207*
Hung, C. P., 18, 19, *33*, 172, *182*
Hutchin, P. C., 168, *181*, *183*
Hutchinson, M. T., 152, *157*
Hyman, L. H., 138, *156*, 299, 309, 311, 312, *316*

## I

Ibrahim, I. K. A., 39, *49*, 253, *264*
Ichinohe, M., 112, *116*
Inagaki, H., 123, *135*
Inglis, W. G., 248, *265*
Ishibashi, N., 277, *286*
Ishii, M., 153, *157*

## J

Jackson, C. R., 124, *136*
James, G. L., 126, 132, *135*, 175, *182*, 277, *286*
Janzen, G. J., 275, 277, 280, *286*
Jantunen, R., 79, *89*, 308, *315*
Jenkins, W. R., 18, 19, *33*, 85, *89*, 105, *115*, 121, 131, *135*, 152, 155, *157*, 165, 166, 172, *182*, *183*, 190, 222, 223, 224, 225, *234*, 300, 303, 313, *315*, *317*
Jensen, H. J., 187, 188, 189, 196, *208*, *209*
Jha, A., 186, 193, 196, 197, *209*
Jiménez Millan, F., 57, *70*, 204, *209*
Johannsen, W., 68, *70*
John, B., 17, *33*
Johnson, A. W., 122, 124, 132, *135*, *136*
Johnson, D. E., 290, 291, *296*
Johnson, F., 185, *209*
Johnson, H. A., 128, *135*
Johnson, R. N., 26, *33*, 165, 166, 168, 179, *182*, 231, *234*, 251, 252, *265*
Johnson, P. W., 272, 273, *285*, 303, *316*
Johnson, R. W., 245, *246*, 283, 284, *285*
Johnston, T., 237, *246*
Jones, C. A., 299, *316*
Jones, F. G. W., 26, *33*, 55, 58, 64, *70*, 93, 113, 115, *116*, 139, *156*, 248, 251, 252, 253, 257, 260, 261, 262, *264*, *265*, 274, 275, *286*
Jones, N. O., 138, 141, 146, 147, 155, *157*, 190, 204, *209*, 248, *265*
Jones, T. P., 249, *265*
Juhl, M., 273, *286*
Junges, W., 57, *70*

## K

Kaai, C., 60, *70*
Kable, P. F., 170, *182*
Kämpfe, L., 251, 252, *265*, 305, *316*
Kan, S. P., 268, 269, 270, *285*, *286*
Kaplanis, J. N., 228, *233*
Katznelson, H., 250, 253, 263, *265*
Kaul, R., 277, *286*
Kawamura, T., 132, *135*
Kegler, H., 186, 194, *208*
Keilin, D., 297, 310, 314, *316*
Kendall, J., 96, 97, 105, *116*

Kennedy, F. S., 295, *296*
Kerekes, M. G., 281, *285*
Kerkut, G. A., 248, *265*
Kerstan, U., 26, *33*
Ketudat, U., 132, *135*
Khan, A. H., 274, *284*
Khera, S., 148, 152, 155, *156,* 163, 168, *182,* 251, *265*
Kiernan, J., 155, *156,* 161, 162, 178, *181*
Kilpatrick, R. A., 162, 173, *181, 182*
Kim, K. S., 99, 111, *116*
Kirkpatrick, J. D., 309, *317*
Kisiel, M., 124, 132, *135,* 167, 175, *182*
Klingler, J., 139, *156,* 238, *246,* 251, 253, 256, 260, 261, 262, *265,* 299, *316*
Klinkenberg, C. H., 141, 148, 151, 152, 155, *156*
Koehler, B., 185, *209*
Köhler, H., 96, *116*
Kok, M. W. S., 60, *70*
Kommedahl, T., 124, *136*
Konno, I., 170, *181*
Kooistra, G., 187, *207*
Kort, J., 55, 58, 59, 60, 62, 66, *70*
Kostoff, D., 93, 96, 97, 105, *116*
Kostük, N. A., 297, 299, 311, 312, 313, *316*
Kray, W. C., 295, *296*
Krikun, J., 188, *207*
Kröning, F., 23, *33*
Krüger, E., 3, 12, 15, 23, *33*
Krusberg, L. R., 77, 78, 79, 80, 81, 83, 84, 85, 86, 88, *89, 90,* 92, 93, 97, 103, 105, 114, *116,* 148, *157,* 162, 165, 168, 172, 176, 177, 178, *182,* 214, 216, 217, 218, 222, 223, 224, 225, 227, 228, 231, 232, *233, 234,* 242, 243, 245, *246,* 300, 303, 305, 312, *316*
Kühn, H., 113, *116,* 251, 254, 259, *265*
Kuiper, K., 190, 191, *209, 210*
Kushner, V. D., 125, *135*

L

Labruyére, R. E., 122, *136*
Lamberti, F., 193, 194, *209*
Lapp, N. A., 31, *33*
Larbey, D. W., 262, *265*
Larson, R. H., 187, 195, *211*
Lavallee, W. H., 250, 251, 253, *265*
Laviolette, F. A., 197, *207*
Leach, J. G., 128, *136*
Lear, B., 290, 291, *296*
Leathers, C. R., 124, *136*
Lee, C. C., 303, *316*
Lee, D. E., 289, *296*
Lee, D. L., 87, 89, 236, *246,* 248, *265,* 267, 268, 269, 271, *286,* 311, *316*
Lee, K. J., 138, 147, 155, *156,* 190, *208*
Legin, R., 193, *211*
Lehman, P. S., 274, 276, 281, *286*
Lehmann, W., 186, *210*
LeJambre, L. F., 308, *316*
Lengweiler-Rey, V., 299, *316*
Letey, J., 307, *316*
Leutenegger, R., 299, 300, *316*
Libman, G., 128, *136*
Lider, L., 189, *209*
Lima, M. B., 193, *209*
Linderstrom-Lang, K., 244, *246*
Lindvig, P. E., 248, *265*
Linford, M. B., 44, *49,* 80, 85, *89,* 93, 99, 105, 107, 111, 112, *116,* 137, 138, 139, 148, 149, 150, 151, 153, *157,* 250, 251, 259, 262, 263, *265,* 270, *286,* 309, 311, *317*
Lissmann, H. W., 259, 260, *264*
Lister, R. M., 186, 194, 195, 197, 198, 205, *209*
Littrell, R. H., 97, *116,* 122, 124, 132, *135, 136*
Loewenberg, B. F., 171, *182*
Loewenberg, J. R., 250, 251, *265,* 274, 281, *286*
Loof, P. A. A., 57, *70,* 190, *209*
Loos, C. A., 250, 251, *264*
López-Abella, D., 204, *209*
Lopez-Saez, J. F., 101, *116*
Lowe, D., 165, 168, *184*
Lownsbery, B. F., 126, 133, *136,* 162, 164, 165, 168, 172, *182, 183,* 186, 188, 196, 197, 199, *209, 210,* 251, *265,* 272, 274, *286, 287*
Lownsbery, J. W., 162, *182*
Luc, M., 251, 252, 253, 255, *265*
Ludwig, R. A., 275, *287*
Lukens, R. J., 280, *286*
Lundegårdh, H., 253, *265*
Lyons, J. M., 310, *316*

## M

McAllan, J. W., 81, 83, *90*
McBeth, C. W., 290, *296*
McClure, M. A., 25, *33,* 166, 167, 179, *182*
MacDonald, D., 124, *136*
McElroy, F. D., 140, 143, 145, 146, 152, *157,* 172, *182,* 237, *246,* 272, *286,* 307, 308, *316, 317*
McGuire, J. M., 122, *135,* 197, *209*
McGuire, R. J., 107, *115*
McKeen, C. D., 131, *136*
McKeen, W. E., 311, *315*
Mackintosh, G. McD., 45, *49*
McKinney, H. H., 185, *209*
Maat, D. Z., 200, *211*
Mabbott, T. W., 63, *70*
Maggenti, A. R., 45, *49,* 93, 97, 103, 114, *116*
Mai, W. F., 58, 60, 62, *69,* 94, *115,* 124, 132, 133, *135,* 155, *156,* 161, 162, 170, 175, 178, 179, *180, 181, 182,* 190, *207, 208,* 221, *233,* 251, 253, *264*
Malek, R. B., 188, *209*
Mankau, R., 93, 99, 105, 107, 111, *116,* 138, 139, 150, *157,* 304, 312, 313, *316*
Mankau, S. K., 304, 312, 313, *316*
Maramorosch, K., 138, 147, 155, *156,* 163, *182,* 190, *208*
Markham, R., 193, *210*
Marks, C. F., 290, 291, 292, *296*
Marrian, D. H., 278, *286*
Martelli, G. P., 193, 194, *209*
Martin, D. W., 253, *265*
Martin, G. C., 56, *70,* 245, *246,* 283, 284, *285*
Martin, M., 164, *181*
Martin, W. J., 56, 59, *70*
Maupas, E., 28, *33*
Mayoh, H., 87, 88, *90,* 243, *246*
Mayol, P. S., 128, 133, *136,* 171, *182*
Melendéz, P. L., 121, 125, 133, *136*
Merny, G., 56, *70*
Metcalf, H., 164, *182*
Metlitski, O. Z., 54, *70*
Millar, R. L., 82, 83, *89*
Miller, C. R., 123, *136*
Miller, C. W., 80, 85, 86, *89, 90,* 162, 168, *182*
Miller, E. C., 144, *157*
Miller, H. N., 92, *116,* 120, 127, *136*
Miller, J. H., 303, *316*
Miller, L. I., 28, *33,* 56, 58, 62, *70*
Miller, P. M., 277, 280, 283, *286, 287*
Minton, E. B., 121, 122, 133, *136*
Minton, N. A., 56, 60, *70,* 121, 122, 124, 133, *136*
Minz, G., 56, *71*
Mish, L. B., 170, *183*
Mitchell, J. P., 99, *116*
Miyakawa, T., 57, 58, 59, 60, *69*
Mojé, W., 290, 294, *296*
Mollett, J. A., 249, *264*
Moore, J. F., 188, 189, *209*
Mordechai, M., 189, *207*
Moreno, P., 101, *116*
Morgan, G. T., 81, 83, *90*
Moriarty, F., 166, 172, 179, *182*
Morikawa, O., 77, *90,* 251, 252, 261, *265*
Morsink, F., 123, 131, *136*
Moses, V., 217, 218, *234*
Mountain, W. B., 86, *90,* 92, 96, *116,* 124, 131, *135, 136,* 160, 162, 165, 166, 171, 172, 173, 174, 176, 180, *182*
Mowat, W. P., 186, 187, 189, 194, 196, *208, 209*
Mulvey, R. H., 3, 19, 20, 21, 23, 28, *33*
Munnecke, D. E., 128, *136*
Murant, A. F., 189, 195, 196, 198, 200, 201, 205, 206, *209, 210*
Muse, B. D., 61, *70,* 81, 83, 84, 90
Myers, R. F., 76, 81, 82, 84, 85, 86, 87, *89, 90,* 114, *116,* 145, *157,* 167, 168, 173, *183,* 214, 217, 218, 220, 232, *233, 234*
Myuge, S. G., 75, 80, 85, *90,* 95, 96, 114, *116,* 163, *183*

## N

Nakasono, K., 8, 19, *33*
Neal, A. L., 274, 278, *286*
Needham, T., 298, 311, *316*
Nelson, B., 167, *182*
Nelson, P. E., 97, 105, 111, *115*
Nemec, B., 92, 94, 99, 101, 105, 108, *116*
Netscher, C., 20, *33*
Newton, W., 54, *70*
Nicholas, W. L., 79, *89,* 217, 218, *234,*

249, *264*, 298, 308, *315*, *317*
Nickle, W. R., 153, *157*
Nielsen, L. W., 93, 97, 103, 105, *116*
Nigh, E. L., Jr., 128, *136*
Nigon, V., 3, 4, 6, 9, 10, 17, 23, 24, 28, *33*, *34*, 249, *265*
Nirula, K. K., 274, *286*
Nixon, H. L., 195, *208*
Nolte, H. W., 96, *116*
Norgren, R. L., 173, *181*
Norton, D. C., 188, 189, *209*
Nusbaum, C. J., 56, 60, *69*, 123, 133, *136*

## O

O'Bannon, J. H., 124, *136*, 166, 173, *183*, 272, *286*
Oberling, C., 107, *116*
Ohira, K., 170, *181*
Oliveria, J. M., 85, *89*, 93, *116*, 153, *157*
Olthof, T. H. A., 56, 60, *70*, 124, 131, *136*, 174, *183*
Omidvar, A. M., 277, *286*
Oostenbrink, M., 272, 273, *286*
Osche, G., 28, *34*, 306, *316*
Osores-Duran, A., 130, 133, *136*
Osske, G., 277, 284, *286*
Oswald, J. W., 200, *209*
Oteifa, B. A., 92, *115*, 250, 251, 252, *265*
Overgaard Nielsen, C., 236, 237, 239, *246*, 310, *316*
Owen, J. H., 124, *135*
Owens, R. G., 95, 97, 103, 107, 108, 114, *116*, *117*

## P

Paden, W. R., 185, *209*
Palmer, L. T., 124, *136*
Palo, A. V., 60, *70*
Panzer, J. D., 251, *264*
Paracer, S. M., 155, *157*, 168, 172, 173, *181*, *183*
Paramonov, A. A., 11, *34*, 248, *265*
Parrott, D. M., 26, 27, *33*, *34*, 54, 55, 64, *70*, *71*, 113, 115, *116*, 262, *265*
Passey, R. F., 284, *285*, 300, *316*
Pasternak, J., 220, *233*
Patrick, Z. A., 86, *90*, 162, 173, 176, *182*, 292, *296*
Pattullo, W. I., 205, *210*
Paulson, R. E., 105, 109, *116*
Pawelska, K., 55, 58, *70*, 277, *286*
Peachey, J. E., 206, *208*
Peacock, F. C., 139, *157*, 163, 166, 171, 178, *183*, 250, 251, 252, *265*
Pennoit-de Cooman, E., 78, *90*
Percival, J., 185, *209*
Perry, V. G., 57, *70*, 155, *157*, 188, 190, *207*, *209*
Peterson, S. W., 248, *265*
Petrovskaya, E. S., 166, *183*
Pi, C.-L., 86, *90*, 171, 173, 176, *183*
Pillai, J. K., 139, 141, 143, 144, 149, 153, 154, *157*
Pillai, M. J., 274, *287*
Pimekhov, A. F., 166, *183*
Pitcher, R. S., 92, *116*, 120, 123, 127, 128, 129, *135*, *136*, 137, 155, *157*, 162, 169, 172, 173, 174, 175, 176, 180, *181*, *183*, 190, 191, *209*, 250, 251, 253, 258, 259, *265*
Plumb, S. C., 249, *264*
Podger, K. R., 269, 270, *287*
Poinar, G. O., 248, *265*, 299, *316*
Polychronopoulos, A. G., 126, 133, *136*, 164, 172, *183*
Poole, J. B., 309, *317*
Porter, D. M., 122, *136*
Posnette, A. F., 186, 193, 196, 197, *209*
Potter, J. W., 28, *34*, 66, *70*
Poulton, M. E., 249, 255, *264*
Powell, N. T., 92, *116*, 120, 121, 122, 123, 125, 127, 128, 133, *135*, *136*, 175, *183*
Powell, W. M., 124, *135*
Prasad, S. K., 165, 166, 173, 179, *183*
Pratt, H. K., 310, *316*
Price, W. C., 179, *181*

## R

Rackham, R. L., 152, *157*, 308, *317*
Rademacher, B., 278, 281, *286*
Radewald, J. D., 155, *157*, 189, *209*
Raistrick, H., 277, 278, *285*
Ramisz, A., 248, *265*
Rangaswami, G., 95, 114, *115*, 274, *285*
Raski, D. J., 55, 57, *69*, *70*, 138, 139, 141, 146, 147, 155, *156*, *157*, 161, 162, *181*,

185, 186, 188, 189, 190, 191, 193, 196, 197, 198, 199, 200, 201, 203, 204, 205, 206, *207, 208, 209, 210, 211,* 248, *265,* 312, *317*
Rau, G. J., 22, *33*
Ray, H. N., 299, *315*
Read, C. P., 78, 88, *89, 90,* 299, *315*
Rebois, R. V., 92, *117*
Reydon, G. A., 57, *69*
Reyes, A. A., 131, *136*
Reynolds, H. W., 124, *136*
Reynolds, L. I., 170, *184*
Rhoades, H. L., 153, 155, *157,* 270, *286,* 309, 311, *317*
Rich, A. E., 123, 131, *136,* 173, 177, *181, 182,* 250, 251, *264*
Riedel, R. M., 83, 84, *90*
Riggs, R. D., 56, 58, 59, 66, *70, 71,* 99, 111, 112, *116, 117*
Riker, A. J., 162, 165, 168, *181*
Riley, R., 20, *34*
Ritzema Bos, J., 51, *70*
Robbins, W. E., 228, 230, *233*
Roberts, R. N., 80, *89*
Robertson, W. M., 201, 202, 203, 204, *210*
Robinson, T., 274, *286*
Rode, H., 251, 253, 261, *265*
Rodrigues, L., 173, *182*
Rogers, G. E., 45, 46, *49,* 79, 85, 86, *89,* 216, 222, *233*
Rogers, W. P., 237, *246,* 269, 270, 272, *286,* 299, 300, 301, *317*
Roggen, D. R., 138, 141, 146, 147, 155, *157,* 190, 203, 204, *209,* 248, *265,* 306, *317*
Rohde, R. A., 80, 86, 87, *89, 90,* 111, *117,* 155, *157,* 173, 174, 176, 178, *180, 183, 184,* 188, 190, *208, 209,* 238, 239, 246, 241, 245, *245, 246,* 248, *265,* 299, 305, 306, *315*
Roman, J., 18, 19, *34,* 97, *117*
Ross, G. J. S., 26, *33*
Ross, H., 55, 59, 64, *70*
Ross, J. P., 55, *70,* 113, *117,* 122, 132, *136,* 273, *286*
Rothstein, M., 87, 88, *90,* 226, 229, 230, *234,* 243, *246,* 298, *317*
Rubinstein, J. H., 103, 107, *117*
Ruehle, J. L., 92, *117*
Russell, C. G., 155, *157*
Russell, P. B., 277, 278, *286*
Ryder, H. W., 129, 132, *136*

## S

Sachs, H., 306, *317*
Sänger, H. L., 187, 200, *209, 210*
Sandstedt, R., 92, *117,* 166, 177, 178, *183*
Santmyer, P. H., 239, 240, *246,* 310, *317*
Santos, M. S. N. de A., 249, *265*
Sasser, J. N., 56, 59, 60, 66, *70, 71,* 220, *233*
Sauer, N. I., 199, *210*
Saurer, W., 39, 41, *49,* 94, *115,* 142, 150, 151, *156,* 179, *180,* 215, 219, 222, *233*
Savage, H. E., 168, *183*
Sawamura, S., 101, *117*
Sayre, F. W., 164, 169, *181, 182, 183*
Sayre, R. M., 165, 166, 168, 172, *183,* 292, *296,* 309, 310, *317*
Saz, H. J., 236, 241, 243, *246*
Schindler, A. F., 155, *157,* 185, 188, *210*
Schmelzer, K., 186, 194, *208*
Schmidt, H. B., 186, *210*
Schmidt, O., 278, *286*
Schmitt, R. V., 206, *209*
Schrauzer, G. W., 295, *296*
Schreiber, K., 277, 284, *286*
Schroeder, P. H., 165, 166, 168, *183*
Schuster, M. L., 92, 94, 95, *117,* 166, 171, 177, 178, *182, 183,* 250, 251, *265,* 274, 281, *286*
Schwabe, C. W., 301, *317*
Scott, B. I. H., 253, *265*
Sculthorpe, C. D., 237, *246*
Seinhorst, J. W., 60, 62, *70,* 84, *90,* 92, *117,* 122, 132, *136,* 138, 140, *157,* 187, 188, 195, 200, *210, 211*
Sembdner, G., 95, 99, 105, 109, *117,* 277, 284, *286*
Seshadri, A. R., 283, *286*
Setty, K. G. H., 96, *117*
Sharma, R. D., 188, *210*
Shepherd, A. M., 36, *49,* 55, *70,* 139, 140, 141, 142, 150, *156,* 214, 215, 216, 218, 222, 232, *233,* 272, 273, 275, 276, 277, 280, 281, 282, 283, 284, *285, 286, 287*
Sher, S. A., 24, *32,* 57, *69,* 312, *315*
Silverman, P. H., 160, *183,* 269, 270, *287*

Simpson, W. F., 308, 311, *317*
Sinclair, W. A., 131, *135*
Sivapalan, P., 222, 223, 224, 225, *234*, 300, 303, *317*
s'Jacob, J. J., 55, *70*
Skotland, C. B., 123, 133, *135*
Slack, D. A., 272, *287*
Sledge, E. B., 155, *157*
Slootweg, A. F. G., 56, *70*
Small, R. H., 57, 58, 59, 60, *69*
Smart, G. C., Jr., 27, 28, *34*, 53, 54, 55, 57, *70*
Smedley, E. M., 281, *287*
Smith, A. U., 310, *317*
Smith, K. M., 193, *210*
Smith, L., 79, 80, *89*, *90*, 218, *234*, 274, *285*
Sol, H. H., 187, 191, 195, 198, *210*
Sommerville, R. I., 269, 270, *286*, *287*, 299, *317*
Specht, H. N., 95, 97, 103, 108, 114, *116*
Spurr, H. W., 87, *90*
Staar, G., 251, 253, 261, *265*
Stanford, E. H., 56, *69*
Standifer, M. S., 155, *157*
Starr, M. P., 128, *136*
Starr, T. J., 169, *183*
Stauffer, J. F., 244, 245, *246*
Stavely, J. R., 126, *136*
Steele, A. E., 58, *70*, 280, *287*
Steiner, G., 25, 27, *34*, 52, 53, 54, 62, *71*, 148, *157*, 299, 311, *317*
Stephenson, W., 306, *317*
Stessel, G. J., 240, *246*
Stewart, M. A., 305, *317*
Stoddard, E. M., 280, *286*
Stokes, D. E., 57, *71*
Stolzy, L. H., 237, 238, *246*, 307, 308, *316*, *317*
Stone, O. M., 194, *208*
Stotzky, G., 170, *183*
Stover, R. H., 124, *136*
Streu, H. T., 152, *157*
Strich-Harari, D., 56, *71*
Sturhan, D., 27, *34*, 52, 53, 54, 58, 59, 60, 61, 64, 65, *71*, 188, 189, *210*
Sudakova, I. M., 166, 168, *183*
Sullivan, T., 94, 95, *117*, 171, *182*, 250, 251, *265*, 274, 281, *286*
Sutherland, J. R., 148, 154, *157*, 175, 178, *183*
Swarup, G., 274, *287*
Szuszkiewicz, T. E., 307, 308, *316*, *317*

## T

Taconis, P. J., 191, *210*
Tanaka, I., 164, 179, *183*
Taniguchi, R., 253, *265*
Tappel, A. L., 303, *317*
Tarjan, A. C., 193, *210*
Taylor, A. L., 166, *183*
Taylor, C. E., 186, 187, 188, 189, 191, 194, 195, 196, 198, 199, 200, 201, 202, 203, 204, 205, 206, *208*, *209*, *210*, 312, *317*
Taylor, D. P., 22, 32, *33*, 139, 153, 154, *157*, 272, 283, *287*
Taylor, G. S., 277, *286*
Taylor, P. L., 173, *183*
Taylor, R. H., 191, 205, 206, *209*, 312, *317*
Teliz, D., 186, 196, 197, 199, *210*
Thistlethwayte, B., 18, *34*
Thomas, H. A., 152, *157*
Thomas, P. R., 188, 189, 191, 194, 196, 198, 200, 205, *210*
Thomasine, Sister M., 197, *207*
Thomason, I. J., 290, 291, 292, *296*, 302, 309, 313, *316*, *317*
Thompson, J. V., 230, *233*
Thompson, M. J., 228, *233*
Thomson, W. W., 303, *316*
Thorne, G., 299, *317*
Thorpe, H. J., 292, *296*
Tiner, J. D., 162, 163, 164, 165, 166, 180, *183*, *184*
Tischler, G., 101, 105, *117*
Tite, D. J., 206, *211*
Todd, A. R., 277, 278, 280, *285*, *286*
Townshend, J. L., 153, *157*, 162, 169, 173, 176, *184*, 304, 312, *317*
Tracey, M. V., 81, 82, 83, *90*, 222, *234*, 300, *317*
Treub, M., 91, 94, 105, *117*
Trexler, P. C., 170, *184*
Triantaphyllou, A. C., 4, 5, 7, 11, 12, 14, 18, 19, 20, 21, 22, 25, 29, 31, *33*, *34*, 59, 66, *71*

Triffitt, M. J., 99, *117*
Troll, J., 80, *90*, 173, 174, 176, *184*
Trudgill, D. L., 26, *34*, 64, *71*, 114, *117*
Tsao, P. H., 126, *136*
Tyler, J., 47, *49*, 163, 164, 177, *184*

## U

Umbreit, W. W., 244, 245, *246*
Upper, C. D., 113, 114, *115*
Ustinov, A. A., 96, *117*

## V

Valoras, N., 307, *316*
Van Beneden, E., 2, *34*
Van der Laan, P. A., 55, *71*
van der Meer, F. A., 194, *210*
van der Tuin, F., 275, 277, *286*
van der Want, J. P. H., 195, *207*
Van Duuren, A. J., 250, 251, *264*
Van Essen, A., 55, 62, *70*
Van Gundy, S. D., 86, *90*, 126, *136*, 140, 143, 145, 146, 152, *157*, 172, *182*, 217, 222, *234*, 237, 238, *246*, 259, *265*, 272, *286*, 299, 300, 301, 302, 303, 304, 305, 307, 308, 309, 310, 311, 312, 313, *315*, *316*, *317*
Van Hoof, H. A., 57, 58, 60, *71*, 187, 196, 198, 199, 200, 201, *210*, *211*
Veech, J. A., 107, 109, 112, 113, 114, *115*, *117*
Videgård, G., 63, *71*
Viglierchio, D. R., 25, 26, *33*, 96, 114, *117*, 161, 166, 167, 168, 178, 179, *182*, *184*, 231, *234*, 250, 251, 252, 253, *265*, 274, 279, *287*, 306, *315*, *317*
von Brand, T., 213, 222, *234*, 236, 239, 240, *246*, 308, 310, 311, *317*
Vuittenez, A., 193, 194, 206, *211*

## W

Walkinshaw, C. H., 187, 195, *211*
Wallace, H. R., 45, *49*, 52, *71*, 86, *90*, 140, *157*, 217, 222, *234*, 236, 239, 240, 241, *246*, 250, 251, 252, 255, 258, 259, 260, *265*, *266*, 271, 272, 273, 274, 276, 277, 281, 284, *285*, *287*, 299, 300, 301, 302, 303, 304, 305, 306, 307, 309, 310, 313, *317*, *318*
Wallendal, P., 166, *181*
Wålstedt, I., 55, *71*
Walton, A. C., 3, 4, *34*
Waring, W. S., 278, 280, *285*, *286*
Waseem, M., 155, *157*, 172, *183*
Watson, B. D., 268, *287*
Webb, R. E., 112, *115*, 171, *181*
Webber, A. J., Jr., 274, *287*
Webster, J. M., 26, 27, *34*, 53, 54, 55, 63, 64, 65, *71*, 105, 109, *116*, 165, 166, 168, 173, 179, *183*, *184*, 221, *234*
Weiner, A., 189, *211*
Weischer, B., 120, 127, 130, *136*, 188, 189, *211*, 251, 252, 253, 259, *266*
Wheaton, T. A., 310, *316*
Wheeler, A. W., 96, *117*
Whidden, R., 168, *181*
White, P. R., 165, *184*
Whitehead, A. G., 189, 190, 206, *211*
Whitlock, J. H., 308, *316*
Widdowson, E., 166, *184*, 250, 251, 252, 259, *266*, 275, 278, *285*, *287*
Wieser, W., 250, 251, 252, *266*
Wihrheim, S. E., 277, 283, *286*, *287*
Williams, A. S., 61, *70*, 81, 83, 84, *90*
Williams, T. D., 112, *117*
Wilson, E. M., 36, *49*
Wilson, P. A. G., 217, 222, *234*, 300, 301, *318*
Wilski, A., 61, *71*, 86, *90*, 113, *117*, 277, *287*
Windrich, W. A., 60, *71*
Winslow, R. D., 188, 191, 198, 199, 206, *208*, 272, 275, *286*, *287*, 299, 306, *318*
Winstead, N. N., 56, 58, 59, 66, *70*, *71*, 112, *117*
Wisse, E., 39, *49*
Wolfe, R. S., 295, *296*
Wood, J. M., 295, *296*
Woods, R. D., 195, *208*
Wright, K. A., 140, 146, 147, *157*, 204, *211*
Wu, L.-Y., 27, *34*, 54, *71*
Wuest, P. J., 272, *287*
Wyss, U., 57, *71*, 154, *157*

## Y

Yarwood, E. A., 164, 169, *180, 181, 182, 183,* 249, *264, 266,* 268, *287*
Yassin, A. M., 189, 196, *211*
Yeates, G. W., 248, *266*
Younes, T., 22, 24, *33, 34*
Youngson, C., 291, *296*
Yu, P. K., 96, 114, *117,* 231, *234,* 279, *287*
Yuen, P. H., 39, *49,* 145, *157,* 248 *266*

## Z

Zehr, E., 170, *182*
Zinoviev, V. G., 75, 80, 84, 85, *90*
Zuckerman, B. M., 79, 86, *89, 90,* 124, 132, *135,* 148, 152, 155, *156, 157,* 161, 163, 164, 167, 168, 170, 171, 172, 173, 175, 180, *181, 182, 183, 184,* 190, *211,* 243, *246,* 251, *265*

# Subject Index


## A

Absorption of stimuli, 254–255
Acetylcholinesterase, 87, 248
Acid phosphatase role in aging, 301
*Actinolaimus hintoni,* cryptobiosis, 311
Adaptation(s), 47, 258–259
  in Nematodea, 138
  to parasitism, 35–48
Aeration effect on hatching in *Heterodera,* 272
Aerobic environment, 307–310
  respiration, 77–78, 242–243, 301
Aggregation of nematodes, 258–259
Aging, 297–318
  nematodes as models, 298
  pigment and, 303
Amino acids
  changes associated with parasitism, 177
  composition in nematodes, 216–219
  as nematode stimuli, 251
Ammonia in homogenates of *Ditylenchus,* 214
Amphid, 248
  role in molting, 270
Ampulla, 44, 45, 219
Amygdalin, 176
Anaerobic environment, 307–310
  respiration, 77, 235, 242–243, 308
*Angiostrongylus cantonensis,* 309
*Anguina*
  cryptobiosis, 298, 311
  cytology, 14, 18
  oogenesis and fertilization, 5
  respiration, 305–306
*Anguina agrostis,* 241, 306
*Anguina tritici,* 8, 241
  anhydrobiosis, 305–306
  osmotic pressure effects, 306
Anhydrobiosis, 304–307, 314
Anoxybiosis, 307–310, 314
Antibiotics for axenizing nematodes, 161–164
Aphelenchidae, feeding, 155, 156
*Aphelenchoides,* 138, 144, 251
  amino acid composition, 217–219
  cytology, 22
  feeding, 155
  ingestion, 145
  monoxenic culture, 166, 168
  pharynx, 145
  resistance to desiccation, 305–307
  secretory movements, 142
*Aphelenchoides besseyi,* cryptobiosis, 311
*Aphelenchoides fragariae,* 57
  biological races, 57
*Aphelenchoides parietinus,* 138, 250
*Aphelenchoides ritzemabosi,* 57, 238, 239
  amino acid composition, 217
  bacterium interaction, 129, 130, 174–175
  biological races, 57
  carbohydrates, 216
  fatty acid composition, 223
  lipids, 222, 300, 305
  oxygen consumption, 239
  pathogenesis, 172
  respiration, 240
*Aphelenchus*
  cytology, 22
  monoxenic culture, 166, 168
*Aphelenchus avenae*
  aging, 300, 303–305, 312–313
  anhydrobiosis, 304–305
  anoxybiosis, 307
  cryptobiosis, 312–313
  feeding, 138, 155
  fungus interactions, 175
  growth, 267
  hatching behavior, 272, 283

lipid catabolism, 300, 307
molting, 267, 271
osmoregulation, 214
respiration, 301, 307–308, 310
time death studies, 290–291
Arabis mosaic virus, 186, 193, 195, 196, 198, 199, 201
Arousal, 259–263
Arrhenius equation, 239
*Ascaris,* 242
cuticle, 268
exsheathing, 270
growth, 267, 268
hatching, 284
hydrostatic pressure, 143
neurosecretory cells, 270
*Ascaris lumbricoides*
lipid catabolism, 300
sterol composition, 227
Ascorbic acid, role in aging, 303
Attractant dispersal, 249, 255–258
nature of, 249
Attraction
to chemicals, 249
to feeding sites, 249
between individuals, 249
to microoganisms, 251
to plants, 139, 251
to roots, 132, 177–178
between sexes, 249
Automixis, 4, 22–24
Autoradiography, 108
Auxins, 95, 96, 177
Axenic culture, 165–169, 298
Axenization, 161–164
of Dorylaimida, 161
of free-living forms, 164
of Tylenchoidea, 161

## B

Bacteria-nematode interactions, 127–129
in dixenic experiments, 160, 174
Bacteriophagous nematodes, axenization, 164
Basic protein, 44
Behavior, 177–178
hatching, 282–283
in monoxenic culture, 167
*Belonolaimus*
biological races, 57
in disease complex, 122
*Belonolaimus longicaudatus,* 157
Biochemical changes in host
under aseptic conditions, 174, 176, 177
by *Pratylenchus penetrans,* 174
Biological races, 51–71, 178
Biosynthesis
of amino acids, 217–219
of fatty acids, 226
of sterols, 227
Biotype, 68
of nematodes, importance in breeding programs, 115
Brome grass mosaic virus, 186, 194
Buccal stylet, 36
exploration and penetration, 139–141
*Bursaphelenchus fungivorus,* 157

## C

*Caenorhabditis* spp.
aging studies, 300–302, 304–305, 308
cytology, 23, 24
glycogen catabolism, 301
protein content, 220
*Caenorhabditis briggsae,* 243, 249
aging, 301–302
amino acid composition, 217, 219
axenization, 164
culturing, 169
cytology, 23
growth, 268
molting, 268
respiration, 308
sterol requirement, 229
*Caenorhabditis elegans,* 63, 301–302
Calcium hypochlorite, effect on hatch of *Heterodera,* 281
Carbon dioxide, 238–239, 242
attractant, 132, 251, 252, 254, 256, 260, 262, 263
effect on hatching, 271
on molting, 269
on respiration, 238–239
Carnation ringspot virus, 186, 194
Callus tissue, 245
nematodes grown on, 167, 168
rate of reproduction on, 165

Cellulases, 80–82
in various nematodes, 81
Chemical composition, 213–234
amino acids, 216–222
ammonia, 214
auxins, 231
carbohydrates, 215–216
chitin, 215
collagen, 216
cytochromes, 232
enzymes, 220–221
fatty acids, 223–227
glycogen, 216
inorganic, 214–215
lipids, 222
nitrogen, 214
organic acids, 232
phospholipids, 223
plant growth regulators, 231–232
polyphenols, 232
proteins, 216–222
salts, 215
sterols, 227–230
water content, 214
Chemical sterilants, 161–163
Chemical stimuli, 249, 252–254, 263
Chemoreception, 248
Cherry leaf roll virus, 186, 193, 194
Chitinase, role in hatching, 271
Chlorogenic acid, 176, 178
Cholinesterase, 87, 248
*Chromadorina viridis,* 261
Chromatin diminution, 2
Chromosomes, 2, 3, 53, 65
fragmentation, 2
general characteristics, 13–15
induced in host by *Meloidogyne,* 107
numbers, 16, 18–24
in various species, 18–24
Cilia, 248
Citrus nematode, *see Tylenchulus*
Cold tolerance, 309
Control, *see* Nematicides
virus disease, 204–207
Copulation, 263
Criconematidae
feeding, 154
ingestion, 146
*Criconemoides curvatum*
feeding, 154
pathogenesis, 172
*Criconemoides xenoplax,* 283
hatching behavior, 283
Cryobiosis, 309–310, 314
Cryptobiosis, 298–299, 310–315
Culturing
axenic, 164–169, 298
*Caenorhabditis briggsae,* 167, 169
callus tissue, 165, 167, 168
excised root tips, 165, 166
gnotobiotic defined, 160
media, types defined, 160
morphological change in, 167
physiological change in, 167
seedlings, 164
Cuticle, 39, 267–271
chemical composition, 215, 216, 232, 268
enzymes, 268
origin, 216
permeation by nematicides, 292–294
structure, 216, 268
ultrastructure, 40
Cyst nematodes, *see Heterodera*
Cyst wall of *Heterodera*
composition, 214, 215, 305
permeability, 283
Cytochrome system, 238, 243
Cytogenetics, 29–32
Cytology, 2, 4
Cytoplasm, 44
Cytoplasmic condition of syncytia, 113

## D

D-D (dichloropropane-dichloropropene), 206
Dagger nematodes, *see Xiphinema*
Dauer larvae, 249
Degradation of stimuli, 253–255
*Deladenas* sp., 249
Deoxyribonucleic acid (DNA), 6, 95, 99, 107, 108
Desiccation, 241, 304–307
Development in monoxenic culture, 167
Diapause, 272–273
Dibromochloropane, 206, 290
Diffusion of stimuli, 255–256

Digestion, external, 144
*Diplogaster,* cytology, 23
*Diploscapter,* cytology, 24
Dispersal of attractants, 249, 253, 255, 256, 259
*Ditylenchus,* 167
  amino acid composition, 217–221
  carbohydrates, 216
  cytochromes, 232
  cytology, 18, 53
  differentiation by serology, 221
  fatty acid composition, 223–225
  feeding, 150, 151
  gland secretions, 143
  indole growth regulators, 231
  lipids, 222, 300
  monoxenic culture, 166, 168
  pathogenicity, 167
  sterol composition, 227, 228
*Ditylenchus allii,* cryptobiosis, 311
*Ditylenchus destructor,* 54, 151, 167
  biological races, 54
  hatching behavior, 283
  intestinal movements in, 144
  monoxenic culture, 166, 168
*Ditylenchus dipsaci,* 39, 151, 238, 239, 241, 250–253, 258, 260, 261, 263
  anhydrobiosis, 304–307
  biochemical change induced by, 177
  biological races, 52, 54, 59, 63, 178
  chemical composition, 216, 217
  cryptobiosis, 311
  in disease complex, 128
  growth, 267
  ingestion, 144–145
  molting, 267
  monoxenic culture, 168
  osmotic pressure effects, 241
  oxygen consumption, 239
  pathogenicity, 172
  pharynx of, 145
  respiration, 240
  sex ratio, 167, 178
  survival, 305–307, 311
*Ditylenchus intermedius,* 150
*Ditylenchus myceliophagus,* 150
*Ditylenchus radicicola,* 240
  biological races, 55
*Ditylenchus triformis,* 214, 223–225, 232, 242, 243
  anhydrobiosis, 305
"Docking disorder," 206
*Dolichodorus heterocephalus*
  feeding, 157
  monoxenic culture, 168
  pathogenicity, 172
Dormancy, 310–315
Dorsal esophageal gland, 44, 45, 141
Dorylaimida, 36, 237
  adaptation to parasitism, 138
  feeding, 156, 157
*Dorylaimus keilini,* cryptobiosis, 311

## E

Ecdysis, 268
Ecological adaptations, 47–48
EDB, 290–295
Egg, 41, 45, 48
  composition of shell, 214
  fertilization, 219
  hatching, 272–284
Eggsac, *Meloidogyne,* composition, 216
Egg shell
  breakdown in hatching, 272
  composition, 81, 284
  permeability, 272, 283
  structure, 282
EH
  effect on hatching, 271
    on molting, 269
Electrical stimuli, 251, 253, 260, 264
Electron microscope, 36, 39, 42, 44, 46, *see also* Ultrastructure
Electron transport chain, 78, 79
Embden–Meyerhof cycle, 77
Endoplasmic reticulum, 44, 46
Enzyme(s), 61, 73–90, 220, 221
  activity related to aging, 306
  in cuticle, 268
  cytochromes, 78–79, 232
  detection, 74–77
  hydrolases, 80–87
    acetylcholinesterase, 76, 87, 248, 302
    amylase, 75, 81, 84, 85, 302
    cathepsin, 85
    cellulase, 81, 82
    chitinase, 81

cholinesterase, 87, 248
dipeptidase, 85
esterases, 87, 301
β-fructofuranosidase, *see* invertase
β-glucosidase, 86
invertase, 75, 81, 85
leucine aminopeptidase, 85
pectinase, 75, 82, 83
pepsin, 85
phenylalanine deaminase, 86
phosphatase
acid, 86, 302
alkaline, 86
polygalacturonase, 83
prolidase, 85
proteases, 81, 85
saccharase, *see* invertase
sucrase, *see* invertase
tripeptidases, 85
trypsin, 85
urease, 86
inhibition, 79
ligases, 88
citrate synthetase, 88
citrogenase, *see* citrate synthetase
condensing enzyme, *see* citrate synthetase
malate synthetase, 88
lyases, 87–88
aconitase, 88
enolase, *see* phosphopyruvate hydratase
fumarase, *see* fumarate hydratase
fumarate hydratase, 88
isocitrate lyase, 87, 302
pectin methylesterase, 82
pectin transeliminase, 84
phosphopyruvate hydratase, 88
oxidoreductases, 77–80
alcohol dehydrogenase, 80
catecholase, *see* polyphenoloxidase
cresolase, *see* polyphenoloxidase
cytochrome oxidase, 78, 79
diaphorase, 78
diphenol oxidase, *see* polyphenoloxidase
DPNH-cytochrome-c-reductase, *see* NADH-cytochrome-c-reductase
glucose dehydrogenase, 80
lactate dehydrogenase, 77
malic dehydrogenase, 78, 302
NADH-cytochrome-c-reductase, 78
NADP-isocitric dehydrogenase, 78
peroxidase, 79
phenolase, *see* polyphenoloxidase
phosphoglyceraldehyde dehydrogenase, 77
polyphenoloxidase, 79
succinic dehydrogenase, 78
triosephosphate dehydrogenase, 77
tryosinase, *see* polyphenoloxidase
role in hatching, 268, 282, 283
in syncytia, 107
transferases, 80
glutamate-oxaloacetate transaminase, 80
glutamate-pyruvate transaminase, 80
L-glutamine-alanine transaminase, *see* glutamate-pyruvate transaminase
L-glutamine-aspartic transaminase, *see* glutamate oxaloacetate transaminase
hexokinase, 80
phosphoglucomutase, 80
phosphoglyceric mutase, 80
transaminase, 80
Enzyme analysis, 74–77
in exudates, 75
histochemical techniques, 76–77
in homogenates, 75–76
Esophageal glands, 41–45, 94, *see also* Pharyngeal glands
composition, 44
Esophagus
dorsal esophageal gland, 44, 45, 141
in *Trichodorus,* 138
Ethanol excretion, 313
Ethylene dibromide (EDB)
effect on cryptobiotic nematodes, 313
permeation characteristics, 291
sensitivity of nematodes to, 291
Evolution, cytogenetic aspects, 29–32
Excised roots, nematodes cultured axenically, 165, 166
Excretory cell, 45
Exsheathing, 269–270
fluid, 269–270

Exudates, 36, 44, 45
effect on hatching
from microorganisms, 272
from roots, 272–282

F

Fanleaf virus, 185
Fatty acids
saturated, 224
unsaturated, 225
Fatigue, sensory, 254, 259, 263
Fecundity, 249
Feeding, 137–157, 169, 176, 178, 262
apparatus, defined, 138
exploration, 139
ingestion, 143–147
mechanism, 93–94
mycophagous nematodes, 150–157
observations, 138, 169, 178
penetration, 140–141
stimulus, 249
Fermentation, 242
Fermentative metabolism, 308, 310–315
Fertilization
in *Anguina,* 5
in *Meloidogyne,* 7
Foliar nematodes, *see Aphelenchoides*
Food cell, 143
selection of, 139
Food reserve in nematodes, 299–303
Fungus-nematode interactions, 120–127
in dixenic experiments, 178
*Fusarium-Meloidogyne,* 121
*Verticillium*-nematode, 122, 123
*Fusarium*-nematode interactions, 121

G

Gall, 94–96
chemical composition, 95
formation under sterile conditions, 171
mechanisms, 94–96
morphology, 95
stimulation, 94–96
Gametogenesis, 3, 4
Gelatinous matrix, 45, 46
Gene flow, 65
Genetics, 1
of biological races, 64
"Giant cell," 38, 41, 44, 45, 142
formation, 177
in plant tissue, 91–117
Gibberellic acid, 252
Gland
esophageal, 41–45, 141–143, 151–157, 215, 219
rectal, 45–47
secretions, 41–47
Glycogen catabolism, 299–303, 307, 311–314
Glycolysis, 77, 242, 311
Glyoxylic acid, 244
Gnotobiology, 159–184
terminology, 160
Gnotobiotic culture
behavior in, 177
egg hatch, 274
life cycle studies, 177
symptom induction in, 172
Golgi bodies, 44, 46
Gradients, 256–258
chemical, 252–253, 255
heat, 254–255
response, 261
in soil, 262
vapor, 249, 252, 255
Granules, 42, 44, 45
Grape Fanleaf virus, 185, 186, 193, 194–199, 205
yellow mosaic virus, 193
Growing sterile plants, 169–171
Growth
*Aphelenchus avenae,* 267
*Ascaris,* 267, 268
*Caenorhabditis briggsae,* 268
*Ditylenchus dipsaci,* 267
*Meloidogyne javanica,* 268
in nematodes, 267–268
Growth regulators, 95–96
in culture media, 165
indoles in nematodes, 231
Guide sheath, 202

H

*Haemonchus contortus,* 300
Hatching, 271–284
animal parasitic nematodes, 271–272
behavior, 282–283
effect of microorganisms, 274

enzymes, 282–283
*Heterodera,* 140, 272–284
inhibition in, 276–277
stimulation in, 274–282
*Meloidogyne,* 277, 283
nematodes, 271–284
optimum temperature, 272
pH, 273–274
physical factors affecting, 272–274
plant parasitic nematodes, 272–284
seasonal effects, 273
soil
aeration effect, 273
moisture effect, 273
stimulation
chemical, 274–276
in *Hypsoperine,* 274
stimulus, 274–282
Hatching agents, 250, 255, 259
artificial, for *Heterodera,* 278–282
mechanism, 283–284
structure, 278–280
Hatching curve for *Heterodera,* 276
Hatching factor
for *Heterodera,* natural, 277–278
for *H. rostochiensis,* 277
isolation and identification, 277–282
Hatching inhibitors, 250
Hatching mechanism
*Ascaris,* 284
*Heterodera,* 283–284
*Meloidogyne javanica,* 284
Head, defined, 138
Heat stimuli, 251, 253–255
*Helicotylenchus,* 8, 10
cytology, 19
*Helicotylenchus multicinctus,* 251
pathogenicity, 172
*Helicotylenchus nannus,* 157
in disease complex, 127, 128
Hemicellulose, 97
*Hemicriconemoides,* cytology, 22
*Hemicycliophora,* ingestion, 146
*Hemicycliophora arenaria,* 237
feeding, 146, 154
hatching, 272
pathogenicity, 172
stylet penetration, 140
*Hemicycliophora paradoxa,* 251, 252
*Hemicycliophora similis,* 251
feeding, 154
Hermaphroditism, 12
*Heterodera*
amino acid composition, 218–219
carbohydrate composition, 215
cyst wall, permeability, 283–284
cytology, 19, 20
diapause, 272–274
differentiation by serology, 221
effect on hatch
of hydrogen ion concentration, 272, 273–274
of inorganic ions, 281
of organic compounds, 272, 278–282
of root exudates, 272, 274–276, 282, 283
of salts, 281, 283
of season, 272, 273
of soil aeration, 272, 273
of soil moisture, 272, 273
of temperature, 272–273
evolution, 29
fungus interactions, 123, 126
hatching, 272–284
hatching agents, artificial, 278–282
hatching curve, 276
hatching factors, natural, 277–278
hatching mechanism, 283–284
pharyngeal gland ducts of, 142
secretory movements in, 142
sex attraction, 249
stimulation of hatch, 274–282
tanning of cysts, 79, 80
*Heterodera avenae,* 251
biological races, 55, 58, 64
diapause, 273
effect on hatch
of root exudates, 275
of season, 273
hatch in water, 274
*Heterodera carotae*
hatch, 280
in sodium hypochlorite, 281
*Heterodera cruciferae*
action of metacorpus, 142
diapause, 273
effect of season on hatch, 273
feeding, 140, 152

natural hatching factors, 277
organic hatching agents, 280
prefeeding behavior, 139
use of stylet by, 140
*Heterodera fici,* biological races, 56
*Heterodera glycines,* 99, 105, 109, 111, 113, 115, 157
anhydrobiosis, 305
biological races, 55, 58, 60, 62
diapause, 273
in disease complex, 132
effect on hatch
of root exudates, 275
of season, 273
hatch in sodium hypochlorite, 282
hatching agents
inorganic, 281
organic, 280
syncytial development, 99
water hatch, 275
*Heterodera goettingiana*
hatch, 275, 280, 281
in calcium hypochlorite, 281
*Heterodera humuli*
biological races, 56
natural hatching factors, 276
*Heterodera major,* cryptobiosis, 311
*Heterodera rostochiensis,* 36, 39, 45, 95, 105, 109, 113, 114, 233, 248, 251–253, 257–259, 263
amino acid composition, 216
anhydrobiosis, 304
biological races, 54, 55, 58, 61
diapause, 273
eggshell composition, 214–216, 222
effect on hatch
of root exudates, 277, 282
of season, 273
feeding, 140
fungus interaction, 126, 175
growth regulator, 231
hatch in calcium hypochlorite, 281
hatching agents
inorganic, 281
organic, 279–280
hatching behavior, 140, 272–277, 282–283
inhibition of hatch, 132, 284
isolation of hatching factors, 279–280
monoxenic culture, 166
pathogenicity, 95
water hatch, 274
*Heterodera schachtii,* 96, 99, 109, 249–252, 263
biological races, 55, 58
in disease complex, 126
hatch in calcium hypochlorite, 281
hatching agents
inorganic, 281
organic, 279–281
monoxenic culture, 166
natural hatching factors, 277
pathogenicity, 96, 172
water hatch, 274
*Heterodera tabacum,* 53
cyst wall permeability, 283
organic hatching agents, 280
*Heterodera trifolii,* 93, 105, 111, 152
organic hatching agents, 280
penetration, 93
pre-feeding behavior, 139
*Heterodera weissi,* hatch in sodium hypochlorite, 282
Heteroderidae, 39, 91
feeding, 152, 153
Heterozygotes, 66
*Hexatylus*
ingestion, 144
pharynx, 144
secretory movements in, 142
*Hexatylus viviparus,* feeding, 155
Hexose monophosphate shunt, 78, 242
Histochemistry, 76–77, 107, 109, 179
Histones, 44
Hoplolaimidae, feeding, 153
*Hoplolaimus*
cytology, 19
in disease complex, 122, 126
monoxenic culture, 168
Host
finding, 178, 247–266
races, 51
reactions, 60, 143
selection, 250
specificity, 52, 139
Hybridization among nematodes, 27–29
Hydrogen ion concentration effect
on hatch, 271

of *Heterodera,* 272, 274
of *Meloidogyne,* 274
on molting, 269, 271
Hydrostatic pressure controlling secretory movements, 142
host-parasite relationships, 138, 145, 146
in nematodes, 143, 144, 146
of plant cells, 144
Hypersensitivity, 112
*Hypsoperine ottersoni,* 274
stimulation of hatch by root exudates, 274

I

Indoleacetic acid, 177, 252
in nematodes, 177
in plants, 177
Ingestion, 143–147
mechanisms, 144–147
Inheritance of pathogenicity, 64
Inhibition of hatch
in *Heterodera,* 276, 277
in *Meloidogyne,* 277, 283
Inhibitors, 107, 108
Inorganic ions, effect on hatch of *Heterodera,* 281–282
Intestinal movements, 144
Intraspecific variation, 62

K

Kinetin, 165
Klinokinesis, 260, 262
Koch's postulates, 171
Krebs cycle (TCA cycle), 77–78, 242–244

L

Labial papillae, 140
Lance nematodes, *see Hoplolaimus*
Lesion formation in gnotobiotic culture, 176
Lesion nematodes, *see Pratylenchus*
Lethargus, 269
Leucine aminopeptidase, role in molting, 269–270
Light stimuli, 261
Lipase, role in hatching, 272
Lipid, 222–230
catabolism, 299–303, 311
content, 299–303, 305, 310
Lips as sensory receptors, 139
*Longidorus,* 186, 187, 190, 191, 193, 194, 198, 201
life cycle, 189
pathogenicity, 188
relation to soil type, 192
vertical distribution, 191
virus vectors, 185–211
*Longidorus attenuatus,* 186, 189, 192, 199, 206
*Longidorus elongatus,* 186, 188, 189, 191, 192, 194, 195, 198–200, 202, 205, 206
biological races, 57
feeding, 156
*Longidorus macrosoma,* 186, 188, 189, 191, 192, 194, 198, 199, 201
*Longidorus profundorum,* 191
Loss of stimuli, 254–255
Lysosome, 303

M

Male attractant, 249, 257
Meadow nematodes, *see Pratylenchus*
Meiosis, 2, 30, 31
*Meloidogyne,* 249, 251, 252, 262
amino acid composition, 216–219
cuticle, 216–218, 268
cytology, 21, 22
effect on hatch
of hydrogen ion concentration, 274
of inorganic ions, 281
of root exudates, 274
of soil moisture, 273
evolution, 30
feeding, 152–153
fungus interactions, 121, 124
hatching in organic compounds, 280
inhibition of hatch, 277, 283
monoxenic culture, 166–168
pathogenicity, 171
penetration, 93, 95, 179
prefeeding behavior, 93, 94
secretory movements in, 93–94, 142
virus interactions, 129, 130
*Meloidogyne arenaria,* 249
biological races, 56
*Meloidogyne hapla,* 39, 248, 250, 251, 252
biological races, 56

monoxenic culture, 168
oogenesis and fertilization, 7
*Meloidogyne incognita,* 94, 242, 250–252, 261
*acrita,* 126, 251, 252
attraction to host, 139
biological races, 56
culturing, 165
in disease complex, 122, 125–127
monoxenic culture, 166, 168
penetration, 179
*Meloidogyne javanica,* 36, 39, 41, 44, 45, 47, 95, 103, 153, 238, 243, 248, 250–253
aging, 302–303
biological races, 56
cold tolerance, 309
constituents of pharyngeal gland secretions, 142, 215
cuticle, 39–41, 216
in disease complex, 123
effect on hatch
of soil aeration, 273
of soil moisture, 273
of temperature, 273
esophageal glands, 41–45, 215, 219
growth, 47
growth regulators, 231
hatching behavior, 272, 282
lipid content, 222, 300
males, 47–48
molting, 268
muscle, 39, 41
protein content, 220
rectal glands, 45
respiration rate, 301, 310
stylet, 36
survival, 302, 309
time death studies, 290–291
ultrastructure, 36, 38, 41, 43
*Mesorhabditis,* cytology, 24
Metabolism of sterols, 228–230
Metacorpus of pharynx, 144, 145
secretions in, 142
Methionine, role in aging, 303
Methodology
axenizing nematodes, 161–164
culturing sterile nematodes, 164–169
determining chemical composition, 214–233
enzyme detection, 74–77
feeding observations, 138
growing sterile plants, 169–171
separating proteins (electrophoresis), 63, 220–221
serology, 63, 221–222
Methyl bromide, 206
Michaelis constant, 238
Microaerobic environment, 307
Migration of larvae, 93
Mitosis
in nematodes, 2–9, 11, 12
in syncytia, 101, 107
Moisture gradients, 258
Molting, 267–271
animal parasitic nematodes, 267–270
*Aphelenchus avenae,* 267, 271
*Ascaris,* 268
*Caenorhabditis briggsae,* 268
*Ditylenchus dipsaci,* 267
effect of size, 271
function, 267–268
*Meloidogyne javanica,* 268
*Paratylenchus,* 270
plant parasitic nematodes, 267–271
*Rotylenchus buxophilus,* 271
sequence of events, 269
stimulus, 269–271
Monoxenic culture
definition, 160
host reactions, 171–176
propagation of nematodes, 164–169, 298
Morphological variation in culture, 167
Movement of nematodes, 259
distances moved to roots, 253–254
Moving sources of attractant, 258
Mucopolysaccharides, 44, 47
Multinucleation in syncytia, 101
Multivesicular lamellar bodies, 46
Muscle, 39, 41
of pharynx, 146
spear protractor, 145–147
Mutation, 65
in culture, 167
nematode growth factor, 167, 168


## N


*Nacobbus serendipiticus*
feeding, 153

monoxenic culture, 166
pathogenicity, 173
Needle nematodes, *see Longidorus*
Nematicides
control of virus diseases, 204–207
evaluation, 170
gross effects, 290–291
on cryptobiotic nematodes, 313
influence on environment, 291–292
intoxication by, 289–290, 295
mode of action, 289–296
hypotheses, 295–296
model systems, 294–295
permeation characteristics, 292–294
resistance to, 292
time death studies, 290–291
Nematode growth, 267–268
Nematode-bacteria interactions, 127–129
Nematode-fungus interactions, 120–127
Nematode-virus interactions, 129–131, 185–211
Nematode physiology, change associated with parasitism, 179
Nematode tracks, 178
Nematodea, adaptations of, 138
*Nematodirus filicollis,* cold tolerance, 309
*Neoaplectana carpocapsae,* 300
Neotylenchidae, feeding, 155
*Neotylenchus,* intestinal movements, 144
*Neotylenchus linfordi,* feeding, 155
NEPO viruses, 186, 193–195, 199, 202, 205
NETU viruses, 187, 195–196
Neurosecretory cells
*Ascaris,* 270
*Phocanema decipiens,* 270
*Nippostrongylus*
cuticle, 268
rate of lipid utilization, 300
*Nippostrongylus brasiliensis* lipids, 222, 300
*Nippostrongylus muris,* respiration, 301
*Nitella,* 143
hydrostatic pressure in, 143
Nitrogen, 253
Nuclear morphology and cytochemistry, 95
Nuclei, number in syncytia, 103
Nucleic acid, 6, 44, 95, 99, 107–109
Nucleolar condition, 103
Nucleolus, 45
Nucleus, 45
Nutrition in culture, 179

## O

Odontostyle, 188, 190
Onchiostyle, 36
Oogenesis, 4–9
in *Anguina,* 5
in *Meloidogyne,* 7
Organic compounds, effect on hatch of *Heterodera,* 272, 278–282
Orientation, 258–263
Orthokinesis, 259
Osborne's cyst nematode, hatch in sodium hypochlorite, 282
Osmoregulation
in *A. avenae,* 214
in *P. redivivus,* 214
Osmotic potential of plant hosts, 143, 144
Osmotic pressure, 240–241, 306
effect on hatch, 284
on respiration, 240–241, 306–307
*Ostertagia,* survival, 309
Oxidative metabolism, 307
Oxygen, 251, 313
Oxygen uptake, 238, 310
consumption, 306, 308

## P

*Panagrellus*
fatty acid composition, 223–225
lipid content, 222–223
sterol requirements, 229
*Panagrellus redivivus,* 239, 240, 243, 261, 263
axenizing, 164
lipid utilization, 300
osmoregulation, 214
protein content, 220
respiration, 240, 310
water content, 214
*Panagrolaimus,* cytology, 24
*Panagrolaimus rigidus,* 249
*Paraphelenchus acontioides*
feeding, 156
prefeeding behavior, 139
*Paraphelenchus myceliophthorus,* feeding, 156
*Parascaris,* chromatin diminution, 2

Parasitism, monoxenic study of, 176
changes associated with in *Meloidogyne,* 39, 41–45
*Paratylenchus curvitatus*
feeding, 155
molting, 270
survival, 270
*Paratylenchus dianthus, see Paratylenchus curvitatus*
*Paratylenchus elachistus,* feeding, 155
*Paratylenchus minutus, see Paratylenchus elachistus*
*Paratylenchus nanus,* molting, 270
*Paratylenchus projectus*
feeding, 155
molting stimulus, 270
survival, 309
Parthenogenesis
meoitic, 10–11
mitotic, 11–12
Parthenogenetic species
biological races, 66
*Meloidogyne,* 171
PAS technique, 43, 215
Pathogenicity
biochemical changes in host, 79, 80, 176–178
in monoxenic studies, 167, 171–176
morphological changes in host, 151–157
syncytia, 91–117
Pathotypes, 67–68
Pea early browning virus, 187, 195, 200
Pectinases, 82–84
classification, 83
Pectins in giant cell, 97
*Pelodera,* cytology, 23
*Pelodera teres,* 249
Penetration, 92–93, 111, 140–141, 151–157, 179
Pentose shunt, 78
Permeability of plant cells, 144
pH, 253
effect on hatch, 273–274
on nematicides, 292
Pharyngeal gland, 94, 141–143
activity during hatching, 282–284
pump in Tylenchida, 145, 146
Pharynx, 138, 143, 144–147
flow of food through, 144, 145
hydrostatic pressure of, 141
in *Xiphinema,* 146
musculature of in *Xiphinema index,* 146
*Trichodorus,* 138
Phasmids, 248
Phenolic reactions in gnotobiotic experiments, 176
Phenols as hatch inhibitors, 277
*Phocanema decipiens*
cuticle, 268
exsheathing, 270
neurosecretory cells, 270
Physiologic aging, 298, 303
*Phytopthora-Meloidogyne* interactions, 123
Pin nematodes, *see Paratylenchus*
Plastic film isolator, 170
*Plectus,* cryptobiosis, 311–312
Poiseuille's law, 146
Polyploidy, 16, 29
of syncytia nuclei, 105
Population(s)
behavioral variations, 54–59
effects of pathogenic fungi on, 131–132
Population control, 248–249
*Pratylenchus* 251
biochemical changes induced by, 80, 86, 174, 176
cytology, 18, 19, 32
in disease complexes, 122–124, 131, 133
evolution, 32
host finding, 177
monoxenic culture, 165–168
oxygen composition, 238, 306
pathogenicity, 167, 173–174, 179
phenolic host reaction, 86, 176, 178
races, 174
symptoms, 174, 175
*Pratylenchus brachyurus,* monoxenic culture, 166, 168
*Pratylenchus crenatus,* 157
*Pratylenchus loosi,* biological races, 57
*Pratylenchus minyus,* 251
monoxenic culture, 165–166
*Pratylenchus neglectus,* biological races, 57
*Pratylenchus penetrans,* 201, 238, 239, 241, 250, 251, 253

amino acid composition, 217
anhydrobiosis, 305
biological races, 56
effect of $CO_2$, 238
fatty acid composition, 223–225
fungus interaction, 175
lipid content, 222, 300
monoxenic culture, 166, 168
respiration, 340
*Pratylenchus pratensis,* 250, 251
*Pratylenchus scribneri,* 259, 263
freeze susceptibility, 309
*Pratylenchus thornei,* anhydrobiosis, 305
*Pratylenchus vulnus,* monoxenic culture, 168
*Pratylenchus zeae,* monoxenic culture, 168
Preconditioning, 253
Precursors of DNA and RNA, 107, 109
Prefeeding behavior of nematodes, 139
Procorpus of pharynx, 145
Proprioreception, 140
Protease, role in hatching, 272
exudates, 85
Protein, 39, 44, 46
composition in nematodes, 219–222
Pseudogamy, 12
*Pseudomonas*-nematode interaction, 127
Pump
defined, 138
metacorpal, 145
pharyngeal, 145
in Criconematidae, 146
operation in Tylenchidae, 145, 146
Pyruvate, 242, 243
*Pythium*-ectoparasitic nematode interactions, 124

Q

Quiescence, 263, 304–310

R

Race, 51–71
biological, *see* Biological races
geographical, 51
identification, 62
variation in pathogenicity, 63
Radioisotopes, labeling, 217, 226
sterol incorporation, 228–230
tracers, 283
*Radopholus similis*
biological races, 57
fungus interactions, 124, 175
monoxenic culture, 166, 168
pathogenicity, 173
Raspberry ringspot virus, 186, 193–196, 198–200, 202, 205, 206
Receptors, 251, 254, 258
Recombination, genetic, 66
Rectal glands, 45–47
Redox potential, 252, 269, 271, *see also* EH
Reducing agents, 251, 252
Reproduction, 3–13
amphimixis, 4
automixis, 4
in callus tissue, 167
hermaphroditism, 12–13
parthenogenesis, 10, 11
pseudogamy, 3, 4, 12
Repulsion, 250, 260
Resistance, 178
definition, 112–113
to Heteroderidae, 111–115
Respiration, 235–245
influence of carbon dioxide, 238–239
osmotic pressure, 240–241
oxygen, 237–238
reduced oxygen levels, 237–238
temperature, 239–240
measurement of, 244–245
rates, 236, 284, 301
Respiratory quotient, 240, 301
Responses, 258–259, 263
Retention by attraction, 262
*Rhabditis*
axenizing, 164
cytology, 23
gynogenesis, 3
osmotic pressure tolerance, 306
*Rhabditis anomala,* 244
*Rhabditis belari,* 249
*Rhabditis oxycerca,* 253
*Rhizoctonia-Pratylenchus* interactions, 124
Ribonucleic acid (RNA), 6, 95, 109
Ribosomes, 44, 46

Ring nematode, *see Criconemoides*
Ringspot viruses, 136
Root attraction, 177, 178
  zones of, 250–252
Root exudates
  effect on hatch
    of *Heterodera,* 272, 274–276, 282
    of *Hypsoperine ottersoni,* 274
    of *Meloidogyne,* 274, 275
Root-knot nematodes, *see Meloidogyne*
Root stimuli, 250–253, 254–255
*Rotylenchulus,* cytology, 19
*Rotylenchulus reniformis*
  biological races, 57, 65
  in disease complex, 126
*Rotylenchus,* cytology, 19
*Rotylenchus buxophilus,* molting, 271
*Rotylenchus robustus,* feeding, 153
*Rotylenchus uniformis, see Rotylenchus robustus*

S

Salts
  effect on hatch, 271
    of *Heterodera,* 272, 281, 283
    on molting, 269
Saturated fatty acid composition, 224
Season, effect on hatch of *Heterodera,* 272, 273
    of *H. avenae,* 273
    of *H. cruciferae,* 273
    of *H. glycines,* 273
    of *H. rostochiensis,* 273
Secretions, 44, 45, 75, 80–82, 85, 93, 94, 248–249, 256
Seedgall nematodes, *see Anguina*
*Seinura*
  cytology, 22
  evolution, 32
Selection, 66
Senescence, 299–303, 312, 315
Sensory receptors, 248, 258
Serological differentiation, 62–63
Serology, species differentiation in *Heterodera* and *Ditylenchus,* 62–63, 221
Sex attractants, 249, 257
Sex determination, 3, 16–17
  related to syncytia, 113, 114
Sex ratio, 167, 178
  underproduction of males, 47
Sheath nematode, *see Hemicycliophora*
Shock responses, 261–262
Sibling species, 53–54
Sodium hypochlorite, effect on hatch of *Heterodera,* 281
Soil aeration, effect on hatch of *Heterodera,* 273
  of *Meloidogyne javanica,* 273
Soil moisture, effect on hatch of *Heterodera,* 273
  of *Meloidogyne,* 273
Somatic muscles, 39, 41, 48
Spermatogenesis, 9–10
Spiral nematodes, *see Helicotylenchus, Rotylenchus*
Stem nematodes, *see Ditylenchus*
Sterile root chamber, 169–171
Stimulation
  of galls, 94–96
  of hatch in *Heterodera,* 272, 274–284
Stimuli
  causing accumulation, 250–253, 258
  chemical, 139, 251
  disappearance and dispersal, 254–258
  hatching, 250, 274–276
  molting, 269–271
  responses to, 258–263
  from roots, 250–252
  tactile, 139
Stimulus
  for hatch in animal parasitic nematodes, 271–272
  for molting, 269–271
Sting nematodes, *see Belonolaimus*
Stomatostyle, 36
Strawberry, latent ringspot virus, 156, 172–173, 175, 186, 193–196, 198, 199, 201, 202
Stress, 47
*Strongyloides ratti*
  lipid utilization, 300
  respiration, 301
*Strongylus edentatus,* 309
*Strongylus vulgaris,* 309
Stubby root nematode, *see Trichodorus*
Stylet, 139, 140, 144

exudate, 36, 215
thrusting in hatching, 282, 283
Subventral esophageal gland, 42, 44
Swarming, 304
Syncytium, syncytia, 1–2
composition, 97
cytoplasm, 108–111
definition, 92
development in plant, 97–101

## T

Tactile stimuli, 248
Taxis, 260–261
TCA (Tricarboxylic acid cycle), 242–244
Temperature
cold effect on survival, 309–310, 312
effect on hatch of *Heterodera*, 272–273
of *Meloidogyne javanica*, 273
on molting, 269
on nematicides, 291–292
on respiration, 239–240
on survival, 305
gradients, 254
Terminal oxidation, 238, 243
*Tetylenchus joctus*, feeding, 150
Thigmokinesis, 248
Thigmotaxis, 248
Time-death studies, 290–291
Tobacco rattle virus, 187, 195–201
mosaic virus, 130
ringspot virus, 129, 186, 193, 197–199
Tomato black ring virus, 186, 194–196, 198, 199, 201, 205
Tomato ringspot virus, 186, 193, 196, 197, 199, 200
Trehalose, release in hatching, 272
Trichodoridae, feeding, 157
*Trichodorus*, 186, 187, 189–191, 195, 200, 206
host range, 190
ingestion, 144, 146
life cycle, 190
pathogenicity, 94, 190
pharyngeal gland ducts of, 141
pharynx, 144, 201–204
relation to soil type, 192
stylet, 36–39, 147
in feeding, 93–94, 139–141, 178
in virus transmission, 200–204
vertical distribution, 191
virus vectors, 60, 186–211
*Trichodorus allius*, 187, 190, 196–200
feeding, 157
*Trichodorus anemones*, 187
*Trichodorus christiei*, 187, 190, 195
axenization, 161
biological races, 58
in disease complex, 126
feeding behavior, 147, 157, 190
freeze susceptibility, 309
mechanism of, 147
physiologically differing populations, 62
secretions, 94
*Trichodorus cylindricus*, 187
*Trichodorus nanus*, 187
*Trichodorus pachydermus*, 187, 190–192, 195, 196, 198–200, 203
biological races, 57–58
ultrastructure of esophagus, 202–204
*Trichodorus porosus*, 187, 196
*Trichodorus primitivus*, 187, 191, 195
*Trichodorus proximus*, feeding, 157
*Trichodorus similis*, 187
*Trichodorus teres*, 187, 190–192, 196
*Trichodorus viruliferus*, 187, 190, 191, 250, 251, 258
feeding, 157
*Trichonema*, 260
*Trichostrongylus axei*, survival, 305
*Trichostrongylus colubriformis*, 259
survival, 305
Trophotype, definition, 68
Tropotaxis, 261
*Turbatrix*
aging studies, 301
fatty acid composition, 224–226
sterol, 227–228
*Turbatrix aceti*, 228, 243
life span, 301
lipid content, 222
organic acid content, 232
Turgor, *see* Hydrostatic pressure
Tylenchida, 36, 150–156
adaptation to parasitism, 138
pharyngeal form, 145

pumping, 145, 146
secretions, 142
stylet retraction, 141
Tylenchidae, feeding, 150–151
Tylenchoidea, feeding, 150–155
Tylencholaiminae, feeding, 156–157
*Tylenchorhynchus*
cytology, 18
monoxenic culture, 168
*Tylenchorhynchus claytoni*
anhydrobiosis, 305
fatty acid composition, 223–225
feeding, 150, 251
fungus interactions, 124, 126, 131, 175
lipid content, 222, 300
monoxenic culture, 168
pathogenicity, 173
*Tylenchorhynchus dubius*
feeding, 150
gland secretions, 141
*Tylenchorhynchus icarus,* 241
osmotic pressure effects, 241
survival, 305
*Tylenchorhynchus martini,* 237, 253
anoxybiosis, 237
*Tylenchulus semipenetrans,* 39
biological races, 57–60
in disease complex, 124
freeze susceptibility, 309
hatching, 272
lipid content, 222, 300
respiration rate, 301, 310
survival, 302
time-death studies, 291
*Tylenchus agricola,* 157
fungus interactions, 175
monoxenic culture, 168
pathogenicity, 173
*Tylenchus bryophilus,* 157
*Tylenchus emarginatus,* feeding, 150
*Tylenchus hexalineatus,* monoxenic culture, 168
*Tylenchus polyhypnus,* cryptobiosis, 311

## U

Unsaturated fatty acid composition, 225
Ultrastructure
*Meloidogyne javanica,* 43
cuticle, 41
subventral esophageal glands, 42, 215
of syncytia
*Heterodera* induced, 99, 114
*Meloidogyne,* 97–100
*Trichodorus,* esophageal region, 147, 201–204
Uterine wall, 45

## V

Valve
apparatus, 138, 145
defined, 138
Vapor gradients, 249, 252, 255
Variation
in host-range, *see* Biological races
in morphology, 62
in pathogenicity and physiology, *see* Biological races
*Verticillium-Pratylenchus* interactions, 122
Viruses, 185–207
acquisition, 197–198
control of diseases, 204–207
diseases, 192–196
distribution in soil, 191–192
hosts, 189
nematode vectors, 186
NEPO, 193–195
NETU, 195–196
persistence, 198–199
relationships with nematodes, 129–131
specificity, 199–200
transmission, 57, 58, 196–197
mechanism, 200–204
Vitamin E, in aging, 303
Vitelline membrane, 8

## W

Wall formation, dissolution in syncytia, 97–101
"Wirrspuren," 261

## X

*Xiphinema,* 186, 191
host range, 188, 189

life cycle, 188–189
pathogenicity, 188
relation to soil type, 192
stylet retraction in, 141
vertical distribution, 191
virus vectors, 185–211
*Xiphinema americanum,* 186, 188, 189, 193, 197–200
species complex, 193
*Xiphinema coxi,* 186, 194, 201
*Xiphinema diversicaudatum,* 186, 188, 189, 191–194, 197–199, 201
feeding, 157
pharynx, 146
*Xiphinema index,* 185, 186, 188, 191, 193, 196–199, 201, 203, 204
axenization, 161
feeding mechanism, 146, 157
osmotic pressure effects, 306
*Xiphinema mediterraneum,* 192, 193
*Xiphinema vuittenezi,* 189, 191